Practical Control Engineering

About the Author

David M. Koenig had a 27 year career in process control and analysis for Corning, Inc., retiring as an Engineering Associate. His education started at the University of Chicago in chemistry, leading to a PhD in chemical engineering at The Ohio State University. He resides in upstate New York where his main job is providing day care for his six month old grandson.

Practical Control Engineering

A Guide for Engineers, Managers, and Practitioners

David M. Koenig

New York Chicago San Francisco
Lisbon London Madrid Mexico City
Milan New Delhi San Juan
Seoul Singapore Sydney Toronto

The McGraw·Hill Companies

Library of Congress Cataloging-in-Publication Data

Koenig, David M.
Practical control engineering : a guide for engineers, managers, and practitioners / David M. Koenig.
p. cm.
Includes bibliographical references and index.
ISBN 978-0-07-160613-4 (alk. paper)
1. Automatic control. 2. Control theory. I. Title.
TJ213.K578 2009
629.8—dc22 2008037488

McGraw-Hill books are available at special quantity discounts to use as premiums and sales promotions, or for use in corporate training programs. To contact a special sales representative, please visit the Contact Us page at www.mhprofessional.com.

Practical Control Engineering

1 2 3 4 5 6 7 8 9 0 DOC/DOC 0 1 4 3 2 1 0 9

ISBN 978-0-07-160613-4
MHID 0-07-160613-0

The pages within this book were printed on acid-free paper containing 15% post-consumer fiber.

Sponsoring Editor
Taisuke Soda

Acquisitions Coordinator
Michael Mulcahy

Editorial Supervisor
David E. Fogarty

Project Manager
Preeti Longia Sinha

Copy Editor
Ragini Pandy

Proofreader
Bhavna Gupta

Indexer
David M. Koenig

Production Supervisor
Richard C. Ruzycka

Composition
International Typesetting and Composition

Art Director, Cover
Jeff Weeks

**To Joshua Lucas, Ryan, Jennifer, Denise,
Julie and Bertha and
in memory of Wilda and Rudy.**

Contents

Preface

You may be an engineering student, a practicing engineer working with control engineers, or even a control engineer. But I am going to assume that you are a manager.

Managers of control engineers sometimes have a difficult challenge. Many companies promote top managerial prospects laterally into unfamiliar technical areas to broaden their outlook. A manager in this situation often will have several process control engineers reporting directly to her and she needs an appreciation for their craft. Alternatively, technical project managers frequently supervise the work of process control engineers on loan from a department specializing in the field. This book is designed to give these managers insight into the work of the process control engineers working for them. It can also give the student of control engineering an alternative and complementary perspective.

Consider the following scenario. A sharp control engineer, who either works for you or is working on a project that you are managing, has just started an oral presentation about his sophisticated approach to solving a knotty control problem. What do you do? If you are a successful manager, you have clearly convinced (perhaps without foundation) many people of your technical competence so you can probably ride through this presentation without jeopardizing your managerial prestige. However, you will likely want to actually critique his presentation carefully. This could be a problem since, being a successful manager, you are juggling several technically diverse balls in the air and haven't the time to research the technological underpinnings of each. Furthermore, your formal educational background may not be in control engineering. The above-mentioned control engineer, embarking on his presentation, is probably quite competent but perhaps he has been somewhat enthralled by the elegance of his approach and has missed the forest for the trees (it certainly happened to me many times over the years). You should be able to ask some penetrating questions or make some key suggestions that will get him on track and make him (and you) more successful. Hopefully, you will pick up a few hints on the kind of questions to ask while reading this book.

The Curse of Control Engineering

The fundamental stumbling block in understanding process control engineering is its language—applied mathematics. I could attempt to skirt the issue with a qualitative book on control engineering. Not only is this difficult to do but it would not really equip the manager to effectively interact with and supervise the process control engineer. To do this, the manager simply has to understand (and speak) the language.

If terms like $\frac{dy}{dt}$ or $\int_0^a dt\, e^{st}$ strike fear in your heart then you should consider looking first at the appendices which are elementary but detailed reviews of the applied mathematics that I will refer to in the main part of the text and that control engineers use in their work. Otherwise, start at the beginning of the book. As you progress through it, I will often show only the results of applying math to the problem at hand. In each case you will be able to go to an appendix and find the pertinent math in much more detail but presented at an introductory level. The chapters are the forest; the appendices are the trees and the leaves.

You may wonder why much of the math is not inserted into the body of the text as each new topic is discussed—it's a valid concern because most books do this. I am assuming that you will read over parts of this book many times and will not need to wade through the math more than once, if that. After all, you are a manager, looking at a somewhat bigger picture than the control engineer.

Also, you may wonder why there are so many appendices, some of them quite long, and relatively few chapters. You might ask, "Are you writing an engineering book or an applied mathematics book?" To those who would ask such an "or" question I will simply pause for a moment and then quietly say, "yes."

Style

The book's style is conversational. I do not expect you to "study" this book. You simply do not have the time or energy to hunker down and wade through a technical tome, given all the other demands of your job. There are no exercises at the ends of the chapters. Rather, I foresee you delving into this book during your relaxation or down time; perhaps it will be a bedtime read...well, maybe a little tougher than that. Perhaps you could spend some time reading it while waiting in an airport. As we progress through the book I will pose occasional questions and sometimes present an answer immediately in small print. You will have the choice of thinking deeply about the question or just reading my response—or perhaps both!

On the other hand, if this book is used in a college level course, the students will likely have access to Matlab and the instructor can easily assign homework having the students reproduce or modify the figures containing simulation and control exercises. I will, upon request, supply you with a set of Matlab scripts or m-files that will generate all the mathematically based figures in the book. Send me an e-mail and convince me you are not a student in a class using this book.

References

There aren't any. That's a little blunt but I don't see you as a control theory scholar—for one thing, you don't have time. However, if you are a college-level engineering student then you already have an arsenal of supporting textbooks at your beck and call.

A Thumbnail Sketch of the Book

The first chapter presents a brief qualitative introduction to many aspects of control engineering and process analysis. The emphasis is on insight rather than specific quantitative techniques.

The second chapter continues the qualitative approach (but not for long). It will spend some serious time dealing with how the engineer should approach the control problem. It will suggest a lot of upfront time be spent on analyzing the process to be controlled. If the approaches advocated here are followed, your control engineer may be able to bypass up the development of a control algorithm altogether.

Since the second chapter emphasized process analysis, the third chapter picks up on this theme and delves into the subject in detail. This chapter will be the first to use mathematics extensively. My basic approach here and throughout the book will be to develop most of the concepts carefully and slowly for simple first-order systems (to be defined later) since the math is so much friendlier. Extensions to more complicated systems will sometimes be done either inductively without proof or by demonstration or with support in the appendices. I think it is sufficient to fully understand the concepts when applied to first-order situations and then to merely feel comfortable about those concepts in other more sophisticated environments.

The third chapter covers a wide range of subjects. It starts with an elementary but thorough mathematical time-domain description of the first-order process. This will require a little bit of calculus which is reviewed in Appendix A. The proportional and proportional-integral control algorithms will be applied to the first-order process and some simple mathematics will be used to study the system. We then will move directly to the s-domain via the Laplace transform (supported in Appendix F). This is an important subject for control engineers and can be a bit scary. It will be my challenge to present it logically, straightforwardly, and clearly.

Just when you might start to feel comfortable in this new domain we will leave Chapter Three and I will kick you into the frequency domain. Chapter Four also adds two more process models to the reader's toolkit—the pure dead-time process and the first-order with dead-time process.

Chapter Five expands the first-order process into a third-order process. This process will be studied in the time and frequency domains. A new mathematical tool, matrices, will be introduced to handle the higher dimensionality. Matrices will also provide a means of looking at processes from the state-space approach which will be applied to the third-order process.

Chapter Six is devoted to the next new process—the mass/spring/dashpot process that has underdamped behavior on its own. This process is studied in the time, Laplace, frequency and state-space domains. Proportional-integral control is shown to be lacking so an extra term containing the derivative is added to the controller. The chapter concludes with an alternative approach, using state feedback, which produces a modified process that does not have underdamped behavior and is easier to control.

Chapter Seven moves on to yet another new process—the distributed process, epitomized by a tubular heat exchanger. To study this process model, a new mathematical tool is introduced—partial differential equations. As before, this new process model will be studied in the time, Laplace, and frequency domains.

At this point we will have studied five different process models: first-order, third-order, pure dead-time, first-order with dead-time, underdamped, and distributed. This set of models covers quite a bit of territory and will be sufficient for our purposes.

We need control algorithms because processes and process signals are exposed to disturbances and noise. To properly analyze the process we must learn how to characterize disturbances and noise. So, Chapter Eight will open a whole new can of worms, stochastic processes, that often is bypassed in introductory control engineering texts but which, if ignored, can be your control engineer's downfall.

Chapters Eight and Nine deal with the discrete time domain, which also has its associated transform—the Z-transform, which is introduced in the latter chapter. As we move into these two new domains I will introduce alternative mathematical structures for our set of process models which usually require more sophisticated mathematics.

In Chapter Five, I started frequently referring to the state of the process or system. Chapter Ten comes to grips with the estimation of the state using the Kalman filter. A state-space based approach to process control using the Kalman filter is presented and applied to several example processes.

Although the simple proportional-integral-derivative control algorithm is used in the development of concepts in Chapters Three through Nine, the eleventh chapter revisits control algorithms using a slightly different approach. It starts with the simple integral-only algorithm and progresses to PI and the PID. The widely used concept of cascade control is presented with an example. Controlling processes subject to white noise has often been a controversial subject, especially when statisticians get involved. To stir the pot, I spend a section on this subject.

This completes the book but it certainly does not cover the complete field of process control. However, it should provide you with a starting point, a reference point and a tool for dealing with those who do process control engineering as a profession.

If you feel the urge, let me know your thoughts via

dmkoenig@alumni.uchicago.edu.

Good luck while you are sitting in the airports!

Practical Control Engineering

CHAPTER 1

Qualitative Concepts in Control Engineering and Process Analysis

This will be the easiest chapter in the book. There will be no mathematics but several qualitative concepts will be introduced. First, the cornerstone of control engineering, the feedback controller is discussed. Its infrequent partner, the feedforward controller is presented. The significant but often misunderstood differences between feedback and feedforward control are examined. The disconcerting truth about the difficulty of implementing error-free feedback control is illustrated with an industrial example. Both kinds of controllers are designed to respond to disturbances, which are discussed briefly. Finally, we spend a few moments on the question of what a control engineer is.

1-1 What Is a Feedback Controller?

Consider the simple process shown in Fig. 1-1. The level in the tank is to be maintained "near" a target value by manipulating the valve on the inlet stream. Now, place the "as-yet-undefined" controller in Fig. 1-2. The controller must sense the level and decide how to adjust the valve. Notice that for the controller to work properly

1. There must be a way of measuring the tank level (the "level sensor") and a way of transmitting the measured signal to the controller.
2. Equally important there must be a way of transmitting the controller decision or controller output to the valve.
3. At the valve there must be a way of converting the controller output signal into a mechanical movement to either close or open the valve (the "actuator").

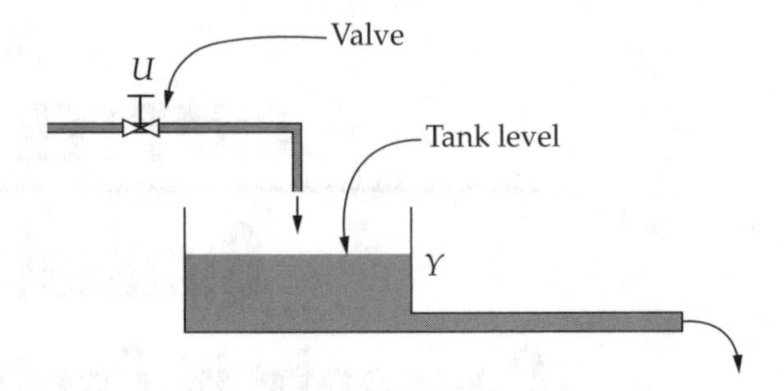

FIGURE 1-1 A tank of liquid (a process).

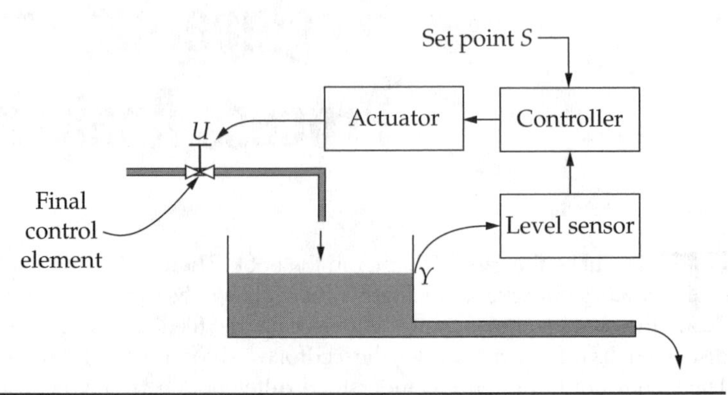

FIGURE 1-2 A tank of liquid with a controller added.

An abstract generalization of the above example is shown in Fig. 1-3, which is a schematic block diagram. The lower box represents the process (the tank of liquid). The input to the process is U (the valve position on the inlet pipe). The output is Y (the tank level). The process

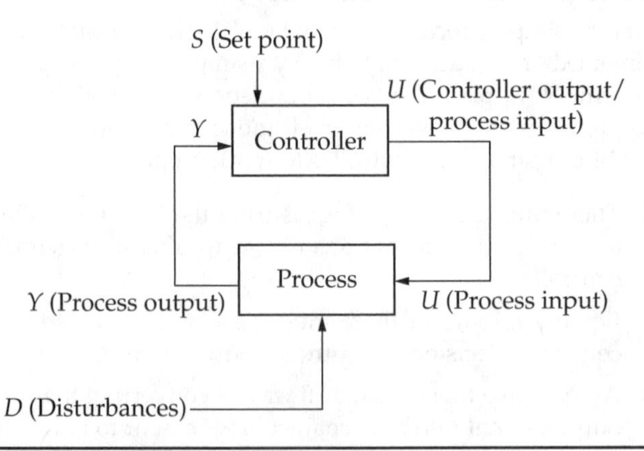

FIGURE 1-3 Block diagram of a control system.

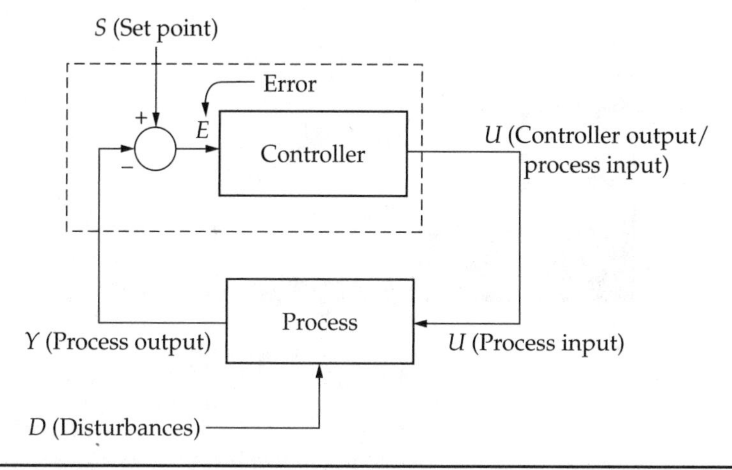

FIGURE 1-4 Block diagram of a control system showing the error.

is subject to disturbances represented by D. The process is therefore an engine that transforms an input U and disturbances D into an output Y. The inputs to the controller are the process output Y (the tank level) and the set point S or target. The controller puts out a signal U (the valve position) designed to cause the process output Y to be "satisfactorily close" to the set point S. You need to memorize this nomenclature because Y, U, S, and D, among some others soon to be introduced, will occur repeatedly.

A more specific form of the controller is shown in Fig. 1-4. The process output is subtracted from the set point to form the controller error E, which is then fed to another box containing the rest of the control algorithm. The controller must drive the controller error to a satisfactorily small value. Note that the controller cannot "see" the disturbances. It can only *react* to the error between the set point and the current measurement of the process output—more about this later. Also note that there will be no control actions unless there are controller errors. Therefore one must reason that an active feedback controller (meaning one where the control output is continually changing) may not keep the process output exactly on set point because control activity means there are errors.

1-2 What Is a Feedforward Controller?

Before getting into a deep discussion of a feedforward controller, let's develop a slightly modified version of our tank of liquid. Consider Fig. 1-5, which shows a large tank, full of water, sitting on top of a large hotel (use your imagination here, please). This tank is filled in the same manner as the one in the previous figures. However, this tank supplies water to the sinks, toilets, and showers in

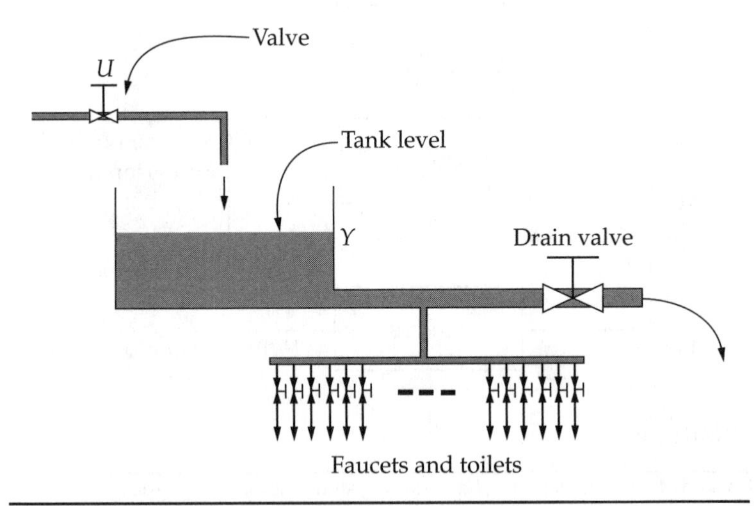

FIGURE 1-5 Large hotel water tank.

the hotel's many rooms. At any moment the faucet or toilet usage could disturb the level in the tank. Moreover, this usage is *unpredictable* (later on we will use the word "stochastic"). There is also a drain valve on the tank which, let's say, the hotel manager occasionally opens to fill the swimming pool. Opening the drain valve would also be a disturbance to the tank level but, unlike the faucet usage, it could probably be considered "deterministic" in the sense that the hotel manager knows when and approximately how much the adjustment to the valve would be. We will spend a fair amount of time discussing stochastic and deterministic disturbances in subsequent sections.

A feedforward controller might be designed to control this latter kind of disturbance. Figure 1-6 shows how one might construct such a controller. Again, the reader must use her imagination here, but assume there is some way to measure the drain valve position and that there is some sort of algorithm in the feedforward controller that adjusts the inlet pipe valve appropriately whenever there is a change in the drain valve.

As before we need to generalize and abstract the concept so Fig. 1-7 shows a block diagram of the feedforward concept. The input to the feedforward controller is the measurement of the disturbance D. The output of the feedforward controller is signal U designed to somehow counteract the disturbance and keep the process output Y satisfactorily near the set point. Unlike the feedback controller, the feedforward controller does "see" the disturbance. However, it does not "see" the effect of the control output U on the process output Y. It is, in effect, operating blindly with regard to the consequences of its actions.

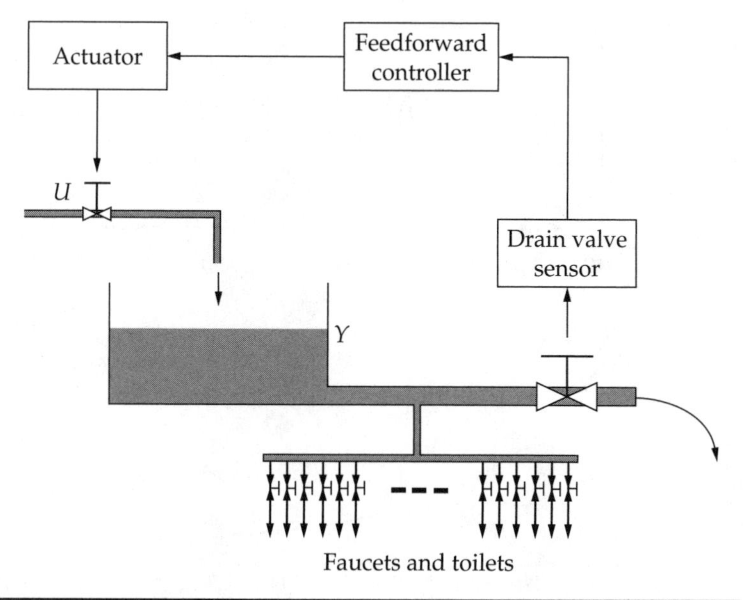

FIGURE 1-6 A feedforward controller.

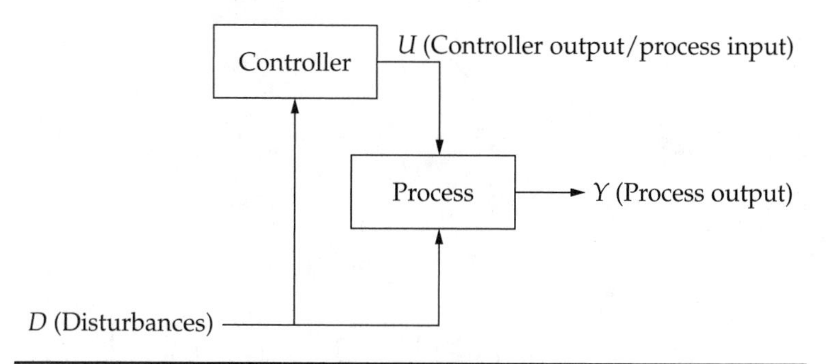

FIGURE 1-7 Feedforward controller block diagram.

1-3 Process Disturbances

Referring back to Fig. 1-5, the tank on the hotel roof, let's spend some time discussing the impact of the faucets, the toilet flushings, and the drain valve on the tank level. First, consider the response of the tank level to a step change in the drain valve position. That is, we suddenly crank the drain valve from its initial constant position to a new, say more open, position and hold it there indefinitely. Figure 1-8 shows the response. This kind of a disturbance is considered *deterministic* because one would usually know the exact time and amount of the valve adjustment.

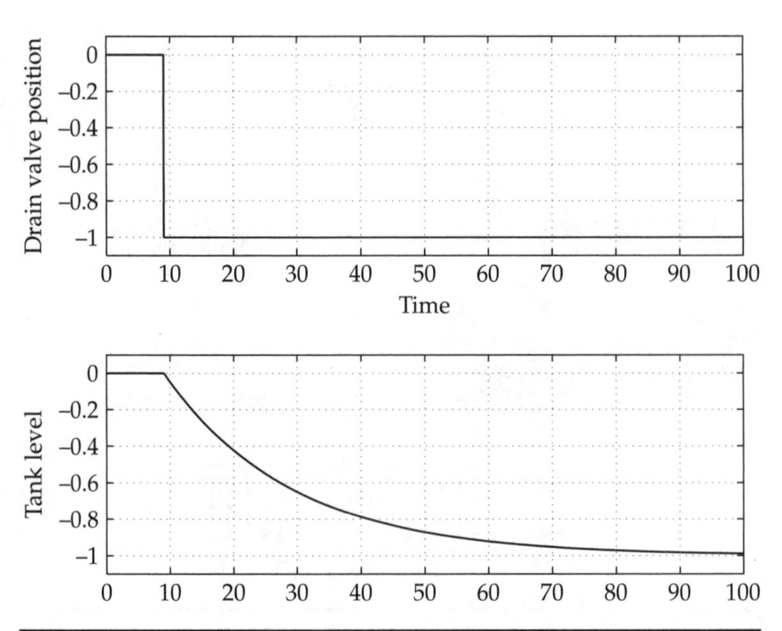

FIGURE 1-8 Response to a drain valve disturbance.

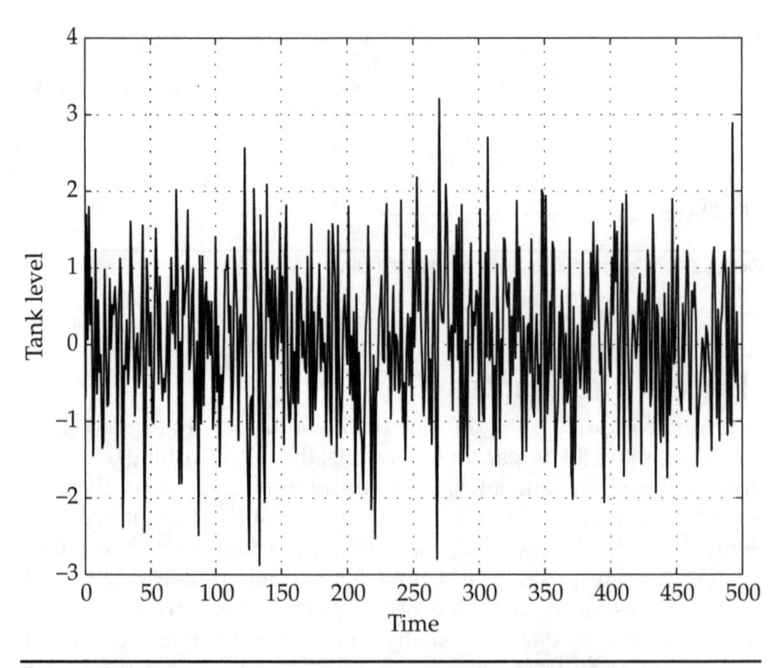

FIGURE 1-9 Variation of tank level due to unpredictable actions.

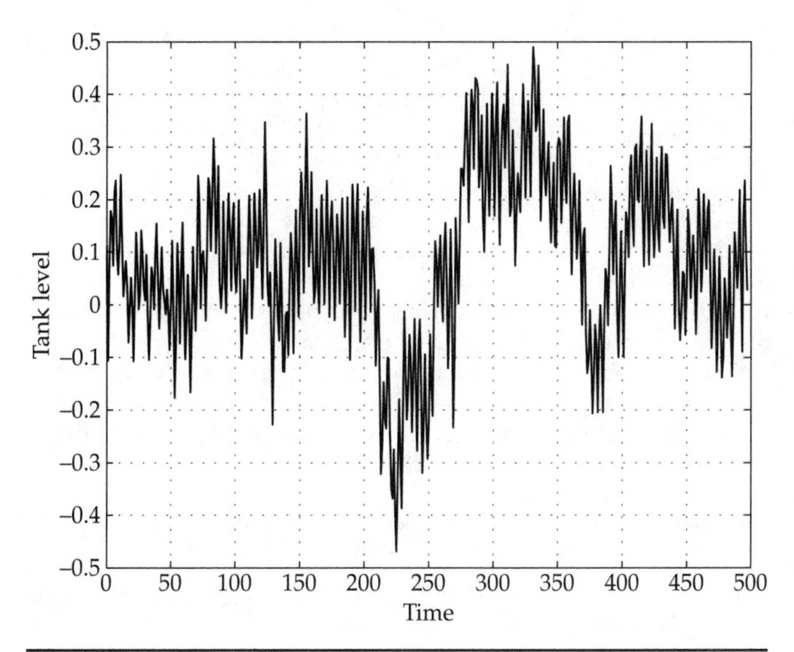

FIGURE 1-10 Autocorrelated stochastic variation of the hotel tank level.

When the flushing of the toilets and the usage of the faucets in the rooms is completely unpredictable and independent of each other, the tank level variation might look like Fig. 1-9. For the time being we will refer to these kinds of fluctuations as *unautocorrelated stochastic disturbances* where the word "stochastic" means conjectural, uncertain, or unpredictable. We will avoid using the word "random" because of the many confusing connotations. Also, we will defer the definition of "unautocorrelated" until a later chapter. If the stochastic variation is *autocorrelated,* the hotel tank level might look like Fig. 1-10.

Later on in Chap. 8, a significantly more quantitative definition will be attached to these two kinds of disturbances and we will find out how to characterize them. For the time being, suffice it to say that unautocorrelated disturbances are stochastic variations with a constant average value while autocorrelated disturbances exhibit drift, sometimes with a constant overall average and sometimes not.

1-4 Comparing Feedforward and Feedback Controllers

The feedforward controller can act on a measured event (such as the drain value position) before it shows up as a disturbance in the process output (such as the tank level). Unfortunately, the feedforward controller has no idea how well it did. Furthermore, it is often rather difficult to measure the disturbance-causing event. Sometimes there will be many disturbance-causing events, some of which cannot be measured. Also, it is not always clear how the algorithm should react to the measured disturbance-causing event. Often, each feedforward control algorithm is a special custom application. Finally, if perchance, the feedforward control algorithm acts mistakenly on a perceived disturbance-causing event it can actually generate a more severe disturbance.

The feedback controller cannot anticipate the disturbance. It can only react "after the damage has been done." If the disturbance is relatively constant there may be a good chance that the feedback controller can slowly compensate for it and perhaps even remove it. As we will show in the next couple of pages, there are some disturbances that simply should be left alone. The feedback controller can tell how well it has been done and it can often react appropriately. Unlike the case with feedforward control algorithms, there are a few well-known, easily applied feedback control algorithms that, under appropriate conditions can deal quite effectively with disturbances.

Question 1-1 Can a set point change be considered as a disturbance? If so, could it be used to easily test a feedback controller?

Answer Yes, to both questions. Changing a set-point is a repeatable test for evaluating the tuning of a feedback controller.

1-5 Combining Feedforward and Feedback Controllers

Figure 1-11 shows how feedforward and feedback controllers can be combined for our hotel example and Fig. 1-12 shows an abstraction of

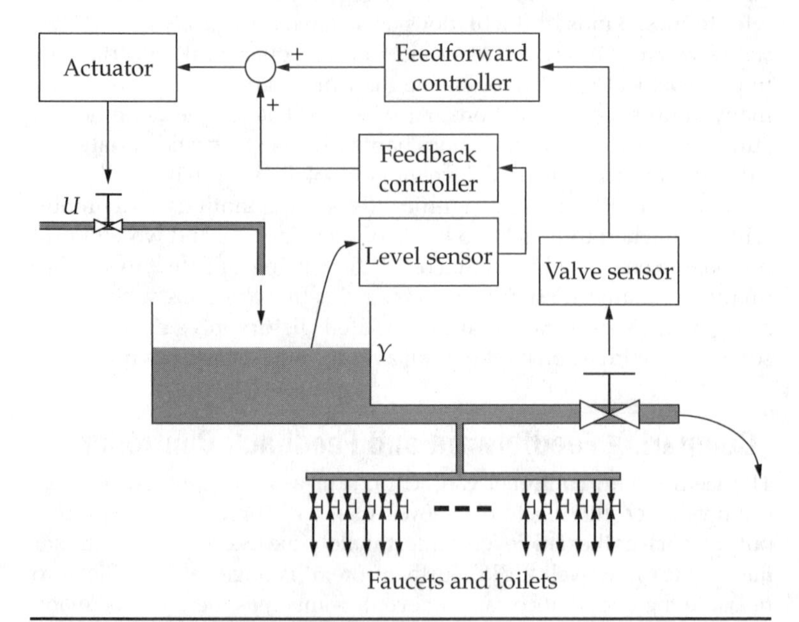

FIGURE 1-11 A feedforward/feedback controller.

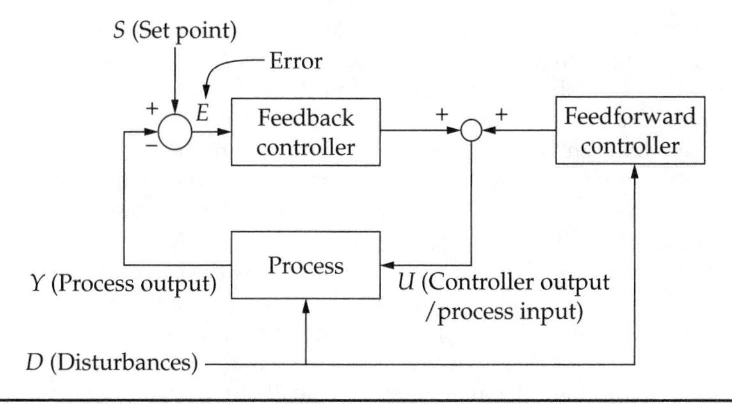

FIGURE 1-12 A feedforward/feedback controller block diagram.

the concept. The outputs of the feedforward and feedback controllers are combined at a summing junction and fed to the valve actuator. This scheme has the advantage of being able to react, via the feedback controller, to any unmeasured and unpredictable (stochastic) disturbances, such as the faucets and toilets, as well as to inaccuracies in the feedforward controller algorithm should it be needed during a swimming pool filling. The feedforward algorithm can provide anticipation for the feedback algorithm while the feedback algorithm can provide a safety net for the feedforward algorithm.

1-6 Why Is Feedback Control Difficult to Carry Out?

Depending on the type of disturbances, feedback control can be difficult to carry out. To illustrate this point, consider the act of driving an automobile. The left-hand side of Fig. 1-13 shows that driving a car is a skillful combination of feedforward and feedback control with a

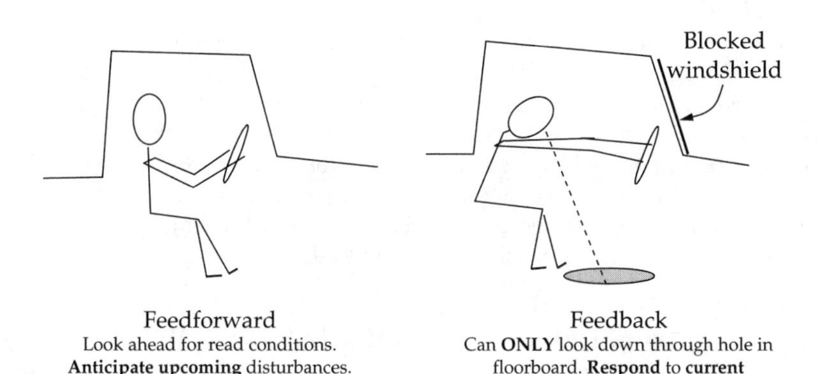

FIGURE 1-13 Comparison of feedback and feedforward control.

strong emphasis on the feedforward component. At the risk of oversimplification, driving a car depends heavily on the driver looking ahead, noting changes in the road and traffic, anticipating disturbances, and making adjustments in the steering wheel, gas pedal, and brake pedal. The actions taken by the driver are the result of many months and sometimes years of training and constitute a human feedforward algorithm. There are human feedback components to these feedforward adjustments but they are mostly corrections for inaccuracies (hopefully small) in the training and experience that constitutes the human feedforward algorithm.

If the automobile were to be driven exclusively by feedback control, the right-hand side of Fig. 1-13 shows that the driver could not look out through the windshield. Instead, the driver must make adjustments based only on information gathered by looking at the road through a hole in the floorboard. This kind of restriction would force the driver to maintain a slow speed. Here the driver is carrying out feedback control and is able to *react* only to current disturbances and has no information on upcoming disturbances.

Consider the case of driving down the center of the road by following the white line as seen through the hole in the floorboard in the face of strong gusting crosswinds. Since this is a hypothetical question, put aside the obvious fact that this activity would be illegal and dangerous. One can surmise that a strategy of reacting *aggressively* to short-term random bursts of wind to keep the white line precisely in the center of the floorboard opening would probably put the car off the road. Instead, because the disturbances are not constant but unpredictable, the driver's best strategy might be to conservatively adjust the steering wheel to keep the white line, *on the average,* "near" the center of the floorboard opening and tolerate a reasonable amount of variation. Therefore, rather than react to *short-term* variations, the driver would have to be content with addressing *long-term* drifts away from the white line. I recently drove from New York to Colorado and back. I found myself reacting to sustained bursts of crosswind in a feedback mode. Therefore, the arguments of this section suggest that the sustained bursts of crosswind might not be classified as unautocorrelated.

Based on this rather extreme example, we can perhaps conclude that using feedback control on a noisy industrial process will probably not produce perfect zero-error control. Since feedforward control is rarely available for industrial processes, if one really wants to decrease the impact of short-term nonpersistent disturbances, he must actually "fix" the process, that is, minimize the disturbances affecting the process.

1-7 An Example of Controlling a Noisy Industrial Process

To illustrate the impact of feedback control on noisy processes, consider a molten glass delivery forehearth shown in Fig. 1-14. Since the reader may not have a glass-manufacturing background, a little

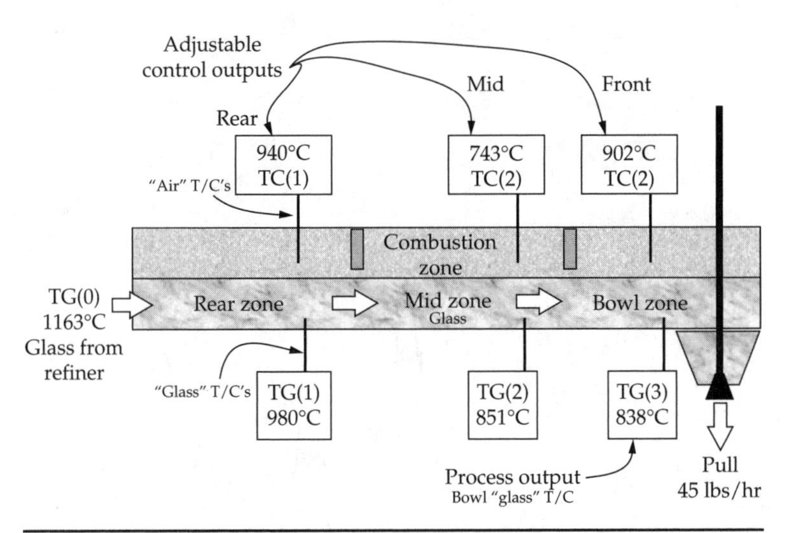

FIGURE 1-14 A molten glass forehearth.

explanation of the process depicted in Fig. 1-14 is necessary. The forehearth is a rectangular duct made of refractory material about 1 ft wide, about 16 ft long, and about 6 in deep. Molten glass at a relatively high temperature, here 1163°C, enters the forehearth from a so-called refiner. The forehearth is designed to cool the glass down to a suitable forming temperature, in this case 838°C. There is a gas combustion zone above the glass where the energy loss from the glass is controlled by maintaining the gas (*not the glass*) temperatures at desired values via controllers, the details of which we will gloss over for the time being.

There are three zones: the rear, mid, and bowl. In each zone, the gas combustion zone temperature above the glass is controlled by manipulating the flow of the air that is mixed with the natural gas before combustion. The amount of gas drawn into the combustion zone depends on the amount of air flow via a ventura valve. In the rear zone, a master control loop measures the TG(1) glass temperature (as measured by a thermocouple inserted into the molten glass) and adjusts the set point for a second loop, called a slave loop, which controls the gas combustion zone temperature TG(2) by in turn manipulating the flow of combustion air. There is a similar pair of control loops in the midzone and the bowl zone. In Chap. 11, we will treat this combination of two control loops, called a *cascade* control structure, in detail.

Therefore, in each zone the control challenge is to adjust the combustion zone temperature set point so as to keep the bowl temperature TG(3) sufficiently close to 838°C. It is a tough task. The incoming glass varies in temperature, the manufacturing environment ambient

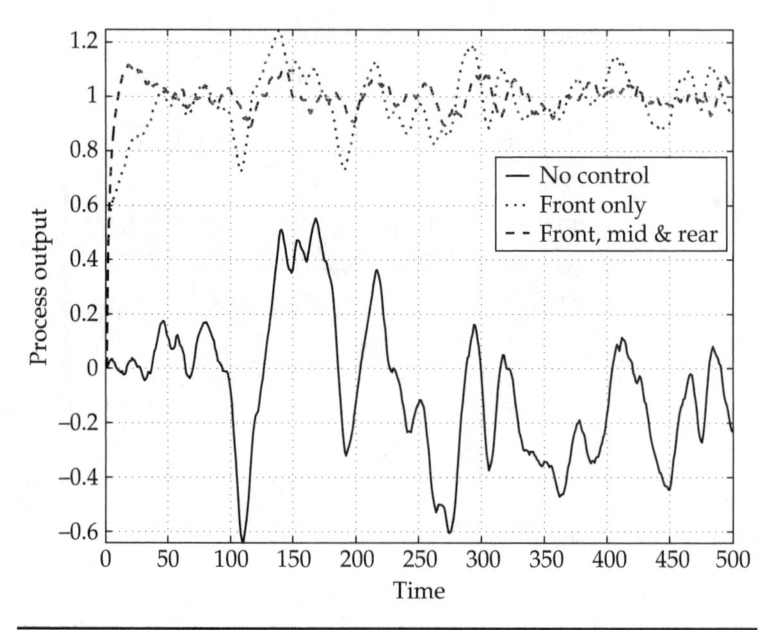

FIGURE 1-15 Control of a molten glass forehearth.

temperature varies because of drafts, and there are variations in the "pull" or glass flow rate. These disturbances are manifested in "noisy" TG(1), TG(2), and TG(3) temperature values.

Figure 1-15 shows a time trace of TG(3) for three cases: (a) no zones under control, (b) front zone–only under control, and (c) all three zones under control. The nominal values of the temperatures have been normalized by subtracting a constant value. A temperature value of 1.0°C in Fig. 1-15 represents the desired 838°C. A temperature value of 1.5°C in Fig. 1-15 represents 838.5°C.

Satisfactory glass forming requires that the bowl temperature varies no more than about 0.3°C. For no control, the TG(3) temperature in Fig. 1-15 shows significant excursions beyond the desired limit and the average value is nowhere near the desired value of 1.0°C. Figure 1-16 shows a closer view of the TG(3) temperature when under the two control schemes. Having all three zones under control is better than having only one but, even with all zones in control, the TG(3) trace still exhibits noise or disturbances.

To further remove variation, the emphasis probably should be placed on decreasing the variation of the glass entering the forehearth from the refiner and on environmental variation. To illustrate the idea that the controller could in fact drive the process output to set point if it were not for the noise and disturbances consider Fig. 1-17. Near the middle of the simulation (at time $t = 250$) I have magically removed the disturbances and I have changed to set point to 1.0. Notice that, in

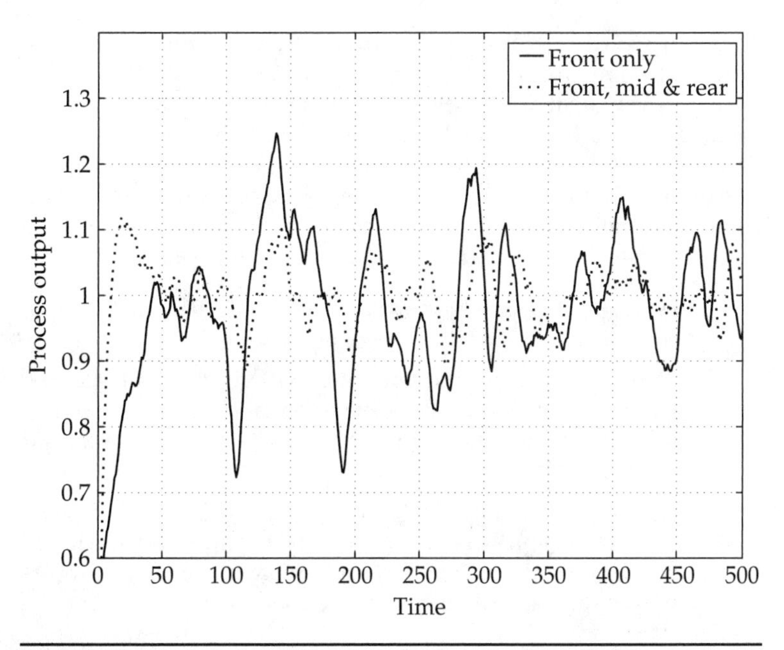

Figure 1-16 Control of a molten glass forehearth, alternative view.

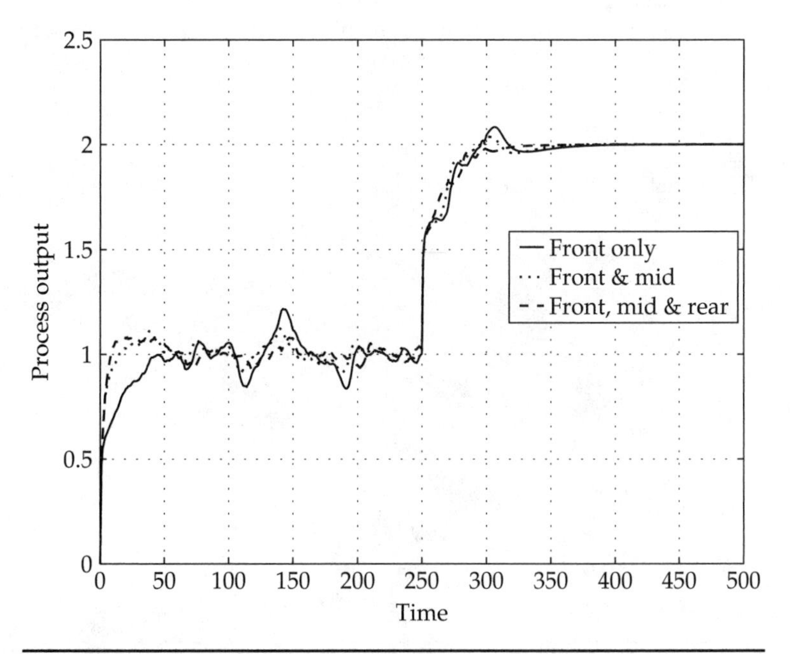

Figure 1-17 Effect of removing disturbances.

the absence of disturbances, the controller drives the process variable to new set point quickly.

Before leaving this example, we should make a few comments on the nomenclature associated with the disturbances discussed. In Chap. 8, we will discuss how to quantitatively characterize these disturbances but for the time being consider the noise riding on the temperature signals in Fig. 1-15 as an example of a stochastic variation.

We perhaps can conclude the following from this example:

1. When process is subject to stochastic disturbances, feedback controllers can *not* "draw straight lines."
2. Although there may be some attenuation, Disturbances In → Disturbances Out. As we shall see later on, the process itself may tend to attenuate input disturbances. Controllers can aid in the attenuation.
3. Controllers can move the process to a neighborhood of a new set point. Controllers may not be able to "draw straight lines" but they may be able to move the *average* value of the process output satisfactorily near a desired set point.

Figure 1-18 gives a pictorial summary of the above comments. When confronted with set-point changes in the face of relatively small stochastic disturbances a feedback controller can be extremely useful. If one is so lucky to have good measurements on incoming streams that represent disturbances to the process, feedforward control coupled with feedback control probably is a good choice.

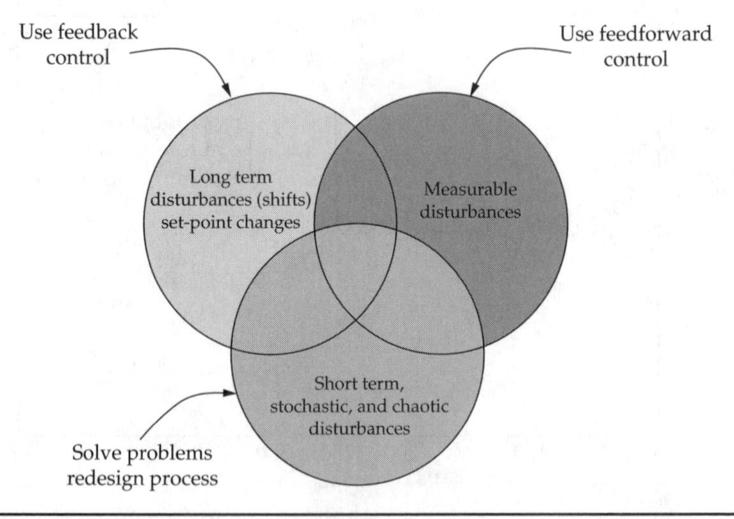

FIGURE 1-18 Different approaches for different problems.

Finally, if the main challenge is trying to maintain a process output satisfactorily near a set point in the face of persistent stochastic disturbances then the best approach probably should be the formation of a problem-solving team to deal with both the process and the environment.

1-8 What Is a Control Engineer?

So far we have implied that a control engineer designs control algorithms. In fact, the title of control engineer can mean many things. The following list, in no particular order, covers many of these "things":

1. Installer of control/instrumentation equipment (sometimes called an "instrumentation engineer"): In my experience this is the most prevalent description of a control engineer's activities. In this case, the actual design of the control algorithm is usually quite straightforward. The engineer usually purchases an off-the-shelf controller, installs it in an instrumentation panel, probably of her design, and then proceeds to make the controller work and get the process under control. This often is not trivial. There may be control input sensor problems. For example, the input signal may come from a thermocouple in an electrically heated bath of some kind and there may be serious common and normal mode voltages riding on the millivolt signal representing the thermocouple value. There may be control output actuator problems. There may be challenging process dynamics problems, which require careful controller tuning. In many ways, instrumentation engineering can be the most challenging aspect of control engineering.
2. Control algorithm designer: When off-the-shelf controllers will not do the job, the scene is often set for the control algorithm designer. The vehicle may be a microprocessor with a higher-level language like BASIC or a lower-level language like assembly language. It may even require firmware. Many control/instrumentation engineers fantasize about opportunities like this. They have to be careful to avoid exotic custom undocumented algorithms and keep it simple.
3. Process improvement team member: Although this person is trained in control engineering, success, as we shall see in Chap. 2, may result from solving process problems rather than installing new control algorithms.
4. Process problem solver: This is just a different name for the previous category although it may be used when the team members have developed a track record of successes.

1-9 Summary

Compared to the rest of the book, this chapter is a piece of cake—no equations and a lot of qualitative concepts. Hopefully, we have laid the foundation for feedback and feedforward control and have shown how difficult they can be to apply especially in the face of process disturbances. The next chapter will retain a qualitative flavor but there will be hints of the more sophisticated things to come. Good luck.

CHAPTER 2

Introduction to Developing Control Algorithms

Before embarking on the quantitative design of a control algorithm it is important to step back and consider some of the softer issues. What kind of approaches might a control engineer take? What kind of up-front work should be done? Is there a difference when dealing with an existing process as compared to bundling a process with the control algorithm and selling the package?

2-1 Approaches to Developing Control Algorithms

Each control/process analysis project is unique but every strategy that I have been involved with has components from the following three approaches.

2-1-1 Style, Massive Intelligence, Luck, and Heroism (SMILH)

In a stylish manner, the engineer speculates on how the process works, cooks up a control approach, and somehow (heroically) makes it work, at least on the short-term. Massive intelligence not only helps but it usually is essential. A massively intelligent person, using the SMILH approach, can, sans substance, exude style and confidence sufficient to overcome any reservations of a project manager. Because this engineer has avoided a couple of methods to be mentioned further, the project will likely experience setbacks and a wide variety of troubles. The successful SMILHer will use these problems as opportunities to show how heroically hard he can work to overcome them. I have always been amazed at the number of managers who can pat the heroic SMILHer on the back for his above-and-beyond-duty hard work and never ask the fundamental question: "Why does this engineer have to resort to such heroics ?" Over the years I have worked with scores of SMILHers. One of the first ones, a great guy named Fred, was the hardware designer while I was the algorithm/software guy. I would program the minicomputer in some combination of FORTRAN and assembly language to (1) act on the inputs served up by Fred's hardware and to (2) send the commands to the output drivers, again provided by Fred. Fred also designed the electrical hardware to connect the operator's panel to the computer. Our trips to the customer's plants had a depressing similarity. We would fire up the system, watch it malfunction, and then I would proceed to find ways to amuse myself, sometimes for days, while Fred dug into the hardware to fix the problems. He was indeed heroic, often putting in "all-nighters." Fred never upset the project manager who thought the world of him . . . actually, as did everyone, including myself. Nobody ever asked "Fred, why don't you do a more thorough job of debugging the system before it goes out to the field or a better job of design in the first place?" Fred went on to be a successful manager.

2-1-2 A Priori First Principles

Some processes invite mathematical modeling up front. The idea, often promulgated by an enlightened (or at least trying to appear enlightened) manager, requires that some mathematically gifted engineer develop a mathematical model of the process based on first principles. Proposed algorithms are then tested via simulation using the mathematical model of the process. This approach is extremely attractive to many people, especially the mathematical modeler who will get a chance to flex his intellectual muscles. Early in my career this was my bag. In retrospect, it makes sense that I would be relatively good at it. I was fresh out of graduate school and knew practically nothing about real-life engineering or manufacturing processes but I did know a little mathematics and I was quite full of myself—a perfect combination.

Success depends mostly upon the style with which the modeler applies himself and presents his results. Many times I have seen beautiful computer graphics generated from modeling efforts that, when stripped of all the fanfare, were absolutely worthless . . . but impressive. Later on, if the algorithm does not work as predicted by the modeling there were always a host of excuses that the modeler could cite.

At least in my experience, a priori mathematical modeling, especially transient time domain modeling, is almost always a waste of time and money. The real goal of this approach should be the gaining of some unexpected insight into how the process works. Unfortunately, mathematical modeling rarely supplies any unexpected performance characteristics because the output is, after all, the result of various postulates and assumptions put together by the modeler at the outset. Often one could just look at the basis for the model, logically conclude how the process was going to behave and develop a control approach based on those conclusions without doing any simulation.

Far more frequently, a priori mathematical modeling simply is not up to the task. Most industrial processes are just too complex and contain too many unknown idiosyncrasies to yield to mathematical modeling. I have more to say on this problem in Sec. 2-4.

2-1-3 A Common Sense, Pedestrian Approach

If the process exists and is accessible, the control engineer adds extensive instrumentation, studies the process using the methods presented next, and, if necessary, develops an algorithm from the process observations.

When the process is not accessible, one makes a heavily instrumented *prototype* of the process and develops a control algorithm around the empirical findings from the prototype.

Alternatively, if it is a new process, yet-to-be-constructed, and a prototype is not practical, the engineer negotiates for added monitoring instrumentation. In addition and, even more difficult, he negotiates for up-front access to the process during which planned disturbances will be carried out so that one can find out how the process actually works dynamically. During this up-front time, many unexpected problems can be discovered and solved. The control algorithm vehicle, usually digitally based, is designed with extensive input/output "hooks" for diagnosis. Finally, the control algorithm is designed around these findings. I have frequently made mathematical models based on the *empirical* evidence gathered during these up-front trials.

This approach is significantly more expensive in the short-term and often violently unpopular with project managers. I have consistently found it to be a bargain in the long-term. There is some style required here; the engineer must convince the management that the extra instrumentation and up-front learning time is required. Junior control engineers usually are not aware of this approach—mostly because they have not yet experienced the disasters associated with SMILH and a priori methods. But, even if they are aware they usually cannot convince a seasoned project manager about the benefits of taking a pedestrian approach simply because they haven't a track record of success in this area.

If the process for which the control algorithm is to be developed already exists then this empirical approach is really the only valid choice IMHO. Since this case is so prevalent and special it will command a whole next section.

2-2 Dealing with the Existing Process

Consider the following scenario. A section supervisor in a manufacturing plant is not satisfied with the performance of the process for which he is responsible. The end-of-line product variance is too high. Thinking that the solution is more or better process control, he calls in the control engineer.

2-2-1 What Is the Problem?

Although most engineers working in a manufacturing environment are formally trained in problem solving, they almost uniformly bypass the most important first step, which is to clearly and exhaustively define the problem. The number of manufacturing-plant section supervisors that I have irritated by persistently and perhaps obnoxiously asking this question seems countless. They often do not want to be bothered by such nonsense. After all, they know more about the process than some staff engineer from headquarters and have already figured out that there is a need for a control upgrade . . . now, just get busy and do it!

Early in my career I obediently plowed ahead and did the project manager's bidding. There were some successes—at least enough to keep me employed—but there were enough failures that I was basically forced to develop the so-called road map for process improvement shown in Fig. 2-1 and discussed in great detail in the following sections.

Before jumping into the approach championed in this chapter, it is critical to convince the project sponsor/manager to develop a *team* containing the control engineer as a member. This team should be diverse, not necessarily in the politically correct ethnic manner, but in the technical strengths of the members. There is little point in fostering competition so only one member of each important discipline should be present. Furthermore, the control engineer need not be the leader; in fact, in my experience it is better to have someone with more leadership skills than technical skills in that position.

2-2-2 The Diamond Road Map

Figure 2-1 shows a diagram containing four corners of a diamond but really consisting of many steps.

Compartmentalization and Requirements Gathering

This is a fancy phrase that simply means, "divide and conquer." Manufacturing processes are almost always complex and consist of many parts, steps, and components. Breaking the process down into *all* of its components or *dynamic modules* is the first step in getting a handle on improving the performance.

Our method will attempt to decrease the variance of the process variables *local* to each module with apparent disregard for the end-of-line performance. Once each module becomes more controllable, the targets for each module can be adjusted more precisely to affect the end-of-line performance beneficially. One usually finds that even without changing the targets of the improved modules, the decreased local variance tends to have a salutary effect on the end-of-line product characteristics. Therefore, there are three benefits to the localization. The *first* benefit is better control, allowing the local set point to be adjusted with the confidence that the module will actually operate

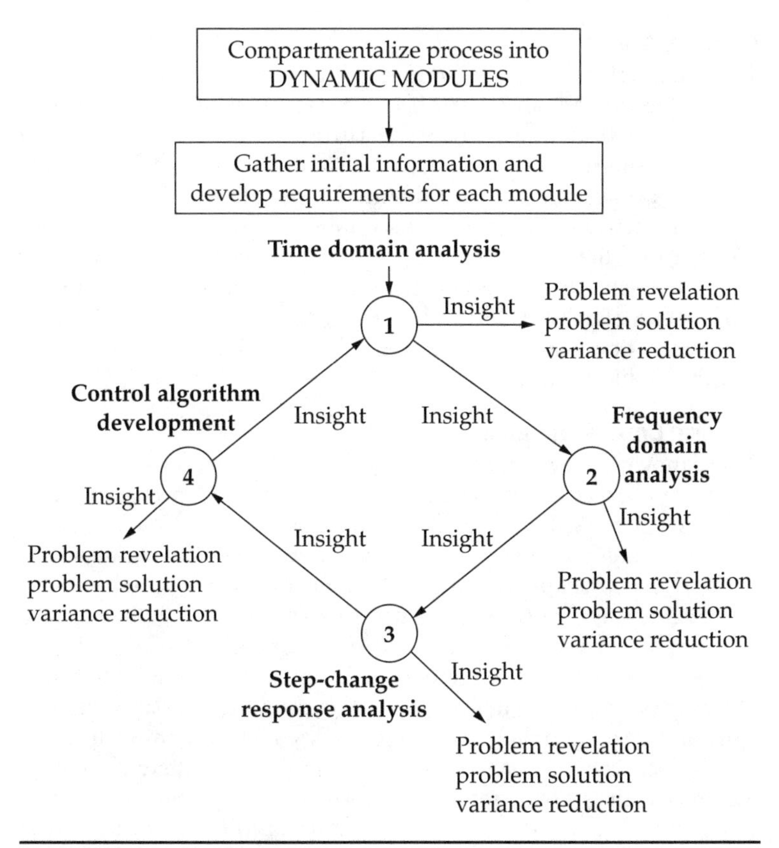

FIGURE 2-1 The diamond road map.

at or satisfactorily near that set point. The *second* benefit is the impact of less variance in that module on the downstream end product. The *third* benefit is that once the module is put under control, the set point can be adjusted to optimize the end-of-line product.

Gathering information about each module, especially its performance requirements, is often the most difficult step. What defines "good performance" for each module? At the end of the manufacturing process where the product emerges, good performance is relatively easy to define. But as you move back into the process this can become quite difficult. There may be no measurements available for many of the "interior" or upstream modules in the process. How do you know if it is performing properly? Could this module be a big player in the observed poor performance at the end of the process? To make sense of these studies one must have a reference point that describes the satisfactory behavior of the module in quantitative terms. In subsequent sections we will discuss in detail methods for studying the performance of a module.

Where to Start?

This is a tough question. Sometimes it is best to start near the product end of the process and work back upstream, especially if analysis suggests that the local variance seems to be coming from the upstream modules. Alternatively, one might start at the most upstream module and work down. In this case the impact of solving problems in an upstream module may not be discernible in the downstream modules because there has been no previous reference point. Finally, it may make sense to start where the hands-on process operators think the most problems are. It's always good practice to include the hands-on process operators in the strategy development, the data review, and the problem-solving activities.

Massive Cross Correlation

Before moving on with the road map, we should make a few comments about an alternative *complementary* and popular approach to process problem solving—the "product correlation approach." Here one cross-correlates the end-of-line performance characteristics with parameters at any and all points upstream in an attempt to find some process variable that might be associated with the undesirable variations in the product. This can be a massive effort and it can be successful. However, I have frequently found that plant noise and unmeasured disturbances throughout the process and its environment will corrupt the correlation calculations and generate many "wild goose chases." Often an analyst will stumble across two variables, located at significantly different points in the process, that, when graphed, appear to move together suggesting a cause and effect. Unfortunately, in a complex process there are almost always going to be variables that move together for short periods of time and that have absolutely no causal relationship.

Figure 2-2 shows a hypothetical block diagram of a complex process. The end-of-line product is the consequence of many steps, each of which can suffer from noise (*N*), disturbances (*D*), and malfunctions (*M*). A massive cross-correlation might easily show several variables

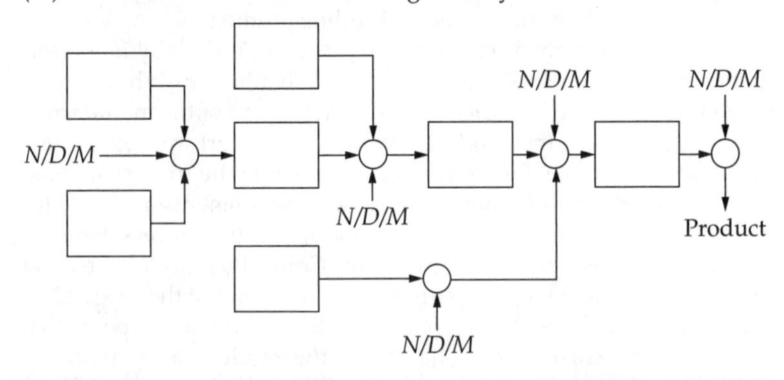

Figure 2-2 A complex process with many sources of noise (*N*), disturbances (*D*), and malfunctions (*M*) .

located at various points in the block diagram that have similar short-term trends due to these disturbances and malfunctions. A good project manager can have both approaches active and complementary.

Time Domain Analysis

Now that a module has been identified and the specifications gathered, it is time to "look" at the process in the simplest most logical way—in the time domain. This means collecting data on selected process variables local to the module and studying how they behave alone and when compared to each other. Before starting to collect the data the team should agree on the key process variables to collect and on what frequency to sample them. This may require installing some new sensors and even installing some data-acquisition equipment.

Decades ago, the only source of data was the chart recorder. Nowadays, most processes have computer-based data-acquisition systems, many of which not only collect and store the data but can also plot it online. These systems can also plot several process variables on the same graph. The opportunities to look at the process dynamics in creative ways are nearly endless. Use your imagination.

Gaining insight and solving problems are the primary goals of the activities associated with each of the four corners of the diamond in Fig. 2-1. The time domain plots will likely reveal problems that should be solved by the team (as soon as possible) thereby reducing variation in the local process variables connected with the module. Reducing variance *locally* is the immediate challenge. Do not worry about the impact of these activities on the end-of-line product variance. That will come later.

Frequency Domain Analysis

Once the time domain analysis/problem revelation/problem solving has begun, it often makes sense to look at the process module in some other domain. The road-map diagram shows a second corner labeled "Frequency domain analysis." Here, without going into too much technical detail, one uses Fast Fourier Transform software to develop line spectra or power spectra for selected variables. Essentially, long strings of time domain process data are transformed to the frequency domain where sometimes one can discover heretofore unknown periodic components lurking in noisy data. Few computer-based data-acquisition/process-monitoring systems have the frequency domain analysis software built in, so the engineer will have to find a way to extract the desired process variables and transfer them to another computer, probably off-line, for this type of analysis.

Figure 2-3 shows a long string of time domain data for a process variable. The variable was sampled at a rate of 1.0 Hz (or every second). In the time domain it simply looks noisy and seems to drift tightly around zero (perhaps after the average has been subtracted). When transformed into the frequency domain, Fig. 2-4 results. Here the

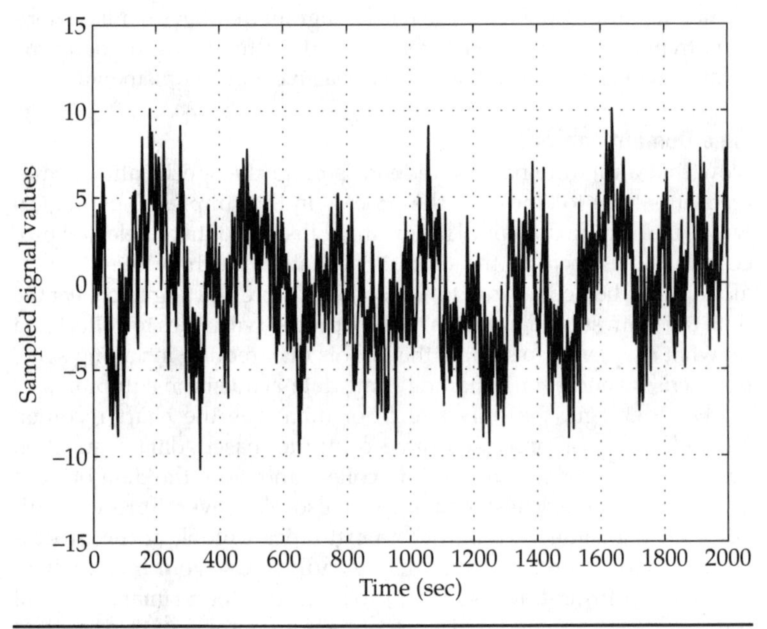

FIGURE 2-3 Sampled noisy signal.

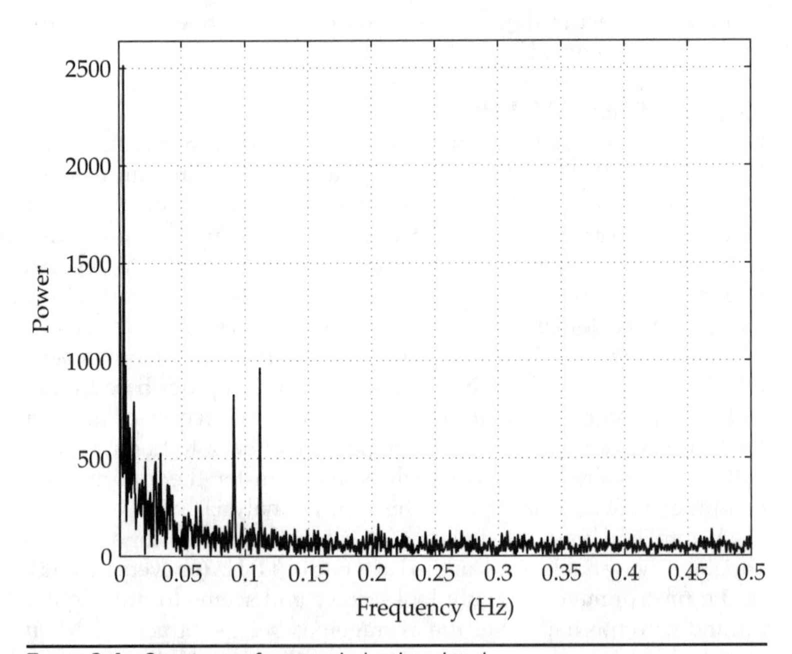

FIGURE 2-4 Spectrum of a sampled noisy signal.

signal power at all the frequencies between 0.0 and 0.5 Hz is plotted versus frequency (see App. C for information on why the frequency is plotted only up to 0.5 Hz). Strong peaks occur at frequencies of 0.091 Hz and 0.1119 Hz, suggesting that buried in the noisy signal are periodic components having a periods of 1/0.091 = 11 sec and 1/0.119 = 8.9 sec. Warning: These two periodic signals could also be *aliases* of higher-frequency signals (see App. C for a discussion of aliasing). Additionally, there is power at low frequencies (less than 0.05 Hz) as a consequence of the stochastic drifting about an average of zero.

If this were real process data it would now be up to the team to collectively figure out where these unexpected periodic components were coming from. Are they logical consequences of some piece of machinery that makes up the manufacturing process or are they symptomatic of some malfunction not immediately obvious but about to blossom into a major problem? In any case they may be significantly contributing to the variance of the local process variable and there may be good reason to remove their source and lower the local variance.

In App. C the power spectrum is discussed in more detail. There, the reader will find that the area under the power spectrum curve is proportional to the total variance of the process variable. Therefore, portions of the frequency spectrum where there is a significant amount of area under the power spectrum curve merit some thought by the process analyst. That appendix will also discuss why only the powers of signals with frequencies between 0.0 and 0.5 Hz (half of the sampling frequency) are plotted.

The data stream should be relatively stable for the frequency domain analysis to be effective. For example, the data analyzed above varies noisily but is reasonably stable about a mean value of zero. Data streams that contain shifts and localized excursions will yield confusing line spectra and may need some extra manipulation before analysis begins.

As with the first corner of the diamond dealing with time domain analysis (Fig. 2-1), the outputs from the frequency domain corner are problem revelation and insight. Should there be problems revealed and then solved, the local variance will be reduced and the module will be more under control.

Step-Change Response Analysis

The first two corner activities provide insight and problem revelation based on noninvasive observation. Sometimes this is not enough. Sometimes, to get enough insight into a process to actually control it, one must intervene. This is where the step-change response analysis comes in.

First of all, the problem-solving team should make a hypothesis regarding what they expect to see as a step response. Then, to carry out the experiment properly, the engineer must turn off any of the

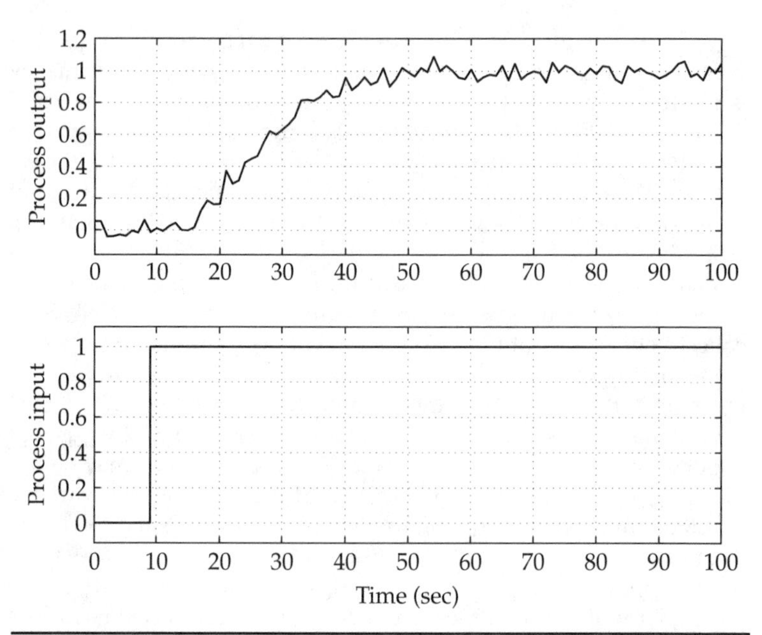

FIGURE 2-5 A typical step-change response.

existing control loops that have any effect on this module. Once the process settles to some acceptable approximation of steady state, the engineer makes an isolated step change to a process input variable. Since this is usually the manipulated variable in a control loop, the step change can often be easily made via the actuator.

Figure 2-5 shows a typical step-change response. Both the process input and output variables have been normalized to lie within the range of 0 to 1 for simplicity. The team should consider if this step response is what they expected. If it is not, then they may have found a problem. Later on, when we talk about the first-order process, the characteristics of this kind of plot will be examined.

Step-change analysis can be useful for at least two reasons. First, as suggested above, it can point out errors in the actuator system and malfunctions in the process. Second, it can give the control engineer valuable information on the dynamic characteristics of the process, which in turn can be used to develop the appropriate control algorithm.

Control Development

This corner of the diamond is where the control engineer has a chance to shine. It is where I used to start when I was inexperienced. Ironically, if the activities associated with the other corners are successful, the variance is often so significantly reduced that there is little need to concoct a sophisticated control algorithm. It has been my experience that a well-constructed process problem-solving

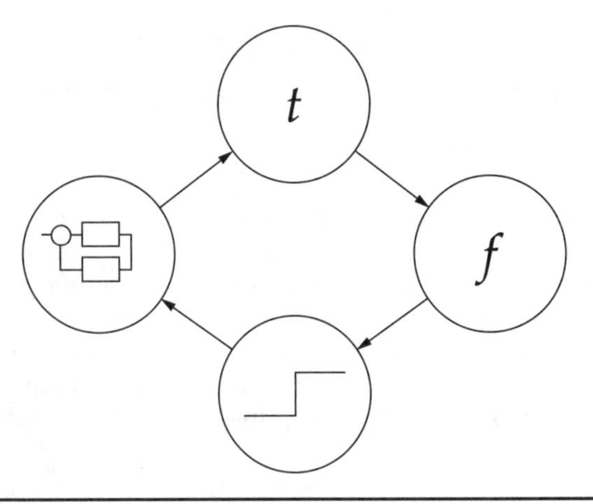

FIGURE 2-6 The diamond road map icon.

team can basically obviate the need for the activities in the "control development corner."

However, should an algorithm be developed and installed, one would move on to the top of the diamond and study the controlled process in the time domain. Having done this, one could continue around the diamond gaining insight and solving problems. For mnemonic purposes the diamond is symbolized in Fig. 2-6.

2-3 Dealing with Control Algorithms Bundled with the Process

What if you are selling a product such as an optical amplifier and you want to augment your product with a controller that will, say, maintain a desired optical output power? Now, you are bundling the process to be controlled with the controller and forming a product that contains two components. This is quite a different situation compared to that covered in Sec. 2-2.

What Is the Problem?

The product now has two components that can have problems: the process and the controller. If the product is constructed so that only the final output, say the optical power in the case of an optical amplifier, can be monitored then how do you diagnose problems? Is it the process or is it the controller?

Separation and Success

The key to success lies in designing the product with ports that will allow the problem solver to tap into internal signals, namely the controller input and the controller output. With these signals available, the problem solver can isolate the controller from the process.

Pay me Now... or... $e^{\text{Pay me Later}}$

FIGURE 2-7 The old Fram oil filter phrase.

How often does this happen? Rarely. It is difficult to convince a project manager to authorize the extra money to design-in the necessary ports. However, it is another example of "pay me now or pay me later" (a phrase from the perhaps famous Fram oil filter advertisements sometimes seen in the 1950s; Fig. 2-7). The money lost in designing the extra ports will pail in comparison to the costs of the engineering time required to diagnose and solve the problem.

Problem Solving with Bundling

The key to problem solving is having access to both the so-called process *and* the controller. If ports are in place then problem solving can be divided immediately into verifying that the process and the controller are performing properly. Without these ports problem solving becomes a guessing game (see Fig. 2-8).

There are ancillary benefits to having the extra ports. During product development, the ability to monitor the process and controller separately can allow for parallel beneficial development (see Fig. 2-9). If the learning about the process and controller are concurrent and interactive, their development can be also be interactive—leading to a synergism and a better final product.

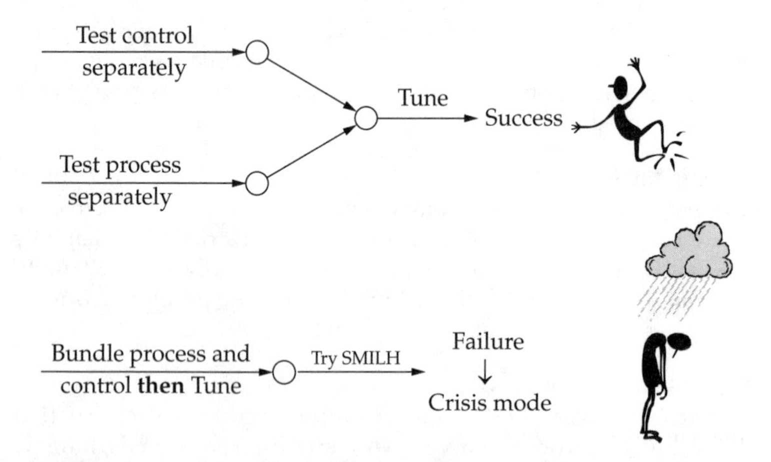

FIGURE 2-8 Bundling process and control as part of a saleable product—testing the process and the algorithm separately.

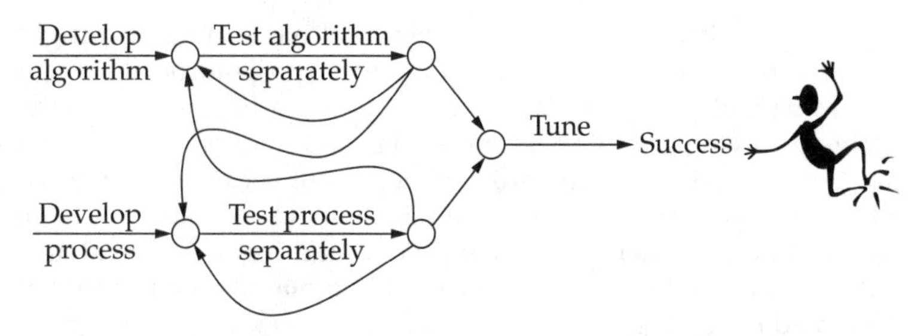

FIGURE 2-9 The benefits of separation: interaction, evolution, synergism, and problem isolation.

2-4 Some General Comments about Debugging Control Algorithms

This is a sore subject with a lot of engineers, yours truly included. Perhaps it's best to simply tell a couple of war stories.

Rookie Fright

I joined a large manufacturing company with only a couple of years of experience after leaving graduate school with a Ph.D. in chemical engineering. Although I had been into a plethora of hobbies and projects before college, my life as a professional student had been reclusive and narrow and I had no hands-on engineering experience, nor much interest in gaining any—hey, I was an applied mathematician (I thought)!

When in Doubt, Simulate—Not!

Given the comments of the above paragraph, I really was good for little other than generating sophisticated mathematical simulations. At my previous job I had been adept at making mathematical models of complicated processes, cooking up complicated algorithms for controlling the model, and using the process model to show how wonderfully the control algorithm would work.

The rationale for using simulation to develop a control algorithm is simple and, in my opinion, quite incorrect. This approach is flawed because the model basically contains the knowledge of the modeler and little else. When it is put through its paces, it will surprise no one. Furthermore, the model will likely not contain any of the subtle idiosyncrasies of the real process—idiosyncrasies that might defeat the control algorithm developed by using the model.

The motivation for using a mathematical model often comes from a manager who has little actual knowledge of mathematical modeling. He has, however, observed that suggesting mathematical modeling as a solution to some difficult process problem often comes across as an enlightened commandment.

The logic behind using a mathematical model to gain insight into a difficult process problem or develop a control algorithm should be faulty on its face. The model is going to be based on the problem-solving group's best knowledge and understanding of the process. Unfortunately, the problem is present because that knowledge and understanding is insufficient. In cases like these, the answer to the problem lies in the process and its discovery almost always requires disturbing the process in a manner similar to that presented in Sec. 2-2.

You, as a manager, should always carefully question the basis for your charges embarking on a mathematical simulation. It can waste an enormous amount of engineering time. I, as a mathematical modeler, never had any problem demonstrating how wonderful and useful my model was and I wasted a lot of the company's money doing it. On the other hand, using a mathematical model to debug a control algorithm is an entirely different matter—see in the following sections.

These last handful of paragraphs have in large part been a regurgitation of earlier material. Sorry, but it's really important.

At Last—Busted!

In my new job, I escaped practical projects for a couple of years but eventually I was given responsibility for putting in a new instrumentation control room for a new process. I was awed by my lack of ability in the hands-on aspect of instrumentation. Consequently I resorted to obsessive planning, preliminary debugging, and extensive use of competent resources (such as the hourly instrument technicians). The instrumentation included a data-logging computer (called minicomputers in those days). To cover myself, I broke all inputs and inserted test voltages to simulate real conditions. I also wrote little subroutines to generate fake signals to the data-logging algorithms. I found many problems and enlisted the aid of talented associates to solve them. The list of debugging tricks could go on but to make a long story short, when it came time for the process to start up, all my stuff worked and there was no opportunity for SMILH—which was a good thing because I simply was not cut out for that kind of thing anyway.

My "customers," the boys who designed and would run the process, were astounded. It was the early days of using computers to monitor industrial processes and they had been conditioned to bring their new processes in without the aid of the data-logging computer for a simple reason: The engineers doing the instrumentation hardware and software were experienced, talented, and confident that they could solve any problems that would come up so they avoided any extensive pre–bring-in debugging and left any problems that would crop up to SMILH.

Although I gained points with my boss over the success of the project, I concluded that large-scale instrumentation installation projects were not fun and did not use my skills properly. I proceeded to avoid them like the plague.

Surprise Sub

After several years of avoiding projects like the one described (and paying for it with an unimpressive rate of promotion), I got involved with a junior engineer, whom I will call Bill, as his mentor. Bill was formally trained as an electrical engineer and had morphed into a software engineer primarily engaged in programming the minicomputer-based data-acquisition and control systems. Without much prior experience, Bill was charged with the instrumentation and control responsibility for another new process. By this time our group had started installing analog output cards and using the minicomputer to close control loops and replace stand-alone analog controllers. I, as a semisenior, semiexperienced associate, was supposed to be available to Bill for questions and counsel.

Bill invited me over to the site once, well into the installation, and showed me around. Since he was programming the computer and since I had minimal knowledge of software systems, the tour was nothing more than a long cup of coffee. I really gained no idea of how he was doing, except that everything looked OK and Bill was cool. I heard no more from him. The process started up without incident until a few days into the operation when a major flaw in the process design was discovered and the boys who owned the process went back to the drawing boards for several months.

When the redesign was finished, three things were apparent. First, nothing topologically had changed with the process so all of Bill's work would supposedly still be applicable. Second, Bill's 5-year company anniversary had arrived and he had planned to spend his extra 3 weeks of vacation (5 weeks total) in Europe with a bunch of like-minded youngsters. So, even though the vacation coincided with the restart of the process he still was allowed to go because of all the trip commitments he had made. Third, because Bill's stuff had apparently worked during the first abortive bring-in, there was no need for his presence and Uncle Dave (*moi*) could simply drop by on the start up day and be available for questions, should there be any (unlikely).

This situation bothered me. I had no idea what Bill had done. So, even though the first process bring-in had been uneventful from an instrumentation point of view, I dropped by during the week before the second bring-in and proceeded to dig into the software. I found that there were several proportional-integral control loops, some connected in a cascade configuration (things we will talk about later).

To check things out I did the obvious. I put test signals into the computer and monitored the data-logging CRTs to see what the control loops did. To make another long story short, there was a plethora of errors, some serious (the cascade loops) and some minor (derived variables based on data-logging inputs). I could correct the FORTRAN source code but since I did not have the software system expertise to link everything together I had to create quite a brouhaha to get the necessary and scarce software resources in time for the bring-in. I was careful not to castigate Bill when he came home blithely exhausted and nearly penniless. Because of the stink I had made about getting software resources, his return was less than pleasant. Ironically, about 15 years later Bill became my boss for a short time—we got along swimmingly.

In retrospect, I was at fault as much as anyone. I should have consulted closely with Bill. Unfortunately, the approach of the software-engineering arm of the instrumentation and control group, of which Bill and I were members, had been to code, compile, link, and turn on the data acquisition and control software. If the computer did not crash at that point then they "shipped it." Unfortunately, this experience did not have a major impact on the managerial approach of the company's software-engineering arm and this kind of nonsense continued for quite a while.

Totally Covering Myself

After several more years of avoiding major installation projects and doing fun process analysis and control design projects in our manufacturing plants I stumbled onto a project in the research wing of the company where I got a chance to design and, with the help of a great team and a great team leader, install a nonlinear dead-time compensation control algorithm that was later patented. It was for a batch process where we started with a "blank" of material, heated it, and formed it into the final product. A batch run would take approximately an hour and consume the blank, which was quite expensive.

After some initial fumbling around we got to the point where I would cook up the latest modifications to the algorithm (it went through some 200 versions before we finally installed it in a manufacturing plant) on my desktop computer terminal as a FORTRAN (and later, C) subroutine. I would link the subroutine to a FORTRAN (and later C) test program of my design, also accessed by my desktop terminal, which simulated the inputs to the process and put the control algorithm through an approximate start up. The process model embedded in my test software was purposely crude and simple but in a qualitative sense it behaved roughly as the real process and it made my algorithm jump through the correct hoops.

After verifying that my algorithm would bring the simulated process up properly, I would e-mail the subroutine to my software team member, located at the research facilities. He would link it to the on-the-floor control minicomputer. Unfortunately, the control computer was used by many other scientists and the software guy always seemed to have trouble with the linking. It appeared that the only way to find out if the linking had gone well was to start the batch process up and see if things worked. (Some alarm bells should be going off in the reader's head about now.) With few exceptions, the linking never went well and we would frequently destroy an expensive blank.

To address this, I hooked up a PC to the analog inputs and outputs of the minicomputer on the research facility's floor and used the PC to crudely simulate the process. I think I did this with an early version of Quick BASIC. With this "process simulator" in place, we could test the linking without worrying about destroying a blank. Again, the process model in the PC was extremely crude but sufficient.

So, any algorithm that I cooked up went through a test simulation on my desktop, a test simulation using the on-the-floor process simulator, and finally, a test on the real process. We ruined no more blanks. It is important to note that the simulation test systems, either on my desk or on the floor, were not particularly sophisticated. I simply wanted to test the control algorithm on a rough approximation of the process to see if, at least, the algorithm was functionally correct. The process (and only the process) itself would tell us if the algorithm was any good at control.

We ended up taking the debugged control system to a manufacturing plant where it was installed without any problems—no SMILH required.

It's Too Complicated—Use the Process for Debugging

This topic is a sore point with many engineers, especially *moi.* Some of my best friends, people who I greatly respect, are staunch supporters of the idea that in some cases one just has to use the process to test out the control system. The decision to use the process to test a control system has taken place repeatedly during my career. Each time I've seen the engineering online debugging costs (never anticipated) overwhelmingly outstrip any cost that would be required to put together a test system that simulates *in some way* the process to which the control system is to be applied.

Whether to simulate or not is a question that will not be settled here. However, I will present my point of view, unpopular though it is. Start with the big picture; the process and the control system with all the interconnections (Fig. 2-10). Let's assume that there are many

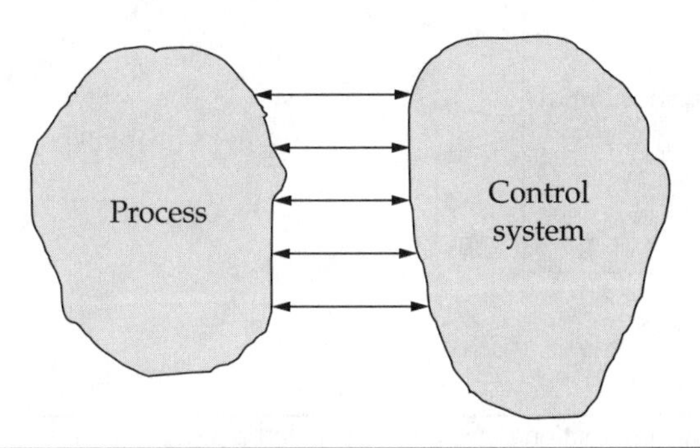

FIGURE 2-10 Interactions between process and control system.

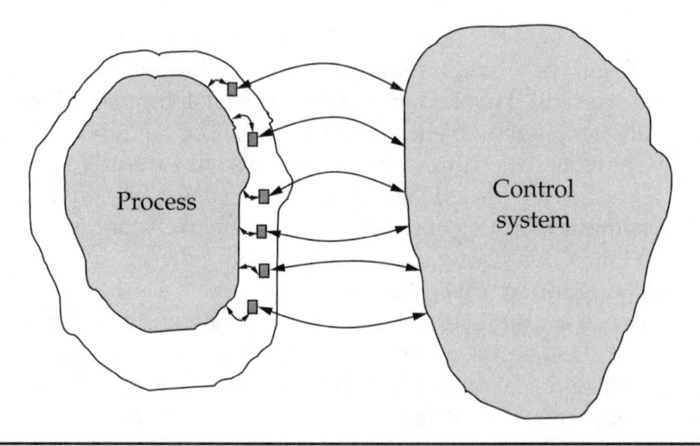

FIGURE 2-11 Interactions between process and control system with interfaces.

connections between the process and the control system: analog inputs and outputs, digital inputs and outputs, and/or serial inputs and outputs. At first glance it may be overwhelming but after the dust has settled, one can look beneath the surface of the process where the lines come from and leave to the control system. At each point there is some sort of conversion of electrical to mechanical, mechanical to thermal, kinetic to electrical, and so on. Some of these conversions can be explicit as in a motor turning a screw but all can be conceptually represented by the small black boxes as shown in Fig. 2-11. It is the inside irregularly shaped object that can be simulated. In Fig. 2-12 a rectangle (a purposely crude approximation to the irregularly shaped object representing the real process) is the simulated process that can be used to test the control system. All of the small black boxes must be simulated, too.

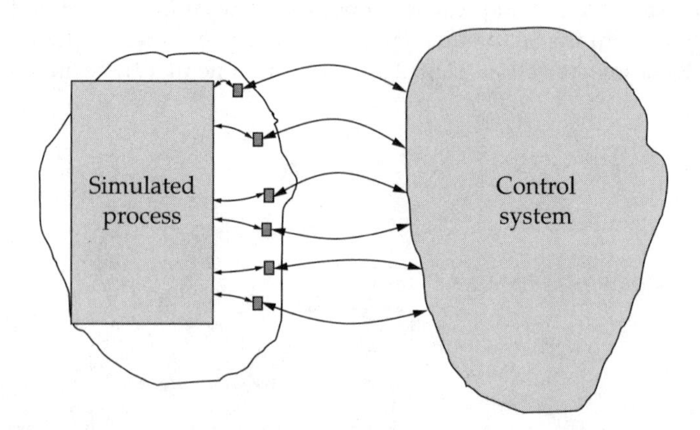

FIGURE 2-12 Interaction between the simulated process and the control system.

This task of simulation at first looks incredibly difficult but it is my position (a minority one, at best) that once the engineer takes the time and trouble to conceptualize the process and control system as shown in Fig. 2-12, she will be able to construct an approximation of the process that is quite useful for testing the control system. It will take time, effort, and resources but the cost will be minuscule compared to the cost of using the process to test the control system.

2-6 Documentation and Indispensability

Consider the following scene. My manager/boss stops by my cubicle/office and says, "Dave, you are doing an absolutely great job on the project. Your contribution is playing a critical part in the success that we are enjoying. You have become indispensable! I will give you a month to thoroughly and clearly document what you have done or I will have to fire you."

That conversation has never taken place with me and my boss nor have I ever heard of it taking place with a coworker. However, I have seen innumerable instances where it should have taken place.

Much folklore and fear has developed around undocumented equipment, algorithms, and methods that at one time, at least, were successful. Many of these "successes" were used with reverence long after they had lost their effectiveness simply because no one knew how or why they really worked. Therefore, they were not improved, adapted, or replaced out of ignorance (and the incompetence of the original manager who did not demand documentation).

Documentation is usually abhorred by engineers until the time comes to solve a problem associated with an undocumented piece of equipment or algorithm (after a call-in at 2:00 A.M.). Many companies have documentation czars who legislate cumbersome structures. This usually extinguishes any creativity or enjoyment on the part of the person best qualified to write the document. Worse yet are the documentation writers who almost always haven't an ounce of technical competence. I have always thought that the person who did the work should document it in any format she liked as long as it had content and clarity. Finally, I have always thought the manager should take a few moments to peruse the document to see if it, in fact, has content and clarity.

I have no other advice other than to ask, "How many managers do you know that lost a promotion because they had their engineers spend too much time on documentation?" . . . or . . . "How many times have you been in a troubleshooting situation where even a snippet of verbiage would have been helpful?"

2-7 Summary

Well, the party is basically over. The next chapter will start with the heavy lifting, the mathematical stuff.

This chapter presented the SMILH concept, which you as a manager probably used (perhaps unaware) to get promoted to your current position. The pedestrian approach, which I advocate will be tough to sell but in the long run, provides measurable benefits.

The road map to process analysis is another tough sell but it can yield great rewards and it is a lot of fun to do.

If, in fact you or your group does develop a control algorithm, it can be a painful experience unless you spend the extra money up front and develop your system to be extremely flexible and accessible for debugging.

OK, on to the math.

CHAPTER 3

Basic Concepts in Process Analysis

Since the basic tenet of this book is to analyze the process before one attempts to control it, we will have to develop some process analysis concepts. Furthermore, since controlling a process inherently deals with transient behavior we will have to deal with process dynamics. Finally, since we need to keep the level of the material in this book reasonable, that is, attractive to a busy manager, we will start with the simplest of constructs—the first-order process. In fact, we will beat it to death. During the beating, the widely used proportional-integral (PI) control algorithm will be introduced. One of the sophisticated tools of control engineering, the Laplace transform, will also be introduced. For technical support, the reader may want to consult App. A (introductory calculus), App. B (complex numbers), App. E (first- and second-order differential equations), and App. F (Laplace transforms).

3-1 The First-Order Process—an Introduction

Let's go back to the tank of water introduced in Chap. 1 (Fig. 3-1). It will be our prototypical first-order process. The dynamic analysis of this tank often consists of studying the step response, which is shown in Fig. 3-2. Here the process input U, the valve, is given a step at time $t = 9$, from an initial value of zero to unity. The process output Y, the tank level, begins to rise and appears to *line out* at a value of 2.0. For convenience, we have chosen the initial value of the valve and the tank level to be zero. In general, these quantities could have almost any initial value but this graph would still apply if the reader is willing to allow us to subtract these nonzero initial values, that is, normalizing the initial values of these quantities to zero.

To proceed we need some nomenclature. First, let the *change* in the process input be signified as ΔU. The symbol Δ usually signifies a change in the quantity following it (or upon which it *operates*). Second, let the resulting change in the tank level be signified as ΔY.

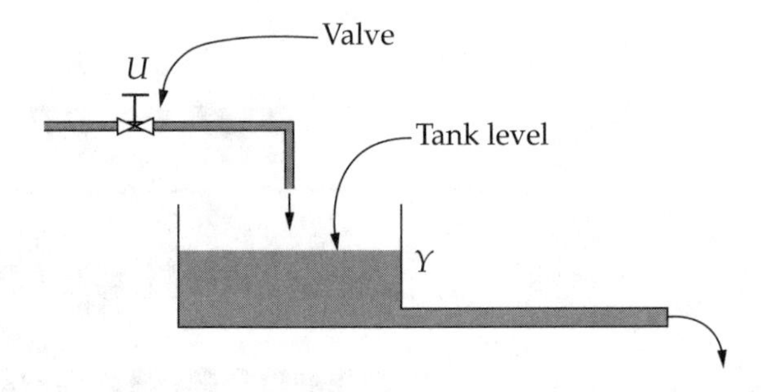

FIGURE 3-1 A tank of liquid (a process).

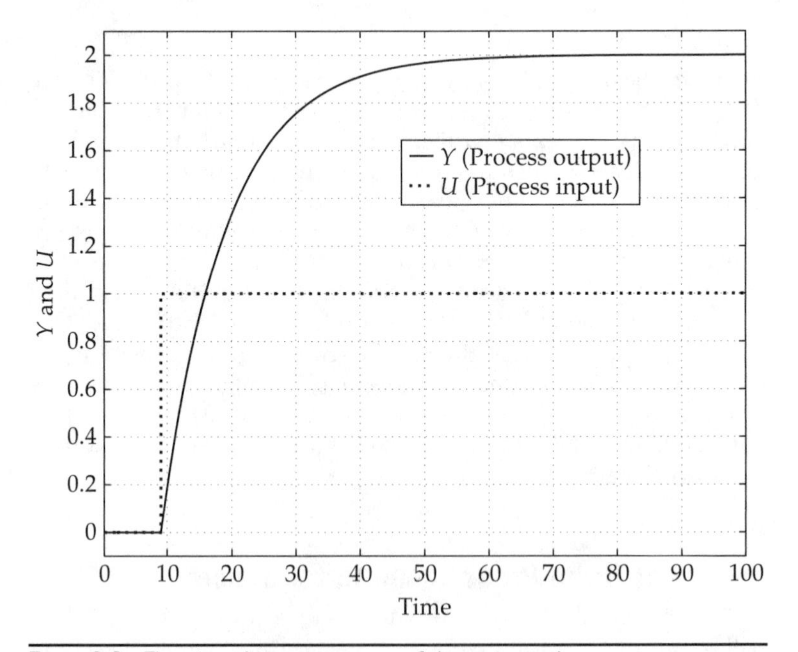

FIGURE 3-2 The step change response of the water tank.

The Process Gain and Time Constant

The transient response for the tank can be characterized by two parameters. The first is the *process gain*, or g, which is the ratio of the change ΔY to the change ΔU:

$$g = \frac{\Delta Y|_{\infty}}{\Delta U} \tag{3-1}$$

The "infinity" subscript in Eq. (3-1) indicates that these changes are those observed after the process has settled in a steady state.

In this case the process gain is 2.0 since $\Delta U = 1.0$ and $\Delta Y|_{\infty} = 2.0$.

The second parameter, called the *process time constant* τ is defined as the time required for the process output to reach 63% of its final value in response to a step change in the process input. Careful examination of Fig. 3-2 will show that the time constant for the water tank is 10.0 time units.

For the time being, these two parameters, the process gain and the process time constant, will suffice for our characterization of the process. Note that the process gain is a static characterization parameter. It can tell the analyst where the process will ultimately settle after a step-change to the input. On the other hand, the time constant is a dynamic characterization parameter for it tells us how the process gets from one state to another.

3-2 Mathematical Descriptions of the First-Order Process

To describe the dynamic behavior of a process we must decide in which domain we will work. The obvious first choice is the continuous *time* domain where for our purposes time will vary continuously from zero to infinity. There are many reasons why one might want to "transform" the domain of analysis into something else, say, the Laplace domain, or the frequency domain, or the discrete time domain. However, first things first! Let us now delve into the continuous time domain. Beware! There will be a lot of math. It will be my challenge to minimize it and keep it simple.

3-2-1 The Continuous Time Domain Model

How can we develop an equation that will describe the behavior of a process? What do we have to work with? Often one has to start with the fundamental conservation laws for mass, momentum, and energy. For our tank of water the apparent fundamental law would be the conservation of mass; in crude terms, what goes in has to either go out or accumulate:

Rate of water in = rate of water out + rate of accumulation of water in the tank

The rate of water flowing in can be represented by F_i, the inlet flow rate in kg/sec. The rate of water leaving the tank can be represented by F_o. The rate of accumulation of water in the tank is the rate of change of the mass in the tank with respect to time. Does this sound familiar? Yes, it is the derivative from first-year calculus (also reviewed in App. A), namely,

$$\frac{d}{dt}(\rho AY) \tag{3-2}$$

where ρ is the water density in kg/m^3, A is the cross-sectional area (which we are assuming is cylindrical with constant diameter) of the tank in m^2, Y is the tank level in m, and the quantity ρAY is the amount of the water in the tank.

Altogether now, one gets

$$F_i = F_o + \rho A \frac{dY}{dt} \tag{3-3}$$

For control purposes, F_i would probably be adjusted to maintain Y on target, so there might be another equation describing the controller's dynamic behavior which we will leave to later. On the other hand, F_o depends on the level of water in the tank, so we need to come up with a "constitutive" equation relating the rate F_o to a potential or tank liquid height Y. A simple relationship is

$$F_o = \frac{Y}{R} \tag{3-4}$$

where R is the resistance associated with the tank's outlet piping. If Y has units of m and F_o has units of kg/sec, then R must have units of m·sec/kg. This is a bit of an idealization because from high school physics the reader probably remembers Torricelli's law, which states that the jet of liquid emerging from a hole in the side of a tank is given by

$$F_o \propto \sqrt{2gY} \tag{3-5}$$

Therefore, Eq. (3-4) is a *linearization* of Eq. (3-5). The concept of linearization is mentioned in App. D in the section about the Taylor's series. Since we expect our controllers to keep the process "near" its nominal values, linearization may well suffice.

Combining Eqs. (3-3) and (3-4) gives

$$F_i = \frac{Y}{R} + \rho A \frac{dY}{dt} \tag{3-6}$$

which, after some simple rearrangement, is

$$\tau \frac{dY}{dt} + Y = RF_i \tag{3-7}$$

where $\tau = R\rho A$. This last quantity is the time constant and the reader should check that the units of the time constant work out to be sec.

For the sake of generality rewrite Eq. (3-7) as

$$\tau \frac{dY}{dt} + Y = gU \tag{3-8}$$

where g is the process gain. The reader should check that the units of the gain are m·sec/kg.

We can verify that g is in fact the process gain by letting time go to infinity assuming that the process input U jumps from zero to a constant value U_c at time zero. At *time equals infinity*, the process has supposedly settled out to a new steady state where the rate of change of all the variables is zero, that is,

$$\frac{dY}{dt} \to 0$$

or

$$\lim_{t\to\infty} \frac{dY}{dt} = 0$$

and

$$\lim_{t\to\infty} Y(t) = Y_\infty$$

which causes Eq. (3-8) to become

$$0 + Y_\infty = gU_c$$

which in turn yields

$$g = \frac{Y_\infty}{U_c} \tag{3-9}$$

The definition of the process gain in Sec. 3-1 shows that the g in Eq. (3-9) is indeed the process gain.

In summary, the model was developed by first applying a conservation law which related flows, that is, F_i and F_o, to the rate of change of the potential, that is, Y. Then, a constitutive equation was used to replace one of the flows with an expression containing a potential. This left us with a model equation that gave the response of the process output Y to the process input U in terms of the process parameters g and τ.

Scaling

As an alternative, the general first-order model equation presented above could be scaled to have unity time constant and unity gain. One would start with

$$\tau \frac{dY}{dt} + Y = gu$$

and make the following changes of variable

$$Y = gy$$

$$t = \tau T$$

$$\tau \frac{d(gy)}{d(\tau t)} + gy = gu$$

$$\frac{dy}{dT} + y = u$$

We will not make much use of this variable change concept until Chap. 5 when it becomes helpful in addressing round-off error.

3-2-2 Solution of the Continuous Time Domain Model

We will solve the model for the first-order process developed in Sec. 3-2-1 for a variety of conditions. First, consider the case where the process input, initially zero, is assigned the nonzero value of U_c at time zero and held there for all time. The reader hopefully will recognize that this situation is the same as giving the process a step change in the process input at time zero. Hence, if the equation can be solved, the step-change response will be obtained and we can compare the results with the graph in Fig. 3-2.

For this case of a constant process input, Eq. (3-8) becomes

$$\tau \frac{dY}{dt} + Y = gU_c \tag{3-10}$$

There are many ways of solving this equation and a couple of them are reviewed in App. E. The solution for the case of a constant input is

$$Y = Y_0 e^{-\frac{t}{\tau}} + gU_c\left(1 - e^{-\frac{t}{\tau}}\right) \tag{3-11}$$

We will refer to this equation repeatedly in this book so the reader should be comfortable with its development before proceeding.

Comments about the Solution

First, the readers should convince themselves that Eq. (3-11) behaves just like the curve in Fig. 3-2 for the case of $Y_0 = 0$, $\tau = 10.0$, $g = 2.0$, and $U_c = 1.0$. To verify the time constant value, look where the process output appears to reach 63% of the final value. It should take τ seconds to get there. Now, look at the final value. It should be two units greater than the initial value of 0.0 because the process gain is 2.0.

With $Y_0 = 0$ there is only one term left in Eq. (3-11), namely,

$$Y = gU_c(1 - e^{-\frac{t}{\tau}}) \tag{3-12}$$

At time zero, this equation yields $Y = 0$ and then, as time increases, Y finally asymptotically approaches gU_c. Second, by taking the derivative of Eq. (3-11), or by solving Eq. (3-10) for the derivative, the rate of change of Y can be obtained as

$$\frac{dY}{dt} = \dot{Y} = -\frac{1}{\tau}(Y_0 - gU_c)$$

Therefore, when $Y_0 = 0$ the rate at time zero (when the step in U is first applied) can be obtained as

$$\dot{Y}(0) = \frac{gU_c}{\tau} \tag{3-13}$$

Since Y and its rate of change were zero before the step in U was applied at time zero, Eq. (3-13) tells us that, at the time of the step in U, the rate of change of Y experiences a *discontinuity*, jumping from zero to gUc/τ. Examination of Fig. 3-2 should support this contention. We know that in real life there are few process quantities that experience true discontinuities, that is, "Mother Nature abhors discontinuities." Therefore, this model is, among other things, an idealization, albeit useful (as we will see).

Equation (3-13) also tells us that the initial rise rate of Y is directly proportional to the strength of the step in U, directly proportional to the gain of the process, and inversely proportional to the process time constant. It is comforting that common sense is supported by simple mathematics, no?

Third, note that when time equals the value of the time constant, that is, when $t = \tau$, and when $Y_0 = 0$, Eq. (3-12) yields

$$Y = gU_c(1 - e^{-\frac{\tau}{\tau}}) = gU_c(1 - e^{-1}) = gU_c(1 - 0.36788) = 0.63212gU_c$$

Therefore, at $t = \tau$, Y equals 63% of its ultimate final value gU_c; hence the basis for the above definition of the time constant.

For future reference, Eq. (3-10) can be rewritten as

$$\begin{aligned} \frac{dY}{dt} &= \left(-\frac{1}{\tau}\right)Y + \frac{g}{\tau}U \\ \frac{dY}{dt} &= aY + bU \end{aligned} \tag{3-14}$$

The simple model equation now has the form that we will use when we get to state space formulations where Y and U will be vectors and a and b will be matrices. We know intuitively that, in the face of steps in U, the response of Y will be bounded or that Y will behave stably. Equation (3-14) suggests that as long as a is negative (which it has to be for our simple example), stability will result. The reader should quickly convince himself that when a is positive (physically unrealistic for this model), the response of Y will be unbounded or unstable.

Question 3-1 Do you understand the comments about "instability"?

Answer Look at Eq. (3-11) with a replacing $-1/\tau$.

$$Y = Y_0 e^{at} + gU_c(1 - e^{at})$$

We know that, by definition, a is negative but if it were not, note that Y would increase without bound in the face of a positive value of U_c, that is, there would be *instability*. For this simple case of a first-order model there is no question about the sign of a but later on when the models get more sophisticated this will not always be the case and the "sign" of whatever replaces a will give us insight into stability.

This temporarily concludes our development of the simple first-order model where we have really beaten a couple of elementary equations to death. As things get more complicated we will repeatedly come back to this model.

3-2-3 The First-Order Model and Proportional Control

Although optional, it would be quite helpful if the reader is able to follow the math in App. E used to arrive at the solution of the differential equation for the first-order model. We will now take a little side trip and see what can be learned from this model from the control point of view.

Let's tack a simple "proportional" controller onto our model and see if we can control the process output to a desired set point. Our starting point is the first-order model for the process to be controlled

$$\tau \frac{dY}{dt} + Y = gU \tag{3-15}$$

Our goal is to try to keep the process output Y "acceptably near" the set point S by adjusting U in some fashion. The simplest "fashion" is to form an error

$$e = S - Y \tag{3-16}$$

and to manipulate U in proportion to the error, that is,

$$U = ke = k(S - Y) \tag{3-17}$$

where k is the proportional control gain.

Before proceeding, let's think about Eq. (3-17) with reference to the water tank. Assume that initially Y is equal to the set point S so that e is initially zero. Also, assume that the nominal initial values have been subtracted from all of the quantities, so Y, S, e, and U are initially zero. If S is stepped up, then e would become nonzero and positive. This would mean that U would increase, assuming that k is positive. An increase in U means more flow into the tank and the level Y should rise. Okay, at least the control algorithm has the correct signs and moves the controller output in the right direction.

Schematically, this feedback control system can be presented as a block diagram (Fig. 3-3). This is a classic schematic that will reappear many times in many forms in the balance of this book. Note how the process output Y is fed back and subtracted from the set point S producing the error E which is fed to the controller which produces the process input U.

Combine Eqs. (3-15) and (3-17)

$$\tau \frac{dY}{dt} + Y = gU$$

$$U = k(S - Y)$$

and get

$$\tau \frac{dY}{dt} + Y = gk(S - Y)$$

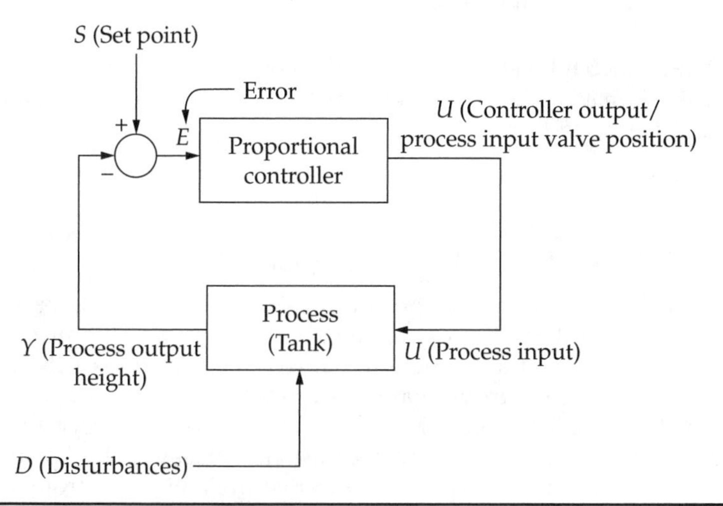

FIGURE 3-3 A feedback controller.

A little rearrangement, which the reader should verify, will yield

$$\left(\frac{\tau}{1+gk}\right)\frac{dY}{dt}+Y=\frac{gk}{1+gk}S \tag{3-18}$$

This is the differential equation that describes the *closed-loop* system containing the process under simple proportional control. It has the same form at Eq. (3-10) except for the following replacements

$$\tau \Rightarrow \frac{\tau}{1+gk} \qquad g \Rightarrow \frac{gk}{1+gk} \qquad U \Rightarrow S$$

Therefore, *by observation*, we can obtain the solution to a process under simple proportional feedback control subject to a step in the set point (from 0 to S_c) at time zero. That is, Eq. (3-11), with the above substitutions, becomes

$$Y = Y_0 e^{-t/\left(\frac{\tau}{1+gk}\right)} + \frac{gk}{1+gk}S_c\left[1-e^{-t/\left(\frac{\tau}{1+gk}\right)}\right] \tag{3-19}$$

Faster Response

Since both g and k are positive, the new *effective* time constant is less than the original one (where there was no control) by a factor of $1/(1+gk)$. As the control gain k increases, the effective time constant decreases. This is something we would hope for since the effect of adding control should be to speed things up.

Offset from Set Point

Look at what Eq. (3-19) yields when $t \to \infty$ (which drives the exponential terms to zero):

$$\lim_{t\to\infty} Y(t) = \frac{gk}{1+gk}S_c$$

So, the proportional control does not ultimately drive the process output all the way to the set point. In fact, the process output settles out at a fraction, namely, $gk/(1+gk)$, of the set point S_c. If the controller gain k is quite large, as in the case of an aggressive controller, this fraction will be nearly unity. Raising the control gain k so as to decrease the offset is risky because our model is an idealization and in real life a high control gain might cause some problems that would lead to instability. Also,

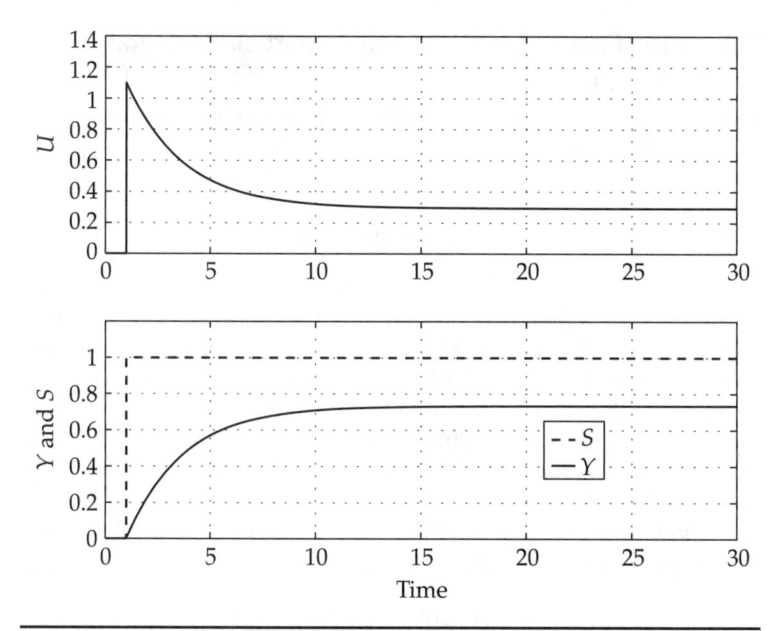

Figure 3-4 Proportional-only control.

high values of k might cause large (and perhaps unacceptable) excursions in the controller output. Figure 3-4 shows the offset that results with proportional-only control.

In case you want to do some simulation yourself, these results were derived from a process that had $\tau = 10$, $g = 2.5$, and the proportional control gain was $k = 1.1$. The Matlab code that generates the results is given in the scripts mentioned in the preface.

Question 3-2 Can you see that no matter how large you make the value of the proportional control gain k, this *idealized* controlled system cannot go unstable?

Answer Let k increase without bound in Eq. (3-19). The result will be

$$Y = Y_0 e^{-t/(0)} + S_c(1 - e^{-t/(0)}) = 0 + S_c = S_c$$

The controller output is proportional to the error and as the error decreases because of the controller action, so does the output. This, in turn, decreases the controller output. This "backing off" of the control output leads to the offset. To obtain a zero error in the face of a step in the set point, we need a controller action that will "keep on going" until the error is removed, even after it stops changing. By the way, we could also have obtained the final value of Y by letting the derivative in Eq. (3-18) go to zero.

3-2-4 The First-Order Model and Proportional-Integral Control

The "keep on going" control feature can be obtained if the controller algorithm is modified to be

$$U(t) = ke(t) + I\int_0^t du\, e(u) \tag{3-20}$$

where we have added a second term that is proportional to the *integral* of the error with I being the proportionality constant. (The integral is reviewed in App. A.) In the face of a step in the set point, assume that the error does not go to zero and remains, say, positive. The second term in Eq. (3-20), because $e(t)$ is positive, will increase and continue to do so. This will cause U to increase until either the error is driven to zero or until U runs out of room.

To get a better feel of how Eq. (3-20) behaves let's simply set the error equal to a constant at time zero, namely,

$$e(t) = 0 \qquad t < 0$$

$$e(t) = C \qquad t \geq 0$$

That is, let's assume that some sort of disturbance is taking place that keeps the process variable away from the set point by a constant amount. Don't worry about how this could actually happen. Eq. (3-20) becomes

$$U(t) = kC + I\int_0^t du\, C$$

$$= kC + ICt$$

which says that, in the face of a constant error C that starts at $t = 0$, the controller output makes an initial jump of kC (the proportional component) at time zero and then ramps up at a rate of IC. (The integral of a constant is reviewed in App. A.) The continual increase in the controller output, due to the term ICt, is the integral action and it is telling us that, as long as the error is constant and as long as the process does not respond to the controller output, the controller output will continue to increase. This is the feature that we needed to reduce the offset between the process output and the set point when proportional-only control was used.

Showing That There Is No Offset

Can we prove this contention of zero offset with our simple mathematics? Well, sort of. Let's combine Eqs. (3-20) and (3-10):

$$\tau \frac{dY}{dt} + Y = gU$$

$$U(t) = ke(t) + I\int_0^t du\, e(u)$$

To combine these two equations, and in the process get rid of the unwieldy integral, we have to take the derivative of each equation and replace e with its definition of $S - Y$. The derivative of the first equation is

$$\tau \frac{d^2Y}{dt^2} + \frac{dY}{dt} = g\frac{dU}{dt}$$

The derivative of the second equation is

$$\frac{dU}{dt} = k\frac{de}{dt} + Ie = k\frac{d(S-Y)}{dt} + I(S-Y)$$

$$= k\frac{dS}{dt} - k\frac{dY}{dt} + IS - IY$$

In taking the derivative of $U(t)$ we used the fact that differentiating the integral simply releases the integrand. For more on this check App. A.

Now, do some minor algebra to eliminate dU/dt between the two equations and get

$$\tau \frac{d^2Y}{dt^2} + (1+gk)\frac{dY}{dt} + gIY = gk\frac{dS}{dt} + gIS \tag{3-21}$$

To avoid some difficulties that we will deal with later on, let's assume that the set point S is constant and has been so for all time, hence $dS/dt = 0$ and

$$\tau \frac{d^2Y}{dt^2} + (1+gk)\frac{dY}{dt} + gIY = gIS \tag{3-22}$$

How could we use Eq. (3-22) to show that Y ultimately goes to Y_{ss}? We could try to solve Eq. (3-22) and then let $t \to \infty$. This would take some effort and at this point it probably is not worth it. Instead, let's just suggest that as $t \to \infty$, things do settle down to a final steady state where

$$\frac{d^2Y}{dt^2} \to 0 \qquad \text{and} \qquad \frac{dy}{dt} \to 0$$

If you can accept this, then Eq. (3-22) immediately yields

$$\lim_{t\to\infty} Y(t) = Y_{ss}$$

However, for future reference, we need to slightly expand the above manipulations and try to find a partial solution to Eq. (3-22). We follow the same path (in App. E) that led to the solution of the first-order differential equation in Eq. (3-10). Assume that the solution consists of a transient part Y_t and a steady-state part Y_{ss}, as in

$$Y = Y_t + Y_{ss}$$

The transient portion of the solution Y_t satisfies

$$\tau \frac{d^2 Y_t}{dt^2} + (1 + gk)\frac{dY_t}{dt} + gIY_t = 0 \tag{3-23}$$

and the steady-state part satisfies

$$gIY_{ss} = gIS$$

(which we already suggested). If we let $t \to \infty$ then the derivatives in Eq. (3-23) will go to zero, as should Y_t, that is,

$$t \to \infty$$

$$\frac{dY_t}{dt} \to 0$$

$$Y_t \to 0$$

If the transient part of the solution goes to zero then all that is left is Y_{ss} which is the same as saying

$$Y = Y_t + Y_{ss} \to Y_{ss}$$

This means that as $t \to \infty$, Eq. (3-22) simplifies to

$$gIY_{ss} = gIS_c$$

so,

$$Y_{ss} = S_c$$

So, when $t \to \infty$, Y_t goes away and the steady-state part remains and it is equal to the set point.

Question 3-3 Does any of this logic based on assuming derivatives go to zero as $t \rightarrow \infty$ bother you?

Answer Actually, it should. What if, somehow, the integral gain was mistakenly set to a negative number? Using the tools of the next section you should be able to show that a negative integral gain will cause instability and that the derivatives will definitely not settle out to zero.

Trying a Partial Solution for the Transient Part

As in App. E, a solution of the form

$$Y_t = Ce^{at} \tag{3-24}$$

is tried. When Eq. (3-24) is inserted into Eq. (3-23), the following quadratic equation results (the reader should try this, verify it, and then perhaps check App. E)

$$\tau Ca^2e^{at} + (1+gk)Cae^{at} + gICe^{at} = 0 \tag{3-25}$$

or, after cancelling Ce^{at}

$$\tau a^2 + (1+gk)a + gI = 0$$

This quadratic equation can be solved for a (the *root* of the equation), yielding two values, a_1 and a_2. The roots of a quadratic equation can be found from the famous quadratic equation root solver (see App. B):

$$a_1, a_2 = -\frac{(1+gk)}{2\tau} \pm \frac{\sqrt{(1+gk)^2 - 4\tau gI}}{2\tau} \tag{3-26}$$

Critical Damping

Eq. (3-26) shows that the roots will have two parts and that if

$$(1+gk)^2 = 4\tau gI$$

or

$$I = \frac{(1+gk)^2}{4\tau g}$$

then both roots will be the same

$$a_1, a_2 = -\frac{(1+gk)}{2\tau}$$

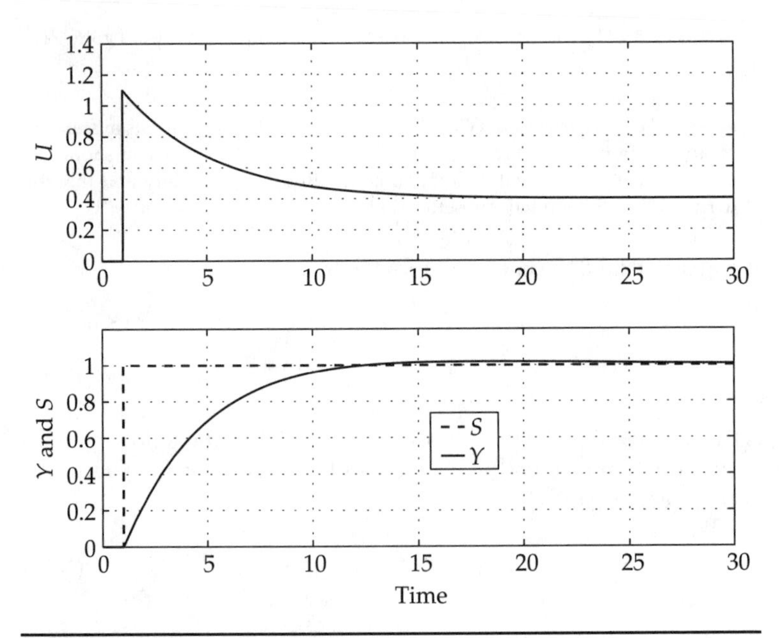

FIGURE 3-5 Critically damped PI control.

and they will be negative. Since these roots appear as a coefficient in the argument of exponential terms, it stands to reason that the transient part of the solution will die away as time increases and this in turn suggests that the process output will settle out at the set point which is the steady-state part of the solution. This situation represents the case where the response is *critically damped*. We will worry about this concept later on.

In the meantime, Fig. 3-5 shows the response of the process variable to a unit step in the set point for the case of critical damping. For this simulation, I used $\tau = 10$, $g = 2.5$, $k = 1.1$. The integral control gain was calculated from

$$(1 + gk)^2 = 4\tau g I$$

$$I = \frac{(1 + gk)^2}{4\tau g} = \frac{(1 + 2.5 \times 1.1)^2}{4 \times 10 \times 2.5} = 0.14$$

Overdamped Response

Note that the two roots, a_1 and a_2, will be real, negative, and *different*, if the argument inside the square root is positive or if

$$(1+gk)^2 > 4\tau g I$$

so, (3-27)

$$I < \frac{(1+gk)^2}{4\tau g}$$

(Note how this integral gain I is less than that for the critically damped case.)

Since there are two roots, the solution will have the form

$$Y = C_1 e^{a_1 t} + C_2 e^{a_2 t} + S_c \tag{3-28}$$

Although we will touch on this later, the response of the process variable for this case will be *overdamped* and might look something like Fig. 3-6. For this simulation I used $\tau = 10$, $g = 2.5$, $k = 1.1$, and $I = 0.1$. There is not much difference between Figs. 3-5 and 3-6.

By the same crude argument given above, you could reason that the transient component of the solution will die away as time increases and the process output will approach the set point.

Question 3-4 Can you support the contention for this last case, namely that the transient part will die away for the overdamped case? While you are at it, can you show that a negative integral gain will cause instability?

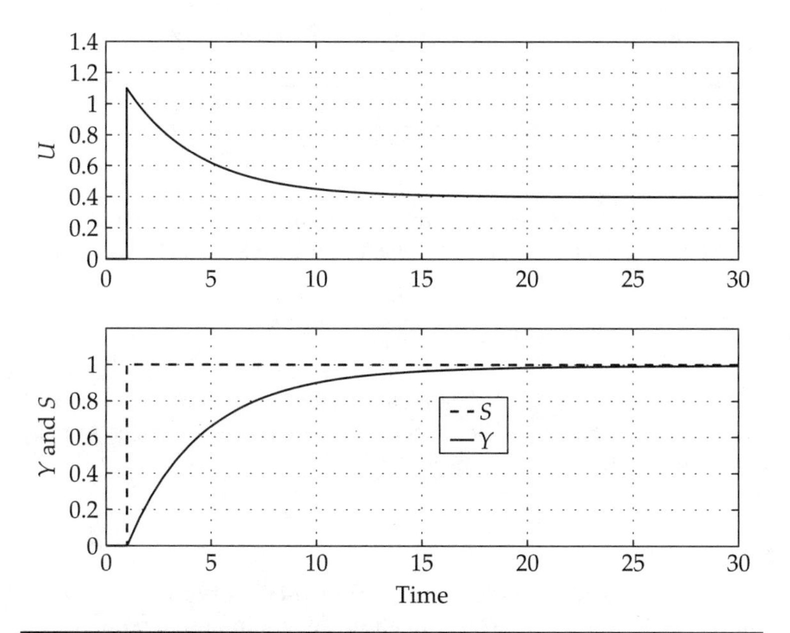

FIGURE 3-6 Overdamped PI control.

Answer Use the quadratic equation root solver

$$a_1, a_2 = -\frac{(1+gk)}{2\tau} \pm \frac{\sqrt{(1+gk)^2 - 4\tau gI}}{2\tau}$$

If Eq. (3-27) holds then the argument of the square root will be positive and the roots will be real. Also, the square root term will be less in magnitude than $(1+gk)$ so the roots cannot be positive. As to the second question, the quadratic equation root solver shows that if $I < 0$ then one of the roots would be positive and in turn would lead to an unbounded response.

Underdamping

Finally, consider the case when

$$(1+gk)^2 < 4\tau gI$$

so,

$$I > \frac{(1+gk)^2}{4\tau g} \tag{3-29}$$

(Note how the integral gain is greater than that for critical damping.) The argument inside the square root is now negative. But we know that

$$\sqrt{-1} = j \qquad \text{and} \qquad \sqrt{-4} = 2j$$

and because of the inequality in Eq. (3-29), the roots are

$$a_1, a_2 = -\frac{(1+gk)}{2\tau} \pm \frac{j\sqrt{4\tau gI - (1+gk)^2}}{2\tau} = \alpha \pm j\beta \tag{3-30}$$

where $\alpha < 0$ and $\beta > 0$ are real numbers. This means that the solution will have exponential terms with imaginary arguments (see App. B) as in

$$e^{(\alpha + j\beta)t} \qquad \text{or} \qquad e^{\alpha t} e^{j\beta t}$$

The $e^{\alpha t}$ term (with $\alpha < 0$) means that the transient response will die away, but what about the other factor? Euler's equation (see App. B) can be useful here.

$$e^{i\beta t} = \cos(\beta t) + j\sin(\beta t)$$

The $e^{j\beta t}$ factor implies sinusoidal or oscillatory behavior while the $e^{\alpha t}$ factor decreases to zero at a rate depending on α. Both factors promise an underdamped behavior where there are oscillations that

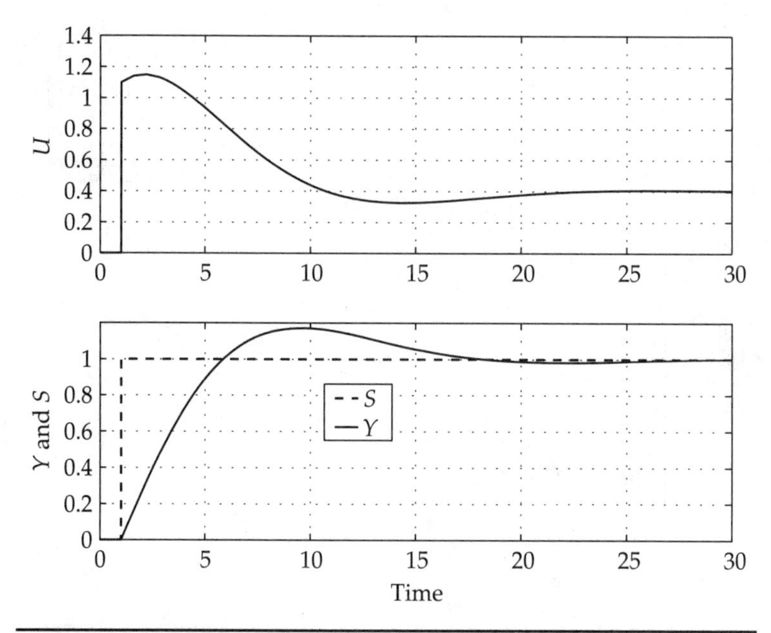

FIGURE 3-7 Underdamped response of the process variable to a unit step in the set point.

damp out with time. Since this condition can result if the integral control gain I is relatively high, overly aggressive control action may lead to underdamped behavior as shown in Fig. 3-7. For this simulation I used $\tau = 10$, $g = 2.5$, $k = 1.1$, and $I = 0.4$.

By applying initial conditions on Y and dY/dt, the two coefficients C_1 and C_2 in Eq. (3-28) could be determined for all of these conditions. Unfortunately, this gets messy quite quickly and we will not proceed in this direction simply because it doesn't add much to our insight. The reader can consult App. E for details.

Question 3-5 What happens to the roots and the system behavior when the control gain I gets really large?

Answer The quadratic root solver equation is

$$a_1, a_2 = -\frac{(1+gk)}{2\tau} \pm \frac{\sqrt{(1+gk)^2 - 4\tau g I}}{2\tau}$$

When I gets really large,

$$a_1, a_2 \Rightarrow -\frac{(1+gk)}{2\tau} \pm \frac{j\sqrt{4\tau g I}}{2\tau} \Rightarrow \pm j\sqrt{\frac{gI}{\tau}}$$

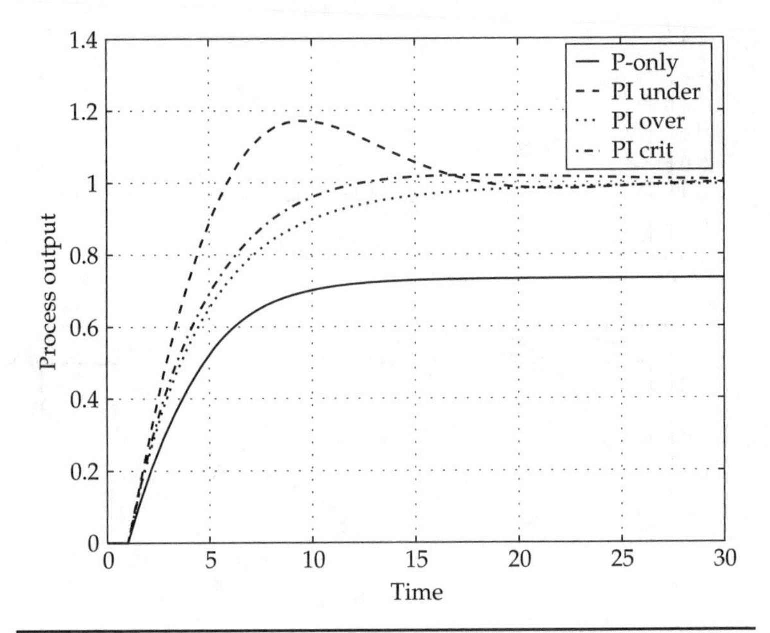

FIGURE 3-8 Comparing the responses.

That is, the square root term dominates and the roots become purely imaginary and the dynamic behavior becomes purely oscillatory with no damping. It is not unstable because the amplitude of the oscillations becomes constant in steady state. This condition is sometimes called marginal stability.

Figure 3-8 shows the process output for the cases covered above. Note that the underdamped response reaches the set point first but overshoots. The overdamped response reaches the set point last and the P-only response does not reach the set point at all.

So What?

This has been the longest section in the book so far and, if you have gotten through it without losing your temper or your patience, there is a good chance that you will make it through the rest of the book—although it will be a little tougher from now on.

I always wonder if going through the mathematics is necessary. Why not just tell about the behavior of the controlled system and let it go at that? There are two reasons, neither of which may be satisfactory to you. First, using the relatively simple mathematics (compared to conventional textbooks on control theory, anyway) *may* help the reader understand the concepts. Second, this *is* the language of the control engineers whom you are working with and it may be to your benefit to be somewhat on the same footing as them.

This section hopefully showed that it is possible to tack a simple "proportional" controller onto a simple first-order process and use it to speed up the response of the process output to a set-point change.

It also showed that proportional control alone will not drive the process variable all the way to set point. The response, although inadequate because of the offset between the process output and set point, was smooth and without oscillations.

When the integral component was added, the process output was driven to set point. Aggressive integral control could cause some overshoot. Excessively aggressive integral control could cause sustained oscillations.

This might be considered a logical point to end the chapter but I choose not to for the simple reason that I need you to quickly move on from the time domain to the Laplace domain before you forget the above results and insights.

3-3 The Laplace Transform

In the last section we had a little trouble with the second-order differential equation. In this section we introduce a tool, the Laplace transform, which will remove some of the problems associated with differential equations but with the cost of having to learn a new concept. The theory of the Laplace transform is dealt with in App. F so we will start with a simple *recipe* for applying the tool to the first-order differential equation.

The first-order model in the time domain is

$$\tau \frac{dY}{dt} + Y = gU \tag{3-31}$$

To move to the Laplace transform domain, the derivative operator is simply replaced by s, the so-called Laplace transform operator, and wiggles are placed over the symbols Y and U since they are in a new domain

$$\tau s\tilde{Y} + \tilde{Y} = g\tilde{U} \tag{3-32}$$

Before dealing with Eq. (3-32) consider some Laplace transform transition rules in the box:

$$\frac{d}{dt} \Rightarrow s$$

$$Y(t) \Rightarrow \tilde{Y}(s)$$

$$U(t) \Rightarrow \tilde{U}(s)$$

$$C \Rightarrow \frac{C}{s}$$

$$\int_0^t Y(u)du \Rightarrow \frac{\tilde{Y}(s)}{s}$$

$$\lim_{s \to 0} s\tilde{Y}(s) = Y(\infty)$$

Most control books have extensive tables giving the transforms for a wide variety of time functions. Note the following comments about the contents of the box given in Sec. 3-3.

1. All initial values must be zero. (Later on, nonzero initial conditions will be covered.
2. The differential operator d/dt is replaced with s.
3. The integral operator $\int_0^t \ldots du$ is replaced with $1/s$.
4. The quantity C is a constant.
5. The last equation in the box is really not a transform rule. Rather it is the *final value theorem* and it shows how one can find the final value in the time domain if one has the Laplace transform. The basis for these rules and the final value theorem are given in App. F.

For the case of the water tank, Y had units of length or m, U had units of volume per unit time or m^3/sec, and time t had unit of sec. In the new domain, s has units of reciprocal time or sec^{-1}, $\tilde{Y}$ has units of m-sec, and $\tilde{U}$ has units of m^3. It's not obvious why $\tilde{Y}$ and $\tilde{U}$ have those units—that should be apparent from the discussion in App. F—but it may make sense that s has units of sec^{-1} by looking at the appearance of τs in Eq. (3-21) and realizing that it would be nice to have this product be unitless.

In any case, Eq. (3-32) does not contain derivatives—in fact, it is an *algebraic* equation and can be solved for $\tilde{Y}$:

$$\tilde{Y} = \frac{g}{\tau s + 1}\tilde{U} = G_p \tilde{U}$$

$$G_p = \frac{g}{\tau s + 1} \tag{3-33}$$

This equation gives the Laplace transform of Y in terms of the Laplace transform of U and a factor G_p that is called the process *transfer function*.

Equation (3-33) will be solved for $Y(t)$ later on in Sec. 3-3-2 but for the time being let's comment on the big picture. We have to take two steps. First, the Laplace transform for U must be found. In the examples so far U has been a step change, so the Laplace transform for the step-change function must be developed. App. F gives the derivation of the Laplace transform of a step change. As a temporary alternative, consider the step in U at time zero as a constant U_c that had zero value for $t < 0$. In this case, using the fourth entry in the box given in Sec. 3-3, the Laplace transform of U is U_c/s.

Second, after replacing $\tilde{U}$ with its transform in terms of s, the modified algebraic equation for $\tilde{Y}$ must be *inverted*, that is, transformed back to the time domain. There are a variety of ways of doing this but the simplest is to break the expression for $\tilde{Y}$ into simple algebraic terms and then go to the mentioned box of Laplace transforms and find the corresponding time domain function. We will get into this soon but, first, a few comments about the transfer function G_p in Eq. (3-33).

3-3-1 The Transfer Function and Block Diagram Algebra

The introduction of the transfer function $G_p(s)$ in Eq. (3-33) is useful because of the block diagram interpretation (Fig. 3-9). The expression in the box multiplies the input to the box to give the box's output.

Alternatively, one can play some games with Eq.(3-33) and get

$$\begin{aligned} \tilde{Y} &= \frac{g}{\tau s+1}\tilde{U} \\ \tilde{Y} + \tau s\tilde{Y} &= g\tilde{U} \\ \tau s\tilde{Y} &= g\tilde{U} - \tilde{Y} \\ \tilde{Y} &= \frac{1}{s}\left(\frac{1}{\tau}\right)\left(g\tilde{U} - \tilde{Y}\right) \end{aligned} \tag{3-34}$$

The last line of Eq. (3-34) suggests that (1) there is some integration going on via the $1/s$ operator and (2) there is some negative feedback since $\tilde{Y}$ is on the right-hand side of the equation with a minus sign. That last line of Eq. (3-34) can be interpreted using block algebra as shown in Fig. 3-10.

The reader should wade through Fig. 3-10 and deduce what each box does. The process output Y is fed back to a summing junction

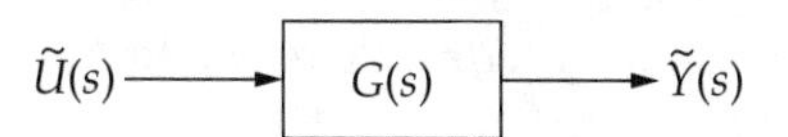

FIGURE 3-9 The transfer function in block form.

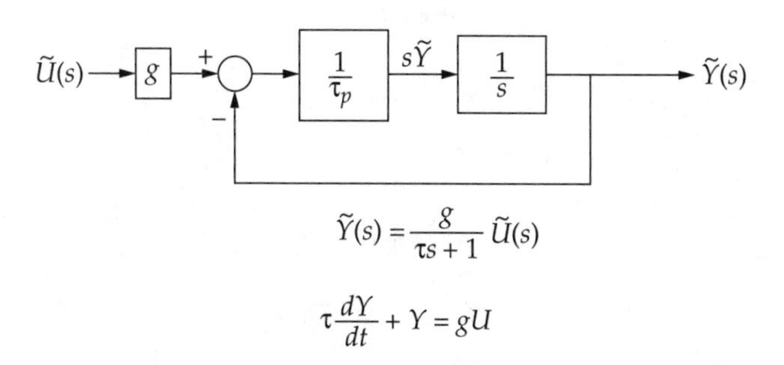

$$\tilde{Y}(s) = \frac{g}{\tau s+1}\tilde{U}(s)$$

$$\tau\frac{dY}{dt} + Y = gU$$

FIGURE 3-10 Block diagram showing integration and negative feedback as part of the process model.

where it is subtracted from the product of the process input U multiplied by the process gain g. This result is multiplied by $1/\tau$. The resulting signal, which is $s\tilde{Y}(s)$ or dY/dt, is then integrated via $1/s$ to form the process output $\tilde{Y}(s)$ or Y, which is fed back, and so on. This structure is similar to the analog computer patchboard of the 1960s. (It is also similar to the block diagrams that make up models in Matlab's Simulink.) This approach to block diagrams will be used in a latter chapter (Chap. 6) when an underdamped process is modified by feedback to present a better face to the outside world.

3-3-2 Applying the New Tool to the First-Order Model

Returning to Eq. (3-33), assume that the time domain function $U(t)$ is a step function having a constant value of U_c. Therefore, it will be treated as a nonzero constant for $t \geq 0$. As with all of our variables, $U(t)$ is assumed to be zero for $t < 0$. The Laplace transform for U_c (see App. F and/or the box given in Sec. 3-3) is

$$\tilde{U} = \frac{U_c}{s}$$

and Eq. (3-33) becomes

$$\tilde{Y} = \frac{g}{\tau s + 1}\frac{U_c}{s} \tag{3-35}$$

To invert this transform to get $Y(t)$, Eq. (3-35) needs to be simplified to a point where we can recognize a familiar form and match it up with a time domain function. *Partial fractions* can be used to split Eq. (3-35) into two simpler terms. Referring to App. F the reader can verify that the new expression for $\tilde{Y}$ is

$$\begin{aligned}\tilde{Y} &= \frac{g}{\tau s + 1}\frac{U_c}{s} \\ &= \frac{gU_c}{s} - \frac{gU_c}{s + \frac{1}{\tau}} \\ &= gU_c\left(\frac{1}{s} - \frac{1}{s + \frac{1}{\tau}}\right)\end{aligned}$$

We already know the time domain functions for the Laplace transforms, namely,

$$\frac{1}{s} \quad \text{and} \quad \frac{1}{s+\frac{1}{\tau}}$$

The first transform is for a step (or a constant) and the second is for an exponential. So, by inspection, we can write the time domain form as

$$Y(t) = gU_c\left(1-e^{-\frac{t}{\tau}}\right)$$

Now, if the reader remembers Eq. (3-11), she will see that a second way has been obtained to solve the differential equation [Eq. (3-10)].

3-3-3 The Laplace Transform of Derivatives

According to the recipe, the derivative in Eq. (3-31) was replaced by the operator s. App. F shows that the basis for this comes directly from the definition of the Laplace transform, which, for a quantity $Y(t)$, is

$$L\{Y(t)\} = \int_0^\infty dt e^{-st} Y(t) \tag{3-36}$$

Note that $e^{-st}Y(t)$ is integrated from $t = 0$ to $t = \infty$. It may seem like a technicality but the integration starts at zero so the value of the quantity $Y(t)$ for $t < 0$ is of no interest and is assumed to be zero. If the quantity has a nonzero initial value, say Y_0, then strictly speaking we have to look at it as

$$Y_0 = \lim_{t\to 0^+} Y(t) = Y(0^+)$$

That is, Y_0 is the initial value of $Y(t)$ when $t = 0$ is approached from the right or from positive values of t. So, effectively, a nonzero initial value corresponds to a step change at $t = 0$ from the Laplace transform point of view. This subtlety comes into play when one evaluates the Laplace transform of the derivative, as in

$$L\left\{\frac{dY}{dt}\right\} = \int_0^\infty dt e^{-st} \frac{dY}{dt}$$

The evaluation of this equation presents a bit of a challenge so I put the gory details in App. F for the reader to check if she wishes. However, after all the dust settles the result is

$$\boxed{L\left\{\frac{dY}{dt}\right\} = s\tilde{Y} - Y(0^+)} \tag{3-37}$$

Thus, the Laplace transform of a derivative of a quantity is equal to s times the Laplace transform of that quantity $\tilde{Y}$, minus that quantity's initial value $Y(0^+)$. In our example and in our recipe box, we stipulated that the initial value was to be zero, so replacing Y by $s\tilde{Y}$ is the correct way to take the transform of the derivative, just as we proposed in the previous section. In most of this book, the initial value of transformed variables will be assumed to be zero.

The Laplace transform of the second-order derivative:

$$\boxed{\begin{aligned} L\left\{\frac{d^2Y}{dt^2}\right\} &= s^2L\{Y\} - sY(0) - \frac{dY}{dt}\Big|_0 \\ &= s^2\tilde{Y} - sY(0^+) - \dot{Y}(0^+) \end{aligned}} \tag{3-38}$$

That is, the Laplace transform of the second derivative of a quantity is s^2 times the Laplace transform of that quantity, $\tilde{Y}$, minus the initial value of that quantity times s, minus the initial value of that quantity's first derivative.

Thus, when the initial conditions are *all* zero, the various derivatives can be transformed by replacing the derivative by Laplace transform of the quantity times the appropriate power of s.

3-3-4 Applying the Laplace Transform to the Case with Proportional plus Integral Control

Equation (3-21) can now easily be transformed. Start with the time domain equation derived earlier

$$\tau\frac{d^2Y}{dt^2} + (1+gk)\frac{dY}{dt} + gIY = gk\frac{dS}{dt} + gIS$$

Apply the Laplace transform rules and get an algebraic equation solvable for $\tilde{Y}$

$$(\tau s^2 + (1+gk)s + gI)\,\tilde{Y} = (gks + gI)\tilde{S}$$

where $\tilde{Y}$ and $\tilde{S}$ have been factored out.

Solving for $\tilde{Y}$ gives

$$\begin{aligned} \tilde{Y} &= \frac{gks + gI}{\tau s^2 + (1+gk)s + gI}\tilde{S} = G\tilde{S} \\ G &= \frac{gks + gI}{\tau s^2 + (1+gk)s + gI} \end{aligned} \tag{3-39}$$

where G represents the *transfer function* from S to Y.

Assume that the set point S is given a step at time zero and that $Y(0)$ is zero. Since for $t \geq 0$, S is a constant, the transform for S is then (remember that for $t < 0, S(t) = 0$).

$$\tilde{S} = \frac{S_c}{s}$$

where S_c is the size of the set-point step.

Equation (3-39) becomes

$$\tilde{Y} = \frac{gks + gI}{\tau s^2 + (1+gk)s + gI} \frac{S_c}{s} \tag{3-40}$$

Question 3-6 What can the final value theorem tell us about whether this controlled process will settle out with no offset?

Answer Applying the final value theorem to Eq. (3-40) gives

$$Y(\infty) = \lim_{s\to 0} s\tilde{Y} = \lim_{s\to 0} s \frac{gks + gI}{\tau s^2 + (1+gk)s + gI} \frac{S_c}{s}$$

$$= \lim_{s\to 0} \frac{(gks + gI)S_c}{\tau s^2 + (1+gk)s + gI} = S_c$$

So, the presence of integral control removes the offset.

Question 3-7 Using the result in App. F for the Laplace transform of the integral, could you arrive at Eq. (3-40) starting with

$$\tau \frac{dY}{dt} + Y = gU$$

$$U(t) = ke(t) + I\int_0^t du e(u)$$

or

$$\tau \frac{dY}{dt} + Y = g\left(ke(t) + I\int_0^t du e(u)\right) \tag{3-41}$$

Answer Applying the Laplace transform to Eq. (3-41) gives

$$\tau s\tilde{Y} + \tilde{Y} = gk\tilde{e} + gI\frac{\tilde{e}}{s}$$

where Eq. (F-19) or the fifth entry in the box was used for the integral. Using the definition of e gives

$$\tau s\tilde{Y} + \tilde{Y} = g\left(k + \frac{I}{s}\right)\tilde{e} = g\left(k + \frac{I}{s}\right)(\tilde{S} - \tilde{Y})$$

$$\tilde{Y}\left(\tau s + 1 + g\left(k + \frac{I}{s}\right)\right) = g\left(k + \frac{I}{s}\right)\tilde{S}$$

If the set point is constant then

$$\tilde{S} = \frac{S_c}{s}$$

and

$$\tilde{Y} = \frac{g\left(k + \frac{I}{s}\right)}{\left(\tau s + 1 + g\left(k + \frac{I}{s}\right)\right)}\frac{S_c}{s} = \frac{gks + I}{\tau s^2 + (1 + gk)s + gI}\frac{S_c}{s}$$

Question 3-8 What can the final value theorem tell us about proportional-only control?

Answer Start with

$$\tau\frac{dY}{dt} + Y = gk(S - Y)$$

and apply the Laplace transform to get

$$\tau\tilde{Y} + \tilde{Y} = gk(\tilde{S} - \tilde{Y})$$

but

$$\tilde{S} = \frac{S_c}{s}$$

so $\tilde{Y} = \dfrac{gkS_c}{s(\tau s + 1 + gk)}$ and $Y(\infty) = \lim_{s\to 0} s\tilde{Y} = \dfrac{gkS_c}{1 + gk}$

Question 3-9 In the development of Eq. (3-22) we set $dS/dt = 0$ and then looked at the dynamic behavior for the case of a constant set point. If we take the Laplace transform of Eq. (3-21) we do not get the Eq. (3-40). Why?

Answer By setting $dS/dt = 0$ we have specified that S has been and forever will be constant. On the other hand, by specifying that S is a step change at time zero we have created an entirely different disturbance to our controlled process, hence the appearance of the different numerator in Eq. (3-40).

3-3-5 More Block Diagram Algebra and Some Useful Transfer Functions

The transfer function $G_p(s)$ for the process in Eq. (3-33) was

$$\tilde{Y} = \frac{g}{\tau s + 1}\tilde{U} = G_p\tilde{U}$$

$$G_p = \frac{g}{\tau s + 1}$$

The transfer function for the control algorithm, G_c, can be developed as follows

$$U(t) = ke(t) + I\int_0^t du e(u)$$

$$\tilde{U}(s) = k\tilde{E}(s) + I\frac{\tilde{E}(s)}{s} = \left(k + \frac{I}{s}\right)\tilde{E}(s)$$

$$\frac{\tilde{U}}{\tilde{E}} = \frac{ks + I}{s} = G_c$$

where G_c is the transfer function for the PI controller. The block diagram for a controlled system can be quickly modified from Fig. 3-3, as in Fig. 3-11.

The overall transfer function relating S to Y under closed-loop control can be derived using the following block diagram algebra:

$$\tilde{Y} = G_p\tilde{U}$$

$$\tilde{U} = G_c\tilde{E}$$

Eliminate $\tilde{U}$ to get

$$\tilde{Y} = G_pG_c\tilde{E}$$

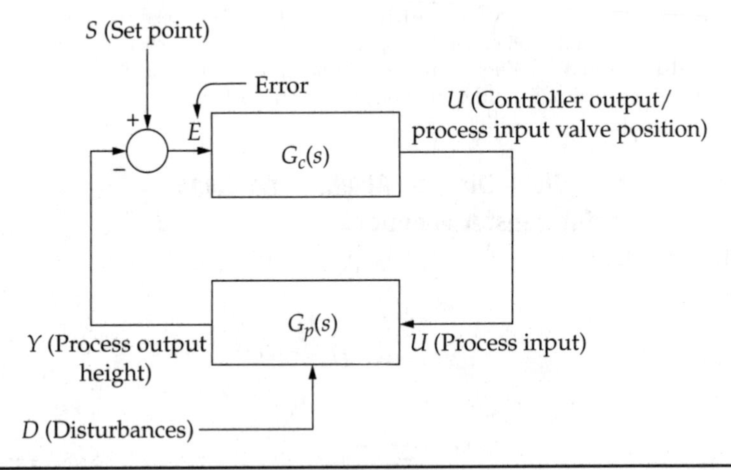

FIGURE 3-11 Block diagram for a controlled system.

Insert the definition of the error

$$\tilde{E} = \tilde{S} - \tilde{Y}$$

$$\tilde{Y} = G_p G_c \left(\tilde{S} - \tilde{Y}\right)$$

Solve for $\tilde{Y}$

$$\tilde{Y} = \frac{G_p G_c}{1 + G_p G_c} \tilde{S} = H\tilde{S}$$

$$\boxed{\frac{\tilde{Y}}{\tilde{S}} = \frac{G_p G_c}{1 + G_p G_c} = H} \tag{3-42}$$

The readers should work through the above steps and make sure she is comfortable with them. The transfer function $\tilde{Y}/\tilde{S}$ or H describes the response of the process output to changes in the set point while under feedback control.

Another, probably more useful, transfer function, which we will call the error transmission function, can be derived using block diagram algebra. Using Fig. 3-12 as a basis, this transfer function, $\tilde{E}/\tilde{N}$, can be developed as follows:

$$\tilde{Y} = G_p \tilde{U}$$

$$\tilde{U} = G_c \tilde{E}$$

$$\tilde{Y} = G_p G_c \tilde{E}$$

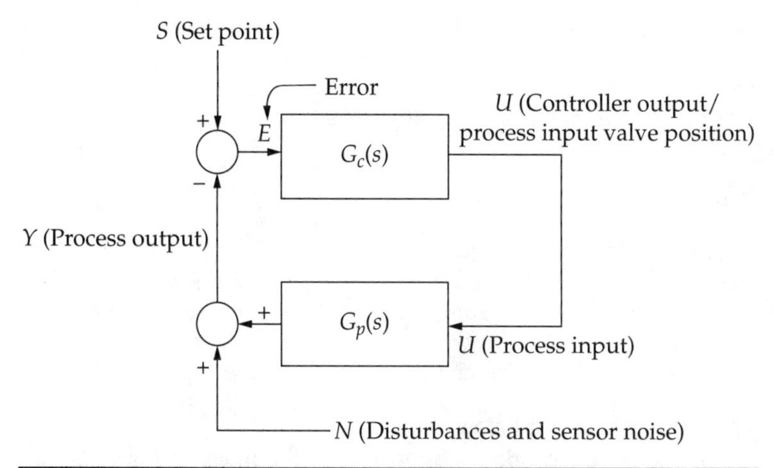

FIGURE 3-12 Block diagram for a controlled system subjected to disturbances and sensor noise.

The error is corrupted by the noise $\tilde{N}$

$$\tilde{E} = \tilde{S} - (\tilde{Y} + \tilde{N})$$

For the time being, ignore the set point

$$\tilde{E} = -(\tilde{Y} + \tilde{N}) = -(G_p G_c \tilde{E} + \tilde{N})$$

Solve for $\tilde{E}$

$$\tilde{E}(1 + G_p G_c) = -\tilde{N}$$

$$\boxed{\frac{\tilde{E}}{\tilde{N}} = -\frac{1}{1 + G_p G_c}} \tag{3-43}$$

During the development of $\tilde{E}/\tilde{N}$ the set point was removed because it is assumed constant at zero. More will be made of $\tilde{E}/\tilde{N}$ when the frequency domain is introduced in the next chapter.

3-3-6 Zeros and Poles

This section will repeatedly refer to Eq. (3-40) which is

$$\tilde{Y} = \frac{gks + gI}{\tau s^2 + (1 + gk)s + gI} \frac{S_c}{s}$$

The numerator in Eq. (3-40), namely, $gks + gI$, has one zero. That is, the value $s = -I/k$ causes this term to be zero, so the zero of this factor is $-I/k$.

The denominator in Eq. (3-40), namely,

$$(\tau s^2 + (1+gk)s + gI)\, s$$

has the same form as the quadratic in Eq. (3-25) with one extra factor. Therefore, the denominator in Eq. (3-40) has three *zeros* (values at which a quantity equals zero). Conventionally, we say that Eq. (3-40) has three *poles* (values at which the quantity becomes infinite) and one zero (the value at which the quantity becomes zero).

Partial Fractions and Poles

Applying the quadratic equation solver, the poles of Eq. (3-40) are found to be

$$-\frac{1+gk}{2\tau} \pm \frac{\sqrt{(1+gk)^2 - 4\tau gI}}{2\tau} \qquad \text{and} \qquad 0.0 \tag{3-44}$$

Two of the roots in Eq. (3-44) are the same as those obtained in Eq. (3-30). Assume for the time being, that the argument of the radical in Eq. (3-44) is positive so that the poles will all be zero or negative real numbers.

To make the following partial fraction algebra a little easier I will factor out τ so that the coefficient of s^2 is unity and Eq. (3-40) becomes

$$\tilde{Y} = \frac{(gks+gI)S_c}{\tau\left(s^2 + \dfrac{(1+gk)}{\tau}s + \dfrac{gI}{\tau}\right)s} = \frac{\dfrac{(gks+gI)S_c}{\tau}}{(s-s_1)(s-s_2)s} \tag{3-45}$$

The resulting quadratic equation for poles is a little different

$$s_1,\ s_2 = -\frac{\dfrac{1+gk}{\tau}}{2} \pm \frac{\sqrt{\left(\dfrac{1+gk}{\tau}\right)^2 - 4\dfrac{gI}{\tau}}}{2}, \qquad s_3 = 0$$

Question 3-10 Is this expression for the poles really different from Eq. (3-30)?

Answer No, a little algebra can show that they are identical.

For the time being, assume that s_1 and s_2 are different and real, that is, assume that

$$\frac{(1+gk)^2}{\tau^2} > 4\frac{gI}{\tau}$$

Expanding Eq. (3-45) using partial fractions gives

$$\tilde{Y} = \frac{(gks+gI)S_c}{\tau\left(s^2+\dfrac{(1+gk)}{\tau}s+\dfrac{gI}{\tau}\right)s} = \frac{\dfrac{(gks+gI)S_c}{\tau}}{(s-s_1)(s-s_2)(s-s_3)} \tag{3-46}$$

The details of the partial fraction expansion and the inversion are carried out in App. F but Eq. (3-46) shows that $Y(t)$ will have three terms: two exponentials from the poles at s_1 and s_2 and one constant from the pole at zero (or at s_3). After the inversion is complete, the result is

$$Y(t) = \frac{gS_c}{\tau}\left(-\frac{k+Is_1}{s_1(s_2-s_1)}e^{s_1t} + \frac{k+Is_2}{s_2(s_2-s_1)}e^{s_2t} + \frac{k}{s_1s_2}\right) \tag{3-47}$$

Therefore, starting with a Laplace transform, partial fractions allowed the transform to be broken down into three simple terms, each of which had a known time domain function as its inverse.

Question 3-11 If Eq. (3-44) had yielded complex poles, how would the development of the partial fraction expansion have changed?

Answer First, One has to remember that s_1 and s_2 are now complex conjugates. Second, one has to figure out how to use Euler's formula to present the result. So, there is no major difference other than a lot more algebra that includes complex numbers. If you are energetic you might try it.

Poles and Time Domain Exponential Terms

The development of Eq. (3-47) suggests that a nonzero pole in the Laplace transform of a quantity relates directly to an exponential term in the time domain. In fact, this is always true and it is a good reason for being so interested in poles. That is, a factor in the Laplace transform having the form showing a pole at $s = p$, as in

$$\frac{1}{s-p}$$

corresponds to a time domain term of

$$e^{pt}$$

Poles can be complex but if so then they must occur in conjugate pairs. Therefore, the factor occurring in a Laplace transform as in

$$\frac{1}{(s-p)(s-p^*)} = \frac{1}{(s-(a+jb))(s-(a-jb))} = \frac{1}{s^2+2as+a^2+b^2}$$

has two complex poles that occur as conjugates. As a consequence, the factor is purely real which you would want because an imaginary process transfer function does not make physical sense.

These complex conjugates also correspond to exponential terms in the time domain except now they occur as

$$C_1e^{(a+jb)t} + C_2e^{(a-jb)t}$$

and end up contributing sinusoidal terms in the time domain.

These pairings suggest several things:

1. A pole at $s = 0$ corresponds to a constant or an offset.
2. When the pole lies on the negative real axis, the corresponding exponential term will also be real and will die away with time.
3. As the pole's location moves to the left on the negative real axis the exponential term will die away more quickly. As the pole moves to the right along the negative real axis in the s-plane it will soon reach $s = 0$ at which point it corresponds to a constant in the time domain. As the pole continues to move into the right-hand side of the s-plane, still along the real axis, the exponential component now increases with time without bound.
4. When the poles appear in the s-plane with components displaced from the real axis then the poles are complex and appear as complex conjugates. The corresponding time domain terms will contain sinusoidal parts and underdamped bounded behavior will result if the poles lie in the left half of the s-plane.
5. If the complex poles are purely imaginary they still appear as conjugates on the imaginary axis and they correspond to undamped sinusoidal behavior that does not dissipate. As the imaginary component of the complex conjugate poles moves away from the real axis (while staying on the imaginary axis) the frequency of the underdamping will increase.
6. If the transfer function has poles that occur in the right-hand side of the s-plane, that is, if the poles have positive real parts, then the process represented by the transfer function will be unstable.

For example, in the development of Eq. (3-33) for the first-order model

$$\tilde{Y} = \frac{g}{\tau s+1}\frac{U_c}{s}$$

there is a pole at $s = -1/\tau$ and at $s = 0$. These poles correspond to an exponential term $e^{-1/\tau}$ and a constant term.

In general, the Laplace transform can be written as a ratio of a numerator $N(s)$ to a denominator $D(s)$

$$\tilde{Y}(s) = G(s) = \frac{N(s)}{D(s)} = \frac{b_0 + b_1 s + \ldots + b_m s^m}{a_0 + a_1 s + \ldots + a_n s^n} = K\frac{(s-z_1)(s-z_2)\ldots(s-z_m)}{(s-p_1)(s-p_2)\ldots(s-p_n)}$$

(3-48)

showing that $G(s)$ has m zeros, $z_1, z_2, \ldots, z_m$ and n poles, $p_1, p_2, \ldots, p_n$, any of which can be real or complex; however, complex poles and zeros must appear as paired complex conjugates so that their product will yield a real quantity.

The inversion of $N(s)/D(s)$ will yield

$$Y(t) = \sum_{k=1}^{n} C_k e^{Pkt} \tag{3-49}$$

Note that if some of the poles are complex they will occur as complex conjugates and the associated exponential terms will contain sinusoidal terms via Euler's formula. Finally, note that to find the poles one usually sets the denominator of the Laplace transform, $D(s)$, to zero and solves for the roots.

The transfer function for the controlled system is

$$G(s) = \frac{G_c G_p}{1 + G_c G_p}$$

To see if this controlled system is stable one could find the values of s (or the poles of $G(s)$) that cause

$$1 + G_c G_p = 0$$

or

$$G_c G_p = -1 \tag{3-50}$$

We will return to this equation many times in subsequent chapters.

The term "pole" may come from the appearance of the magnitude of a Laplace transform when plotted in the s-domain. Consider the first-order Laplace transform

$$G(s) = \frac{g}{\tau s + 1} = \frac{3}{s+3}$$

$$g = 1 \qquad \tau = 0.3333$$

which has a pole at $s = -3$. The magnitude of $G(s)$ can be obtained from its complex conjugate, as explained in App. F, as

$$s = a + jb$$

$$G(s) = \frac{3}{\tau(a + jb) + 3} = \frac{3}{\tau a + 3 + j\tau b}$$

$$|G(s)| = \sqrt{G \cdot G^*} = \sqrt{\frac{3}{\tau a + 3 + j\tau b} \cdot \frac{3}{\tau a + 3 - j\tau b}} = \frac{3}{\sqrt{(\tau a + 3)^2 + \tau^2 b^2}}$$

First, plot the location of the pole in the s-plane where a represents a point on the real axis and b represents a point on the imaginary axis (Fig. 3-13). Next, plot the magnitude of $G(s)$ against the s-plane as in Fig. 3-14. Notice how the magnitude of $G(s)$ looks like a tent that has a tent pole located at $s = -3.0$ which lies on the real axis in the s-plane.

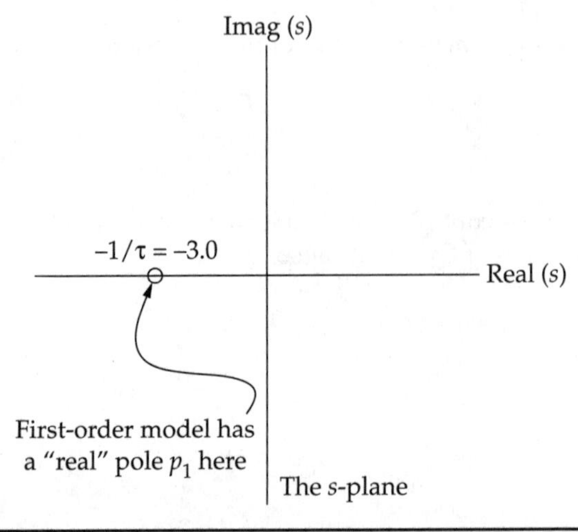

FIGURE 3-13 Location of a pole in the s-plane.

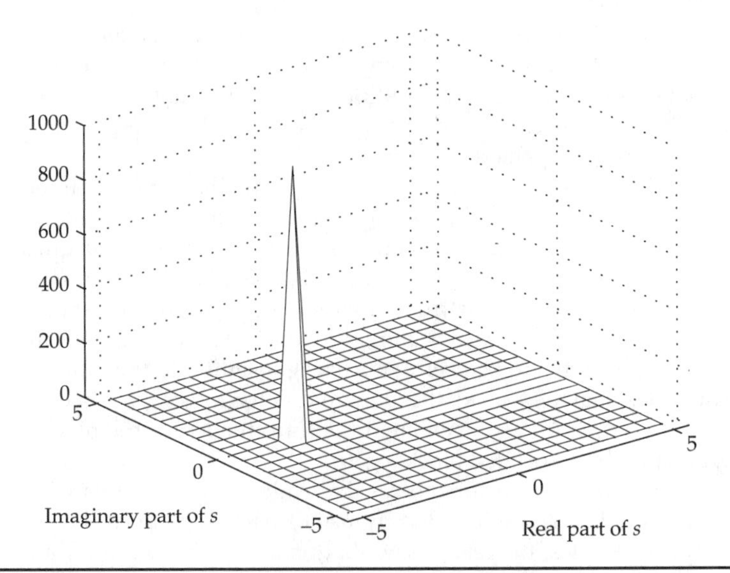

FIGURE 3-14 Magnitude of the first-order Laplace transform.

3-4 Summary

It's time for a break. This chapter has been the first with a lot of mathematics and it probably has been difficult to digest. Hopefully, you haven't lost your motivation to continue (or to reread this chapter along with the appropriate appendices).

We started with an elementary dynamic analysis of a tank filled with liquid. The conservation of mass coupled with a constitutive equation yielded a linear first-order differential equation that described the behavior of an ideal model of the tank. Using elementary methods, the differential equation was solved and the solution was shown to support our intuitive feelings for the tank's dynamics. The important concepts of time constant and gain were introduced.

Simple proportional feedback control was attached to the process, producing another first-order differential equation that was also relatively simple to solve. The failure of proportional control to drive the process output all the way to set point was noted.

Integral control was added. Now, the offset between the process output and the set point could be eliminated. The differential equation that described this situation was second order and required a little more mathematical sophistication to solve. The concepts of critical damping, underdamped behavior, and overdamped behavior were introduced. Although the liquid tank under proportional-integral control could exhibit underdamped behavior, its response to a step change in the set point was shown to always be stable.

The example process was quite simple but the idea that proportional-only control leaves an offset between the set point and the process variable is general. That the addition of integral control can remove the offset but can cause underdamped behavior if applied too aggressively is another general concept.

Perhaps the reader could see that the mathematics required to describe the behavior of anything more complicated than PI control applied to a first-order process was going to get messy quite quickly. This set the scene for the introduction of the Laplace transform which allowed us to move away from differential equations and get back to algebra. The recipe for using the *s* (or Heaviside) operator was introduced and shown to be useful in gaining insight into the differential equations that described the dynamic behavior of processes. The Laplace transform also facilitated the introduction of the block diagram and the associated block algebra.

Coupled with the appendices the reader saw that Laplace transforms could often be inverted by use of partial fractions. From the simple examples, the reader saw that poles of the *s*-domain transfer function are related to exponential terms in the time domain.

At this point it appears as though the Laplace transform is mostly useful in solving differential equations. Later on, we will see that the Laplace transform can be used to gain significant amounts of insight in other ways that do not involve inversion.

We have broken the ice and are ready to dive into the cold, deep water. First, we will move into yet one more domain, the frequency domain. Then a couple of processes more sophisticated than the simple liquid tank will be introduced before we look at controlled systems in the three domains of time, Laplace, and frequency.

CHAPTER 4

A New Domain and More Process Models

Chapter 3 introduced the reader to a relatively sophisticated tool, the Laplace transform. It was shown to be handy for solving the differential equations that describe model processes. It appeared to have some other features that could yield insight into a model's behavior without actually doing an inversion.

In this chapter the Laplace transform will be used as a stepping stone to lead us to the frequency domain where we will learn more tools for gaining insight into dynamic behavior of processes and controlled systems.

Chapter 3 also got us started with a simple process model, the first-order model that behaved approximately as many real processes do. However, this model is not sufficient to cover the wide variety of industrial processes that the control engineer must deal with. So, to the first-order process model we will add a pure dead-time model which will subsequently be combined with the former to produce the first-order with dead-time or FOWDT model. For technical support the reader may want to read App. B (complex numbers), App. D (infinite series), App. E (first- and second-order differential equations), and App. F (Laplace transforms).

As in previous chapters, each new process will be put under control. In this chapter the new tool of frequency domain analysis will be used to augment time domain studies.

4-1 Onward to the Frequency Domain

4-1-1 Sinusoidally Disturbing the First-Order Process

Instead of disturbing our tank of liquid with a step change in the input flow rate, consider an input flow rate that varies as a sinusoid about some nominal value as shown in Fig. 4-1. The figure suggests

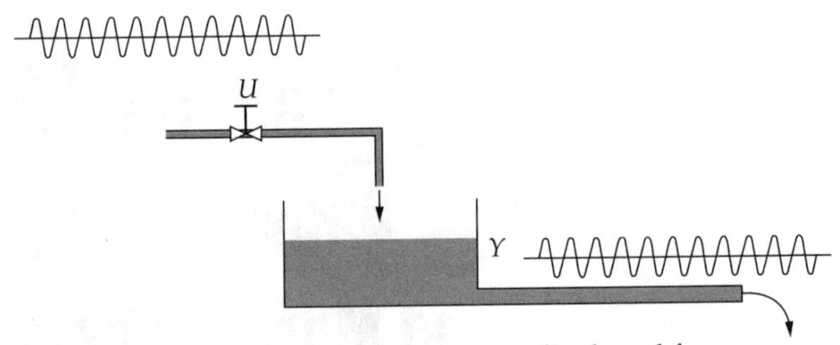

Put in a sinusoidal flow rate U of given amplitude and frequency —what does the output flow rate Y do?

FIGURE 4-1 Frequency response of tank of liquid.

that if the input varies sinusoidally so will the level (and the output flow rate, too). Assume that the input flow rate is described by

$$U(t) = U_c + A_U \sin(2\pi ft)$$

The input flow rate has a nominal value of U_c. The flow rate is varying about the nominal value with an amplitude A_U and a frequency f which often has units of hertz or cycles per second. Another frequency represented by ω is the *radian* frequency, usually having units of radians per second. It is related to the *other* frequency by $\omega = 2\pi f$. Therefore, the input flow rate could also be written as

$$U(t) = U_c + A_U \sin(\omega t)$$

For the time being, consider the output flow rate F_o as the process output. In Chap. 3 the level L was the process output. The simple equations describing these quantities were

$$\tau \frac{dL}{dt} + L = RF_i$$

$$F_o = \frac{L}{R}$$

or, after combining,

$$R\tau \frac{dF_o}{dt} + RF_o = RF_i$$

or

$$\tau \frac{dF_o}{dt} + F_o = F_i \qquad (4\text{-}1)$$

Equation (4-1) shows that, when the output flow rate is the process output and the input flow rate is the process input, the process gain is unity and the time constant is the same as when the level is the process output. Making this choice of process input and output variables will simplify some of the graphs and some of the interpretations. Later on, we can extend the presentation to nonunity gains with ease.

Now, with this simple background in mind, what will the output flow rate look like when the input flow rate oscillates about some nominal value? First, F_o will have a nominal value F_c and it will vary about its nominal value with an amplitude, A_Y, a frequency, f, and a phase (relative to that of F_i), θ, as in

$$F_o(t) = F_c + A_Y \sin(2\pi f t + \theta) \tag{4-2}$$

Note that the frequency of the oscillations in the input flow rate and the tank level is the same. This is an assumption that we will support soon. However, the amplitudes and the phases are different.

Assume that the tank has a time constant of 40 min. Consider Fig. 4-2 where the input flow rate has a period of 100 min or a frequency of 0.01 min^{-1}. Note that the output flow rate lags the input flow rate (has a positive nonzero phase relative to the input flow rate) and has a smaller amplitude.

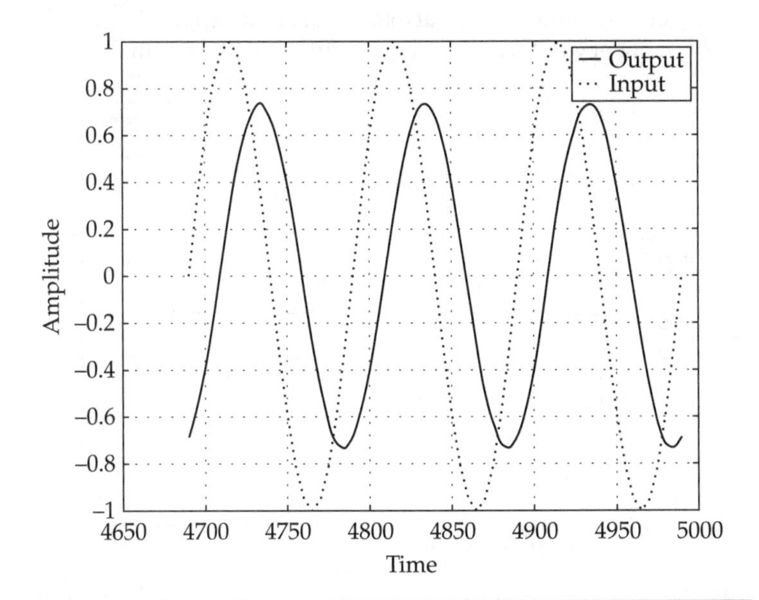

FIGURE 4-2 Input flow rate and level for $f = 0.01$ min^{-1}. Sinusoidal response of tank with period = 100.

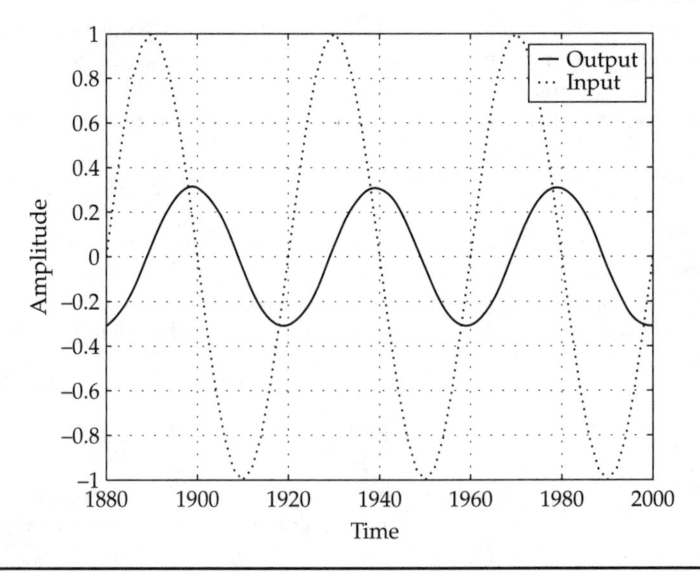

FIGURE 4-3 Input flow rate and level for $f = 0.025\ \text{min}^{-1}$. Sinusoidal response of tank with period = 40.

In Fig. 4-3 the frequency of the input flow rate frequency is increased to $0.025\ \text{min}^{-1}$ (a period of 40 min). Notice that the output flow rate lags the input flow rate even more (greater phase lag) and the ratio of the output amplitude to the input amplitude is smaller (more attenuation) than for the case of the lower frequency.

Figure 4-4 shows the input/output relationship for the case of an input frequency of $1.0\ \text{min}^{-1}$. The amplitude of the output flow rate is

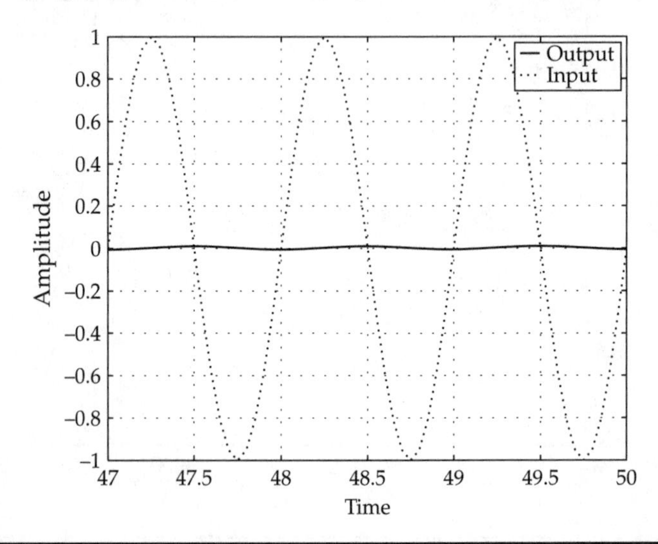

FIGURE 4-4 Input flow rate and level for $f = 1.0\ \text{min}^{-1}$. Sinusoidal response of tank with period = 1.

barely discernable and the lag is almost 90°. Notice that the scales of the time axes on these last three plots are different. The first has a span of 350, the second 120, and the third has a span of 3.0.

What is going on? The inertia associated with the mass of liquid in the tank (characterized by the tank's time constant) causes the output flow rate's response to be attenuated as the frequency of the input flow rate increases. At low input frequencies, in spite of the inertia, the output flow rate is nearly in phase with the input flow rate and there is almost no lag. The slowly varying input flow rate gives the mass of liquid time to respond. As the frequency increases, the mass of the liquid cannot keep up with the input flow rate and the lag increases and the ratio of output amplitude to input amplitude decreases. Note, however, that the frequency of the output flow rate is still identical to that of the input flow rate. As you might expect, and as we will soon show, the phase lag is directly related to the process time constant. Likewise, the attenuation in the amplitude ratio depends on the process time constant.

From the point of view of the flow rates, the tank behaves as a *low pass filter,* that is, it passes low frequency variations almost without attenuation with almost zero phase lag. For high frequency variations it attenuates the amplitude and adds phase lag. Filters as processes or processes as filters will be dealt with later on in this chapter and in Chap. 9.

4-1-2 A Little Mathematical Support in the Time Domain

Let's see if some simple math can "prove" our contentions. Another way of writing Eq. (4-2), ignoring the constant offset value, is

$$U(t) = A_U \sin(2\pi ft) = A_U \operatorname{Re}\{e^{j2\pi ft}\}$$

This makes use of Euler's equation that is presented in App. B. It simply says that a sine function is the real part of a complex exponential function. If this bothers you and you do not want to delve into App. B, then you had best skim the rest of this subsection. If not, then temporarily forget about the "real part" and use

$$U(t) = A_U e^{j2\pi ft} \tag{4-3}$$

This is a common method of control engineers. It says, "make the input flow rate a complex sinusoid (knowing full well that you are only interested in the real part) and use it to solve a problem; then when the solution has been obtained, if it is complex, take the real part of the solution and you're home!" The simple algebra of complex exponentials is often preferable to the sometimes sophisticated complexity of the trigonometric relationships.

With this leap of faith in hand, feed the expression for $U(t)$ given in Eq. (4-3) into the differential equation describing our simple tank of liquid, that is, Eq. (4-1), and assume that the process outlet flow rate, $Y(t)$, will also be a sinusoid with the same frequency but with a phase relative to $U(t)$, namely,

$$Y(t) = Ce^{j(2\pi ft+\theta)} \tag{4-4}$$

Note that the amplitude of the process outlet C is as yet unknown as is the phase θ. As mentioned in our leap of faith statement, assume that the actual process outlet flow rate is the real part of the expression in Eq. (4-4). Using Eq. (4-4) also assumes that after making the process input U a sinusoid, all of the transients associated with that change have died out leaving the outlet flow rate to be a complex sinusoid with a modified amplitude and phase. If the choice of Eq. (4-4) is incorrect it will show up quickly as we turn the crank.

With all these nontrivial preliminaries out of the way, put Eqs. (4-2) and (4-4) into Eq. (4-1) and see if something insightful happens. A lot of the details will be left to App. B. After plugging in the expression for U and Y, Eq. (4-1) becomes

$$\tau(j2\pi f)Ce^{j(2\pi ft+\theta)} + Ce^{j(2\pi ft+\theta)} = A_U e^{j(2\pi ft)}$$

This messy looking expression is really a simple (but complex) equation that has real and imaginary parts. In App. B these real and imaginary parts are collected algebraically and the real parts on the left-hand side of the above equation are equated to the real parts on the right-hand side. The same thing is done with the imaginary parts. This gives two equations that, with the help of some inverse trigonometric identities, can be solved for the unknown amplitude C and the phase θ. The result is

$$\begin{aligned} \theta &= -\tan^{-1}(2\pi f\tau) = -\tan^{-1}(\omega\tau) \\ C &= \frac{A_U}{\sqrt{1+(2\pi f\tau)^2}} = \frac{A_U}{\sqrt{1+(\omega\tau)^2}} \end{aligned} \tag{4-5}$$

This supports the earlier contention that as the frequency increases the phase lag increases (or the phase becomes more negative) as does the amplitude attenuation. It also tells us that as the tank size and its associated time constant increase so does the phase lag and the amplitude attenuation. Furthermore, it supports our contention that the frequency of the process output is the same as that of the input [if this were not the case then the idea of plugging in assumed functions for $U(t)$ and $Y(t)$ would not have worked]. Finally, it suggests that the maximum phase lag for this first-order process is 90°.

Question 4-1 Why is the maximum phase lag of the first-order model 90°?

Answer As ω, in $\theta = -\tan^{-1}(\omega\tau)$, increases without bound to ∞, the arctangent function yields $\tan^{-1}(\infty) = \pi / 2$ or 90°.

4-1-3 A Little Mathematical Support in the Laplace Transform Domain

From Chap. 3 and App. F, the transfer function for the process described by Eq. (4-1) can be obtained directly from Eq. (3-33) by setting $g = 1.0$, resulting in

$$\tilde{Y} = \frac{1}{\tau s + 1}\tilde{U} = G_p \tilde{U}$$

$$G_p = \frac{1}{\tau s + 1} \tag{4-6}$$

Now, another trick! Let $s = j2\pi f$ and find the magnitude and the phase of the result

$$\begin{aligned} G_p(j2\pi f) &= \frac{1}{\tau j2\pi f + 1} \\ &= \frac{1}{\tau j2\pi f + 1} \cdot \frac{-\tau j2\pi f + 1}{-\tau j2\pi f + 1} \\ &= \frac{-\tau j2\pi f + 1}{(\tau 2\pi f)^2 + 1} \\ &= \frac{1}{(\tau 2\pi f)^2 + 1} - j\frac{\tau 2\pi f}{(\tau 2\pi f)^2 + 1} \end{aligned}$$

Here, the numerator and denominator of the complex transfer function were multiplied by the conjugate of the denominator $-\tau j2\pi f + 1$. This got rid of imaginary components in the denominator and allowed us to separate G_p into its real and imaginary parts.

The transfer function is now a complex quantity with a magnitude and a phase, as in

$$G_p = |G_p| e^{j\theta}$$

As shown in App. B, the magnitude of a complex quantity is the square root of the real and imaginary parts

$$|G_p(j2\pi f)| = \frac{1}{\sqrt{(\tau 2\pi f)^2 + 1}} \tag{4-7}$$

and the phase is the angle whose tangent is the ratio of the imaginary to the real part

$$\theta = \tan^{-1}\left(\frac{-\tau 2\pi f}{1}\right) = -\tan^{-1}(\tau 2\pi f) \tag{4-8}$$

The last two equations turn out to be the same as Eq. (4-5). So, we see one more reason why the Laplace transform can be so useful: there is an easy, straightforward path from the Laplace domain to the frequency domain. All you have to do is accept the serendipitous effect of replacing the Laplace operator by $j\omega$. There is one caveat. The result of making the substitution gives the *steady-state* sinusoidal solution after the transients have died out—remember, when you feed a sinusoid to a process, the process output requires some time to evolve toward a sinusoidal function. Refer to App. B where the full solution, including the transient part, is given.

4-1-4 A Little Graphical Support

How can this information be presented more compactly? Try plotting the amplitude ratio and the phase lag versus the frequency. For this example the result, called a Bode plot, would be as shown in Fig. 4-5,

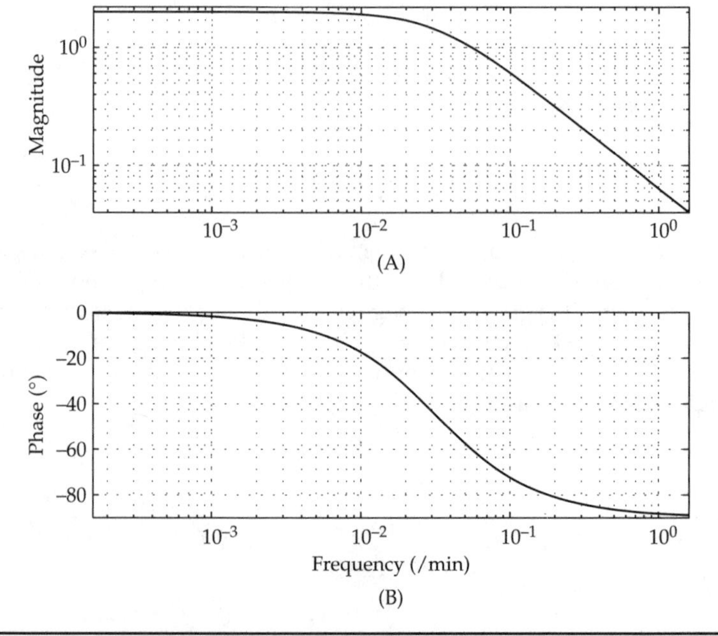

FIGURE 4-5 Bode plot for liquid tank. (A) Ratio of amplitudes: F_o/F_i. (B) Phase of F_o relative to F_i.

which is derived from Eqs. (4-7) and (4-8). Often control engineers use decibels or dB instead of magnitude where

$$dB = 20\log_{10}(\text{magnitude})$$

Replotting Fig. 4-5 in these units gives Fig. 4-6. Note how the amplitude ratio decreases with increasing frequency and how the phase gets more negative as frequency increases. While the amplitude ratio appears to continue to decrease indefinitely as frequency increases, the phase appears to reach an asymptote of –90°, which is consistent with earlier comments.

In the extreme case of an oscillating input flow rate having a nearly zero frequency (or an extremely long period) there would be plenty of time for the output flow rate to respond to any change in the input flow rate and overcome the inertia. Therefore, the amplitude ratio reaches a left-hand asymptote of unity. Likewise, the phase angle between the input and output would essentially be zero. Effectively this asymptote represents a quiescent steady state.

On the other hand, with an extremely high frequency, the inertia of the liquid in the tank would tend to wash out the effect of any

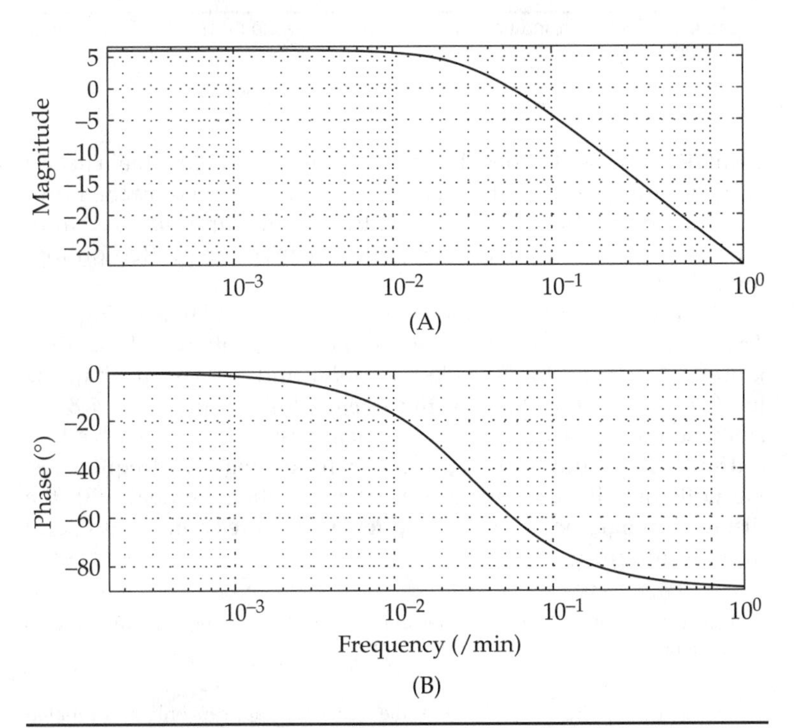

FIGURE 4-6 Bode plot using dB units. (A) Ratio of amplitudes: F_o/F_i. (B) Phase of F_o relative to F_i.

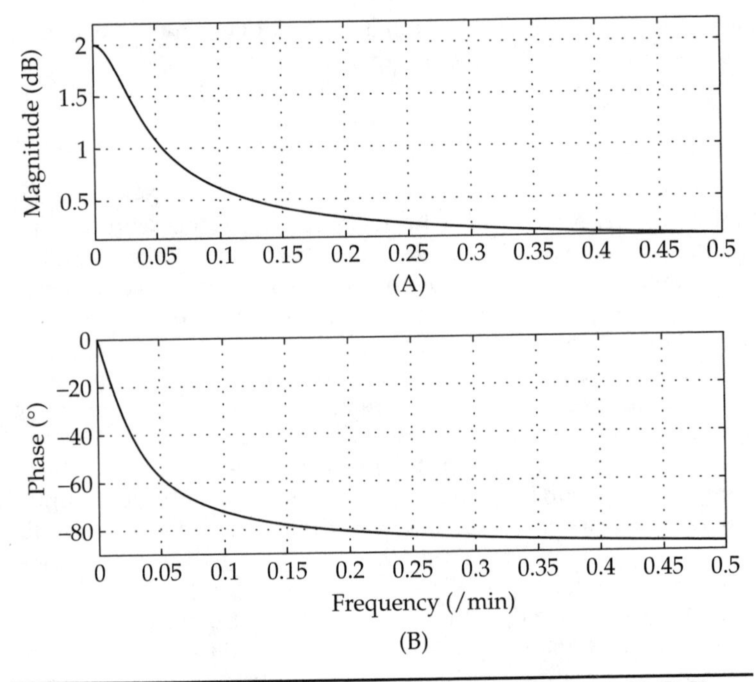

FIGURE 4-7 Bode plot in linear units. (A) Ratio of amplitudes: F_o/F_i. (B) Phase of F_o relative to F_i.

input flow rate oscillations. The output flow rate would remain nearly at its nominal value, almost completely undisturbed by the input flow rate oscillations. The ratio of the outlet amplitude to that of the input would be nearly zero, which cannot be plotted on log-log graphs.

Had we used linear scales the picture would be as in Fig. 4-7. The linear plot tends to compress some of the action and for some kind of design problems is less popular and less useful than the log–log plot. Before leaving the linear plot consider Fig. 4-8, an extension of Fig. 4-7, which has a linear frequency axis with a logarithmic amplitude ratio axis and includes negative frequencies. We will come back to this graph later on when we deal with the discrete time domain. Linear frequency axes are sometimes useful in filter design.

Question 4-2 Why is the magnitude in Fig. 4-8 symmetrical about zero frequency?

Answer Look at Eq. (4-7). Note that the frequency f appears only as a squared quantity. Therefore, the magnitude does not depend on the sign of the frequency.

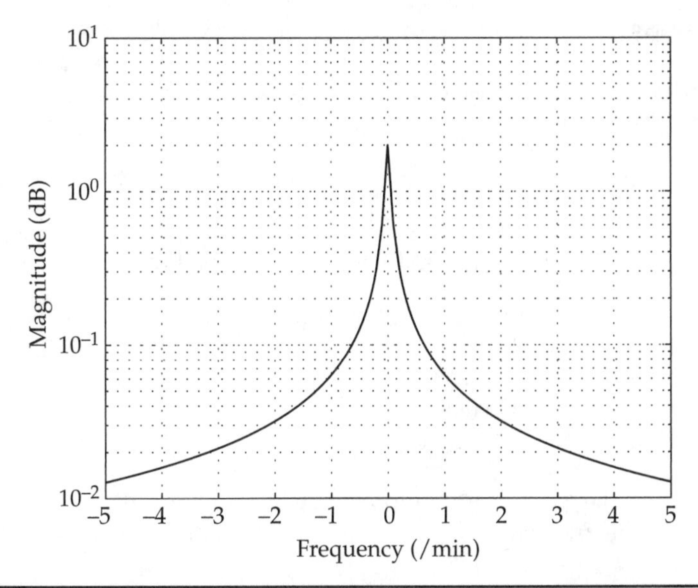

FIGURE 4-8 Linear frequency axis showing negative and positive frequencies.

4-1-5 A Graphing Trick

As we will see, some processes can be modeled with a series of first-order components so it is sometimes handy to be able to sketch the first-order magnitude on the log-log plots.

First, when $f \to 0$ or $\omega \to 0$, Eq. (4-7) shows that the magnitude approaches an asymptote of unity or a value equal to the gain g if it is not unity. On a dB plot with $g = 1$, this asymptote would be at zero.

Second, to see how the magnitude behaves when $\omega \to \infty$, start with Eq. (4-7) and apply some of the rules of logarithms.

$$\log_{10}\left|G_p(j\omega)\right| = \log_{10}\frac{1}{\sqrt{(\tau\omega)^2+1}}$$

$$= \log_{10}((\tau\omega)^2+1)^{-\frac{1}{2}}$$

$$= -\frac{1}{2}\log_{10}((\tau\omega)^2+1)$$

$$\lim_{\omega\to\infty}\left|G_p(j\omega)\right| \cong -\frac{1}{2}\log_{10}((\tau\omega)^2)$$

$$\cong -\log_{10}(\tau\omega)$$

Therefore, if $|G|$ is plotted against $\tau\omega$ on a log-log plot, at large values of $\tau\omega$ the graph will show a straight line with a slope of –1.

Question 4-3 What would the slope at large frequencies be if dB units were used?

Answer For large frequencies $\text{dB} \cong 20\log_{10}(\tau\omega)$ so the slope would be −20 dB per decade change in frequency.

Third, when $\tau\omega = 1$, the magnitude in dB is given by

$$20\log_{10}\left|G_p(j\omega)\right| = 20\log_{10}\frac{1}{\sqrt{(1)^2+1}} = 20\log_{10}\frac{1}{\sqrt{2}}$$

$$= -20\log_{10}\sqrt{2} = -3.0103 \text{ dB}$$

$$\left|G_p(j\omega)\right| = \frac{1}{\sqrt{(1)^2+1}} = 0.7071$$

Thus, the graph of $|G|$ is approximately 0.7071 or 3 dB down when $\tau\omega = 1$. This frequency, $\omega_{cor} = 1/\tau$ or $f_{cor} = 1/(2\pi\tau)$ is called the *corner* frequency.

The shape of the $|G|$ curve can be approximated with two straight lines—one is horizontal from small $\tau\omega$ to the point where $\tau\omega = 1$. The second has a slope of −1 and starts at $\tau\omega = 1$.

The phase in Eq. (4-8) shows a left-hand asymptote at 0° and a right-hand asymptote at 90°. When $\tau\omega = 1$, the phase is 45°.

Figure 4-9 shows a Bode plot for the case of unity gain and a time constant of 10.0. The magnitude is plotted against $\tau\omega$ on a log-log

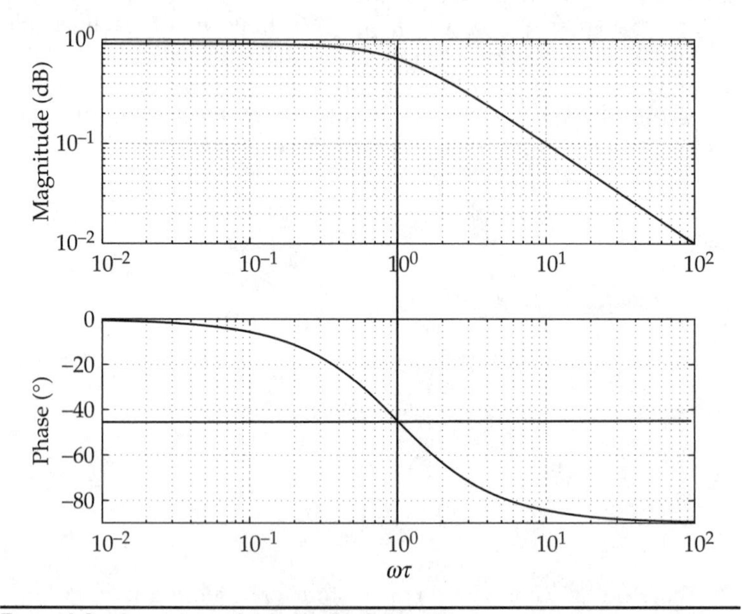

FIGURE 4-9 Asymptotes for first-order Bode plot.

scale and the phase is plotted on a semilog scale. The vertical line indicates $\tau\omega = 1$ and the horizontal line indicates a phase of 45°.

4-2 How Can Sinusoids Help Us with Understanding Feedback Control?

In the Sec. 4-1, the input flow rate was varied sinusoidally and the output flow rate was observed. This was an open-loop disturbance with no control involved. Now, let's dreg up the closed-loop schematic that we talked about in Chap. 3. There is one change, however. For the time being, the process output will be the process outlet flow rate, so Eq. (4-1) with its unity gain describes the behavior of our process. The process input will still be the process input flow rate.

In Fig. 4-10 note that the set point is varied sinusoidally and the feedback loop is cut just before the process output is fed back and subtracted from the set point. We will focus on the output of the cut line as a response to the sinusoidally varying set-point input. The gain and phase of the output at the cut point will be called the open-loop gain and phase.

However, before looking at that input/output relationship, consider what happens at the point where the process output is subtracted from the set point. An equivalent diagram appears in the upper right-hand corner of Fig. 4-10. Here the subtraction is broken up into a negation followed by an addition. What happens to a sinusoid (the process output) that is negated before it is added to the set point? When the sign is changed from positive to negative the process variable immediately experiences a phase lag of 180° or, in other words, negating a quantity causes it to have a phase of –180°. To see this, look at Fig. 4-11.

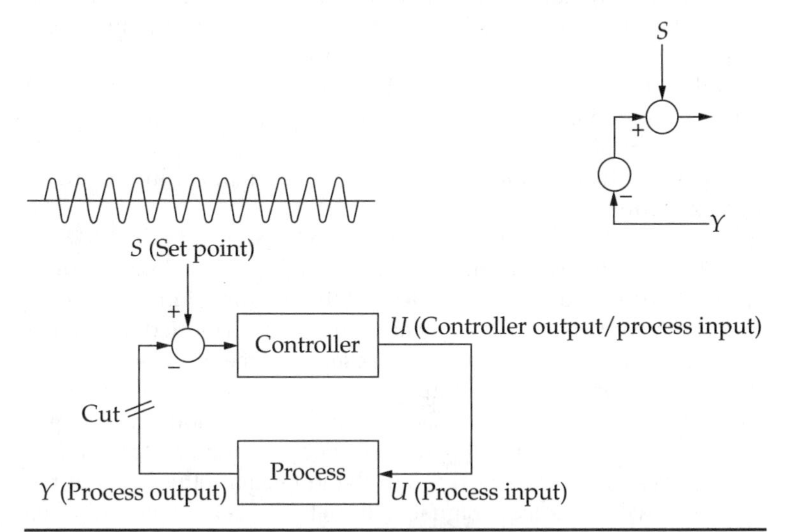

FIGURE 4-10 Varying the set point and cutting the loop.

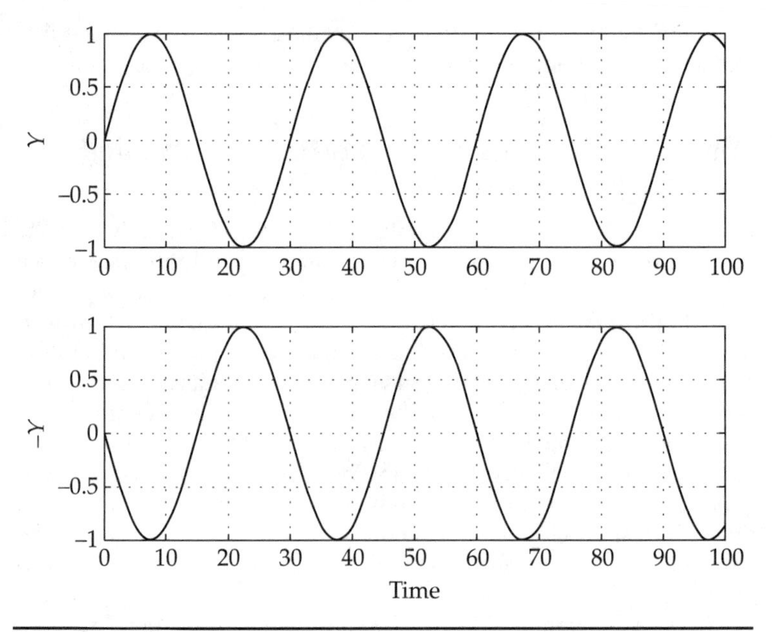

FIGURE 4-11 Negating a quantity.

Now, go back to Fig. 4-10. What can we say about the process output at the cut point? First, we can guess that, relative to the set point (the input) the process output will have less amplitude and more phase lag because of inertial effects in the process (which has unity gain). The controller also might modify the amplitude and phase of this signal. But, *for the time being*, assume that neither the process nor the control adds phase lag and that neither attenuates the amplitude.

When the loop is closed and Y is subtracted from S, the phase lag of Y immediately gains 180°. So, a signal with −180° of phase and unity amplitude is now being *added* to a signal (the set point) having zero phase and unity amplitude. The result is "annihilation" or cancellation.

Now, what if the process and controller were to add −180° of phase to the process output without changing the amplitude from unity? When the addition takes place at the summing junction, the set point with zero phase and unity amplitude will be added to a signal with a phase of −360° (sum of the phase lag from the process/controller *and* the negation) and unity amplitude. So, the result of the summing junction will be to produce a signal with an amplitude that is larger than either the set point alone or the process variable alone. When this enlarged signal passes through the loop again, the result of the summing will produce a signal with yet a larger amplitude. This sequence is shown in Fig. 4-12.

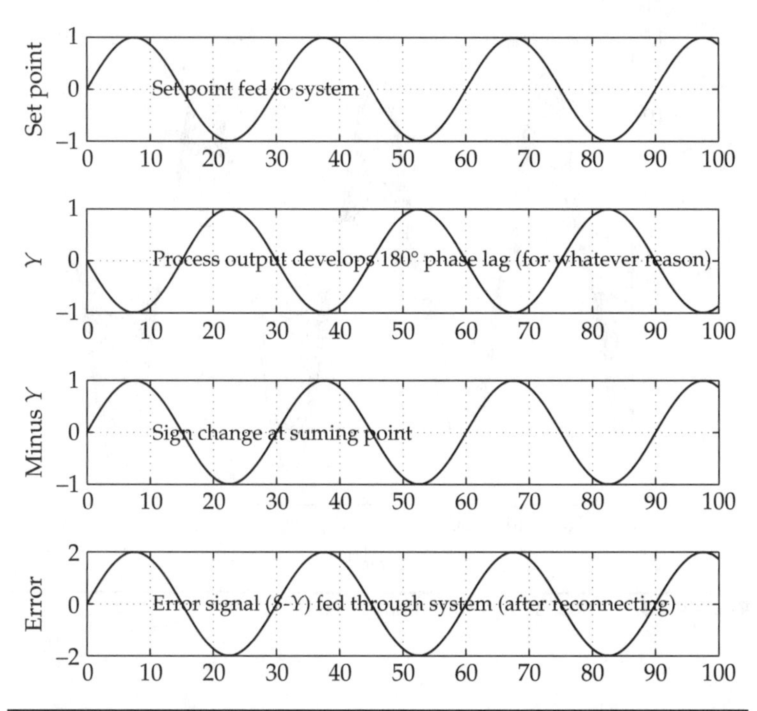

FIGURE 4-12 Signal growth when process/controller provides −180° of phase lag.

At the risk of beating this to death, we repeat the sequence. After one pass through the hypothetical system, the effect of the process/controller causes the process output to maintain the amplitude of unity but to experience a phase lag of 180°. Then when the process output is negated at the summing junction (and becomes the error signal) it gains another 180° of phase lag so that it is now perfectly in phase with the set point and the two signals add. Figure 4-13 shows how the error signal grows without bound (actually after 12 passes through the loop) when there is a phase lag of 180° in the process output. This argument suggests that when a signal in a feedback loop has unity amplitude and −180° of phase just before the subtraction point, there will be unbounded amplification. It further suggests that when the phase lag of the signal being fed back is less than 180° (other things being the same) the unbounded amplification will not occur. Consider the case where the phase lag of the fed-back signal is 170° in Figs. 4-14 and 4-15. Figure 4-15 shows that the signal is amplified as it cycles through the loop but it levels out at a value of about 11.5. So, there is growth but it is bounded.

We leave this section with the thought that we should design a controller such that the open-loop gain (the gain when the loop is cut as in Fig. 4-10) of the total system is less than unity if the phase lag is 180°.

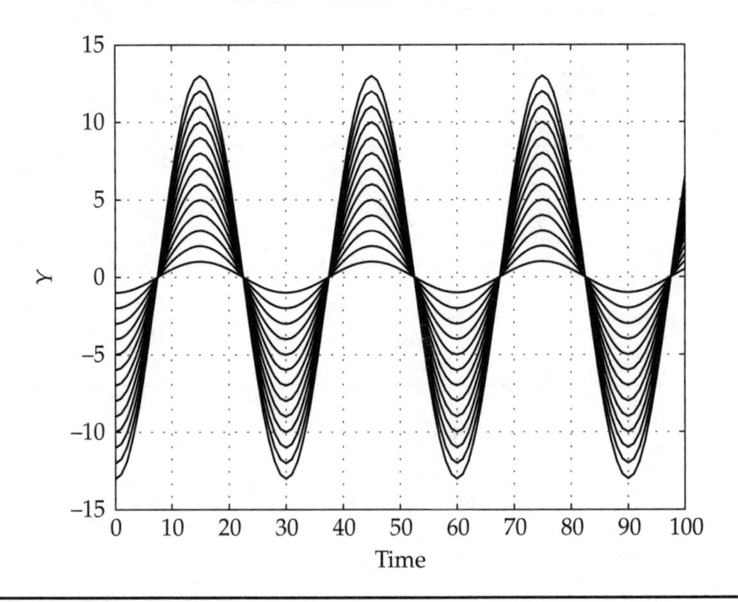

FIGURE 4-13 Unbounded growth when phase lag is 180°.

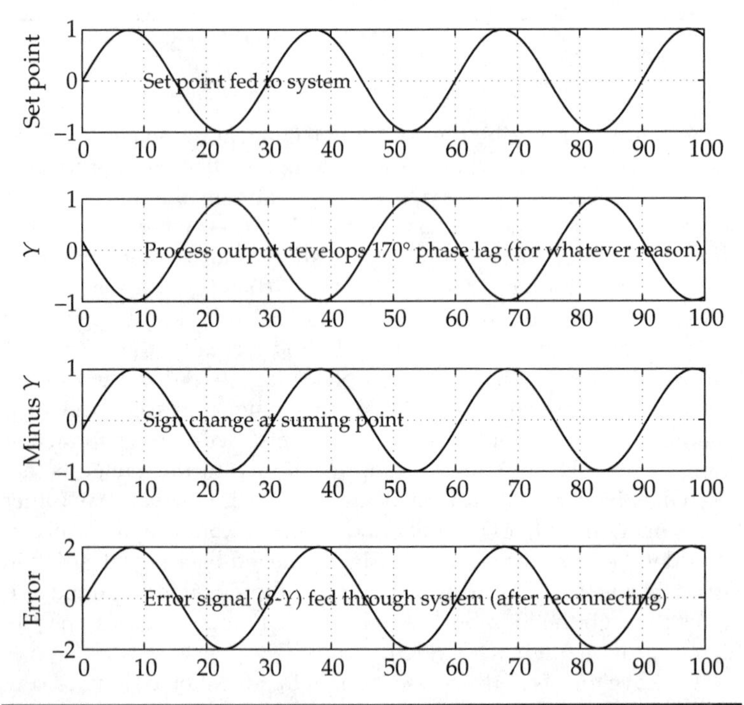

FIGURE 4-14 Signal growth when process/controller provides −170° of phase lag.

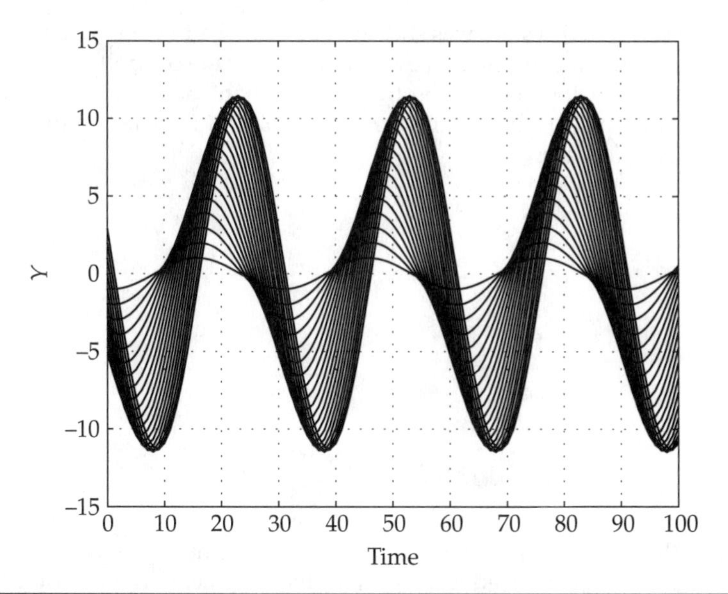

Figure 4-15 Bounded growth when phase lag is 170°.

Alternatively, we should design a controller such that the open-loop phase lag is less than 180° when the open-loop gain is unity. The combination of open-loop amplitudes greater than unity and phase lags greater than 180° spells trouble.

With these concepts in mind, let's take an alternative look at our first-order process under feedback control.

4-3 The First-Order Process with Feedback Control in the Frequency Domain

Back in Sec. 3-2-3 the first-order process under feedback control was studied. With proportional-only control, the differential equation describing the controlled system was first order and suggested that the response was bounded under all conditions. When integral control was added the order of the describing differential equation jumped to two but the response still appeared to be bounded although at high integral control gains there could be underdamped behavior.

Before we can use the results of this section for actual design we need to discuss some process models that have more phase lag than the simple first-order system. These models are not only more complicated but they also are able to describe important characteristics of real processes. But for the time being, we will stick with the first-order process.

When the Laplace transform was applied to this first-order process under PI (proportional-integral) control in Chap. 3, the behavior of the controlled system was shown to be described by the two transfer functions, one for the process G_p and one for the controller G_c. The algebraic development of the transfer function for response of the process output just before the summing point in response to the set point is as follows:

$$
\begin{aligned}
&\tilde{Y} = G_p \tilde{U} \\
&\tilde{U} = G_c \tilde{E} \\
&\tilde{Y} = G_p G_c \tilde{E} \\
&\tilde{Y} = G_p G_c \tilde{S} \\
&\frac{\tilde{Y}}{\tilde{S}}\Big|_{\text{openloop}} = G_p G_c = \frac{g}{\tau s + 1}\frac{ks + I}{s} = G
\end{aligned}
\tag{4-9}
$$

Note that $\tilde{E} = \tilde{S}$ because there is no feedback connection—yet.

Had there been feedback and had the loop actually been *closed,* the algebra would have been carried out as follows:

$$
\begin{aligned}
&\tilde{Y} = G_p \tilde{U} \\
&\tilde{U} = G_c \tilde{E} \\
&\tilde{Y} = G_p G_c \tilde{E} \\
&\tilde{Y} = G_p G_c (\tilde{S} - \tilde{Y}) \\
&\tilde{Y} + G_p G_c \tilde{Y} = G_p G_c \tilde{S} \\
&\frac{\tilde{Y}}{\tilde{S}}\Big|_{\text{closedloop}} = \frac{G_p G_c}{1 + G_p G_c} = \frac{gks + gI}{\tau s^2 + (gk + 1)s + I}
\end{aligned}
$$

which is the closed-loop transfer function. However, since we are not going to close the loop yet we will stick with the result of Eq. (4-9), that is, the open-loop transfer function $G_c G_p$.

To move to the frequency domain we apply the trick of letting $s = j2\pi f = j\omega$, where ω is the frequency in radians per second, while f is the frequency in cycles/sec.

$$
G(j\omega) = G_c G_p = \frac{g}{\tau j\omega + 1}\frac{kj\omega + I}{j\omega} \tag{4-10}
$$

The above transfer function contains four factors: two numerators and two denominators, each a complex quantity with a magnitude and a phase. The simplest way to get the overall magnitude and phase is to look at each factor separately and convert it into a magnitude and a phase. After this conversion, the magnitudes can be multiplied and divided as necessary and the phases can be added or subtracted as necessary:

$$
\begin{aligned}
G(j\omega) &= G_p(j\omega)G_c(j\omega) = \frac{g}{\tau j\omega + 1}\frac{kj\omega + I}{j\omega} \\
&= \frac{g e^{j0}}{\sqrt{(\tau\omega)^2 + 1}\, e^{j\theta_1}} \frac{\sqrt{(k\omega)^2 + I^2}\, e^{j\theta_2}}{1 e^{j\pi/2}} \\
&= \frac{g\sqrt{(k\omega)^2 + I^2}}{\sqrt{(\tau\omega)^2 + 1}} e^{j(\theta_2 - \theta_1 - \pi/2)} \\
&= |G| e^{j\theta}
\end{aligned}
$$

$$
\begin{aligned}
\theta &= \theta_2 - \theta_1 - \pi/2 \\
\theta_1 &= \tan^{-1}(\tau\omega) \\
\theta_2 &= \tan^{-1}\left(\frac{k\omega}{I}\right) \\
|G| &= \frac{g\sqrt{(k\omega)^2 + I^2}}{\sqrt{(\tau\omega)^2 + 1}} \qquad \theta = \tan^{-1}\left(\frac{k\omega}{I}\right) - \tan^{-1}(\tau\omega) - \frac{\pi}{2}
\end{aligned}
\tag{4-11}
$$

The development of these equations used the simple algebra of complex exponentials where the magnitudes multiply and the angles add.

Question 4-4 Could you derive the appropriate equations for the magnitude and angles for the case where $I = 0$?

Answer With $I = 0$ and $k = 1$, Eq. (4-9) simplifies to the equation in Sec. 4-2 for the first-order process without control. See Eqs. (4-7) and (4-8).

Question 4-5 Could you derive the appropriate equations for the magnitude and angles for the case where $k = 0$?

Answer Take the limit as $k \to 0$ in Eq. (4-11) and remember that the angle whose tangent is zero is zero. The result will be

$$
|G| = \frac{gI}{\sqrt{(\tau\omega)^2 + 1}} \qquad \theta = -\tan^{-1}(\tau\omega) - \frac{\pi}{2}
$$

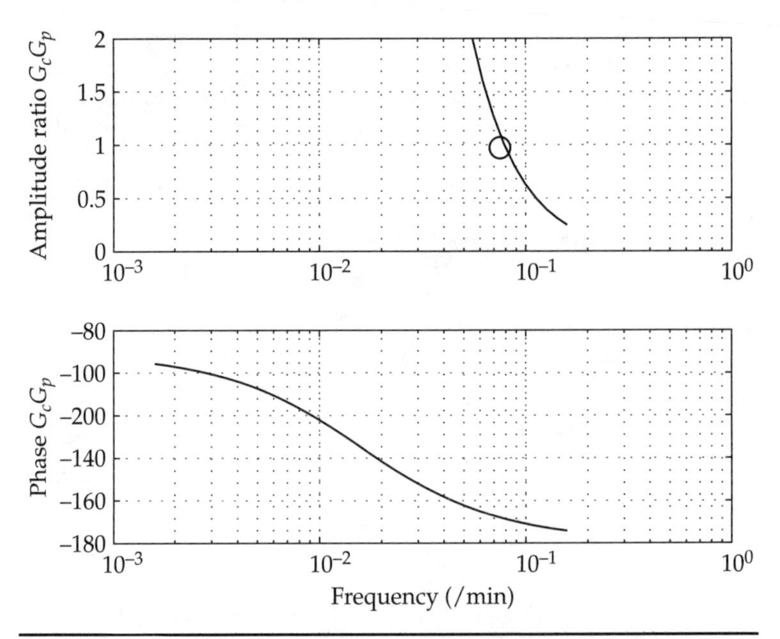

FIGURE 4-16 Open-loop Bode plot for first-order process with I-only.

In Chap. 3 the parameter values used were $\tau = 10$, $g = 2.5$. For the case of proportional-only control, the Bode plot will be basically the same as that shown in Figs. 4-7 and 4-8.

For the case of integral-only control with $I = 1$ the Bode plot is shown in Fig. 4-16. Note that at low frequency the phase is already slightly less than –90° because of the integral controller, which has a constant phase of –90° independent of the frequency (see Sec. 4-3-1). As the frequency increases, the phase lag increases but never exceeds 180°. The small circle indicates the point where the amplitude ratio is unity. So, no matter what the integral gain is, this controlled system cannot become unstable. This can also be seen from Eq. (4-11) by letting $\omega \rightarrow \infty$.

4-3-1 What's This about the Integral?

We may have slipped one past you in the above paragraph where it was quietly stated that "because of the integral controller which has a constant phase of –90° independent of the frequency." To see this take the integral of the sine function:

$$\int \sin u\,du = -\cos u + C$$

The sine and negative cosine functions are plotted in Fig. 4-17. Note that the negative cosine function, which is the integral of the sine function, lags the sine function by a constant 90°. Therefore, the

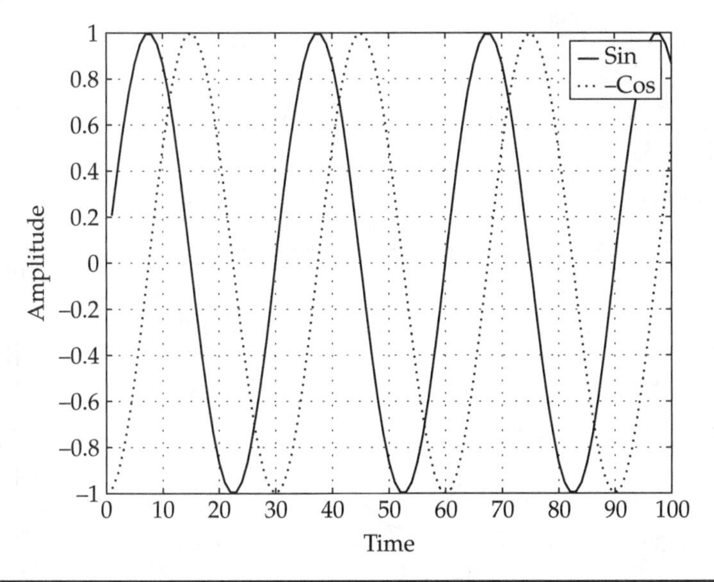

FIGURE 4-17 The sine and negative cosine functions.

presence of an integrator in the controller adds an immediate phase of –90° to the Bode plot.

Question 4-6 If the integral component in a controller adds a constant –90° of phase what does the proportional component do?

Answer If the controller is proportional-only then $G_c = k$ and the open-loop transfer function becomes $G_p k$. The presence of the additional factor k simply modifies the amplitude but has no effect on the phase. In fact, our approach in analyzing controlled processes will be to start with proportional-only control with a control gain of $k = 1$ because the Bode plot is simply that of the process. Then to avoid too much gain when the phase equals –180° we will adjust the control gain appropriately. Since the phase remains independent of the value of k, only the magnitude plot will shift in response to changes in k.

4-3-2 What about Adding P to the I?

The integral gain for the previous I-only controller will be lowered to 0.4 and a proportional component with a control gain k of 1.0 will be added. The impact on the Bode plot is shown in Fig. 4-18. One notices that the phase does not come anywhere near –180°. The time domain behavior for this set of control gains is shown in Fig. 4-19.

Finally, look at the error transmission curve [introduced in Chap. 3, Eq. (3-43)] for this controlled system in Fig. 4-20.

The error transmission curve (sans the phase curve) is replotted with linear frequency axes in Fig. 4-21. This curves shows how

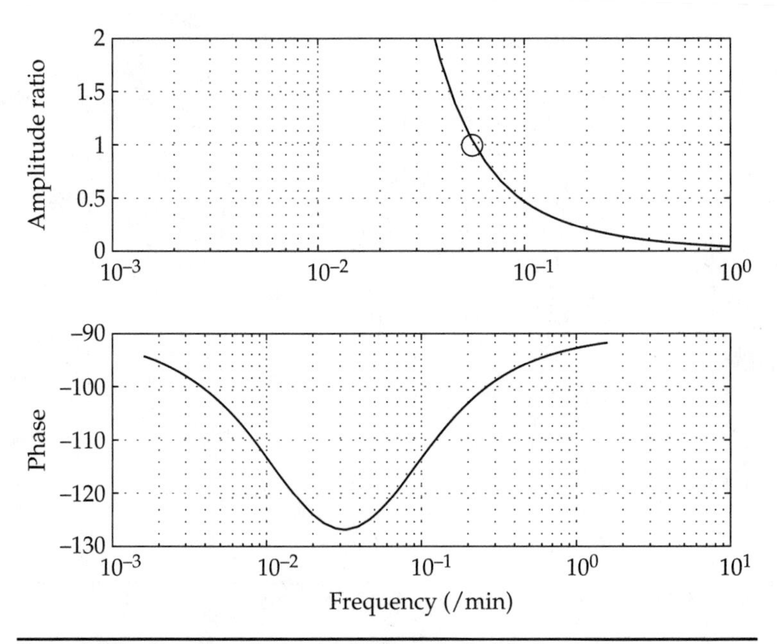

FIGURE 4-18 Open-loop Bode plot for PI controlled first-order process.

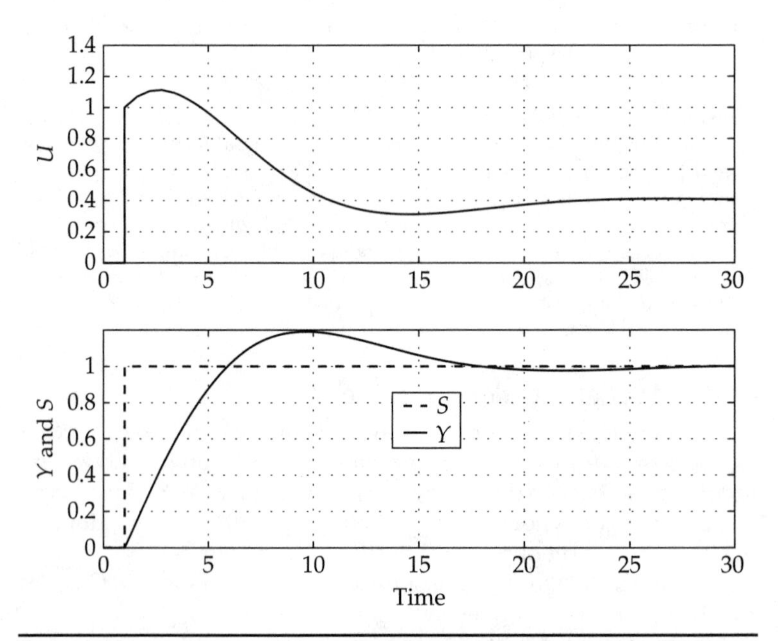

FIGURE 4-19 Response of PI controlled first-order process to unit set-point step.

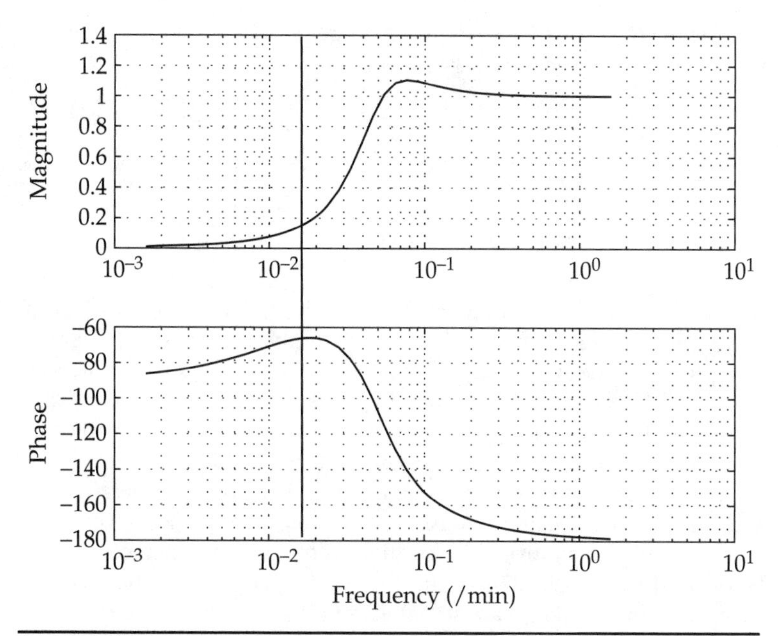

Figure 4-20 Error transmission curve for first-order process with PI control.

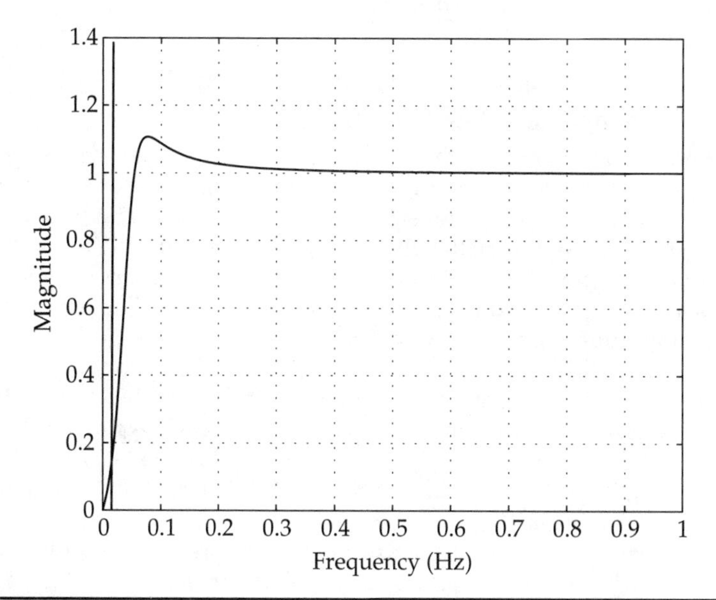

Figure 4-21 Error transmission curve for first-order process with PI control, linear plot.

disturbances of different frequencies are controlled. From Chap. 3, the "corner" frequency of

$$\omega_{cor} = 1/\tau = 1/10 = 0.1 \text{ rad/sec}$$

or

$$f_{cor} = \omega_{cor}/(2\pi) = 0.0159 = 1.5910^{-2} \text{ Hz}$$

denoted a point on the Bode plot where the process magnitude and phase plot showed a change. It is a key variable in the error transmission plot. Disturbances having frequencies below the corner frequency in Fig. 4-20 are attenuated. Figure 4-21 shows that zero frequency disturbances, that is, constant offsets, are completely removed. At the corner frequency things start to change. Disturbances having frequencies above the corner frequency are passed with little effect. Physically, this is to be expected because disturbances with low frequencies would be relatively easy to control whereas high-frequency disturbances would be beyond the capabilities of the controller. Disturbances having frequencies around 0.08 Hz are actually amplified slightly. Sometimes it is difficult for one to grasp the reality that only a small part of the disturbance spectrum is actually controlled when feedback control is applied. This suggests that process improvement and process problem solving is the best way to improve performance.

4-3-3 Partial Summary and a Rule of Thumb Using Phase Margin and Gain Margin

Based on the example in this section it looks like we can gain some insight into the controllability of a process by looking at the Bode plot for the open-loop transfer function G_cG_p. We want to avoid design situations where the phase lag of G_cG_p is near 180° when the amplitude ratio is unity. Experience suggests that a *phase margin* of at least 45° is required for good control performance. That is, when the amplitude ratio is unity we would like the phase lag to be no more than 135°.

Conversely, we want to avoid situations where the amplitude ratio is near unity when the phase lag is 180°. Again, experience has shown that a *gain margin* of at least 6 dB is desirable. That is, when the phase lag is 180° we would like the gain to be less than 0.5 where $20 \log_{10}(.5) = -6.02$ dB.

We can't seem to find a way to make our first-order process go unstable when we put it under PI control. This situation will be changed when we study some new processes in the upcoming sections.

Question 4-7 Why are we having trouble making first-order processes go unstable by adding controllers?

Answer The most phase lag that can come out of a first-order process is 90° and that is only at high frequencies. Proportional-only control adds no phase lag and integral-only adds 90° so the first-order process under integral-only or proportional-integral control can only have 180° phase lag as a limiting case when the frequency is extremely high.

4-4 A Pure Dead-Time Process

Consider the process depicted in Fig. 4-22. Imagine many small buckets nearly contiguous such that when the inlet flow rate is continuous so is the outlet flow rate. With this in mind, Fig. 4-22 suggests that the process output Y will be identical to the process input U except with a shift in time, namely,

$$Y(t) = U(t - D) \tag{4-12}$$

where D is the dead time. If the conveyor belt speed is v and the distance between the filling and dumping points is L then the dead time would be $D = L/v$.

Figure 4-23 shows the step response of a process having a dead time of 8 time units. The process gain is unity—what goes in comes out unattenuated and unamplified. The time constant is zero but there is a dead time between the step in the input and the response of the output.

So much for the time domain. What does the Bode plot for the pure dead-time process look like? Figure 4-24 shows magnitude and phase plotted for linear frequency and Fig. 4-25 shows the same thing plotted with logarithmic frequency. Both figures support our contention that the amplitude ratio of the output to the input is unaffected by frequency. However, the phase lag of the output

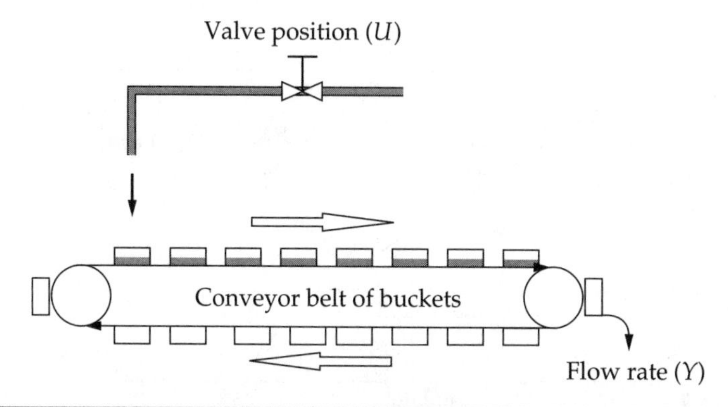

FIGURE 4-22 A dead-time process. Imagine many small buckets together so that the flow is effectively continuous.

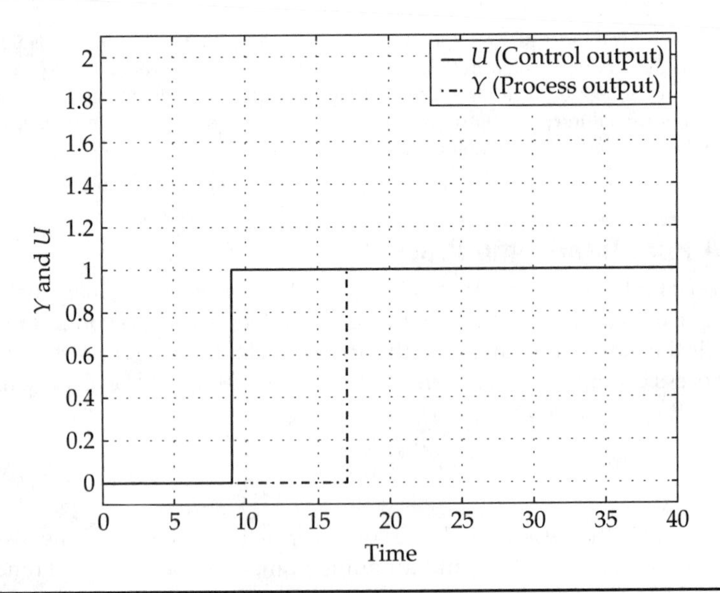

FIGURE 4-23 Step-change response of pure dead-time process ($D = 8$, $g = 2.5$, $\tau = 0$).

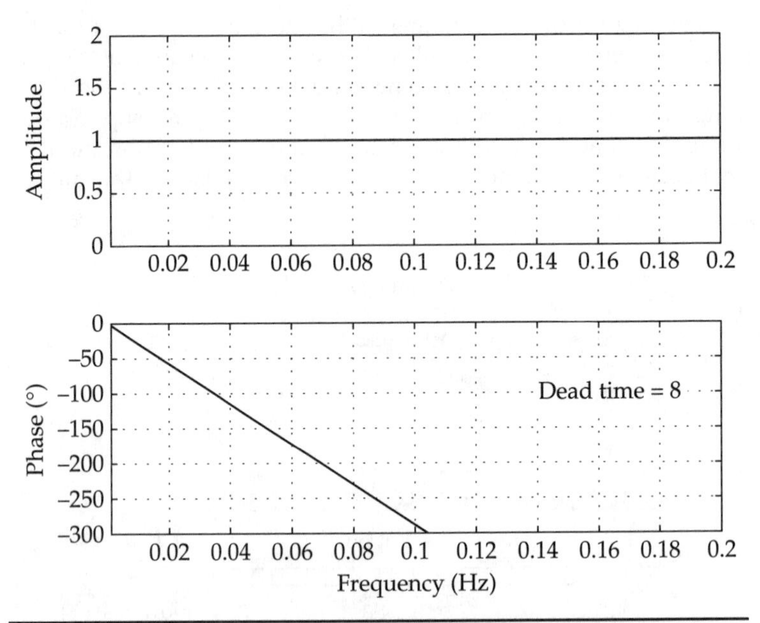

FIGURE 4-24 Bode plot for a pure dead-time process $D = 8$ (linear frequency axis).

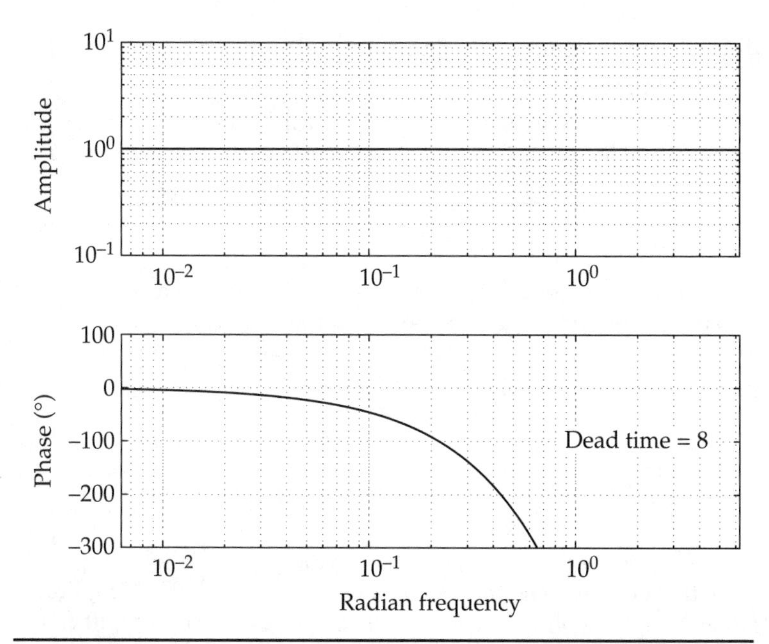

FIGURE 4-25 Bode plot of pure dead-time process $D = 8$ (logarithmic frequency axis).

increases linearly with increasing frequency. This makes sense if the reader can visualize a sine wave entering the pure dead-time process. When it leaves it will still be a sine wave with the same amplitude but it will be shifted in time and the shift will be a function of frequency. At low frequency the outlet will see the beginning of the same cycle that is still entering the process. At high frequency there will be many cycles inside the process and the emerging cycle will be significantly displaced from the entering cycle.

The Laplace transform of a pure dead-time process is

$$L\{Y(t-D)\} = e^{-Ds}\tilde{U}(s) \qquad (4\text{-}13)$$

The basis for Eq. (4-13) is in App. F and the rule is simple. Whenever a quantity $Y(t)$ is delayed in time by an amount D, the Laplace transform of $Y(t-D)$ is the transform of $Y(t)$ multiplied by $\exp(-Ds)$, as in $e^{-Ds}\tilde{Y}(s)$.

Moving to the frequency domain by letting $s = j\omega$ causes the multiplier to become $e^{-j\omega D}$. The magnitude of the delayed quantity is unaffected because the magnitude of the exponential is unity. However, the phase lag is increased by ωD. This linear dependence of the phase on frequency is demonstrated in Fig. 4-24.

4-4-1 Proportional-Only Control of a Pure Dead-Time Process

Based on Chap. 3, we can guess that there will be an offset between the set point and the process output. The open-loop transfer function is

$$G_pG_c(s) = e^{-sD}k$$

where k is the proportional control gain that we wish to estimate. When s is replaced by $j\omega$, the open-loop transfer function becomes

$$G_pG_c(j\omega) = e^{-j\omega D}k = |G|e^{j\theta}$$

$$|G| = k \tag{4-14}$$

$$\theta = -\omega D$$

The Bode plot for $k = 1$ and $D = 8$ is the same as that in Fig. 4-24. Find the point where the phase lag reaches 180°. The amplitude remains unity at all frequencies so a control gain of unity will cause problems. When the control gain k is decreased to 0.7 the whole amplitude line is shifted down to a value of 0.7 and the phase margin requirement is satisfied. Using this control gain, a simulation in the time domain shows the response to a step in the set point (Fig. 4-26).

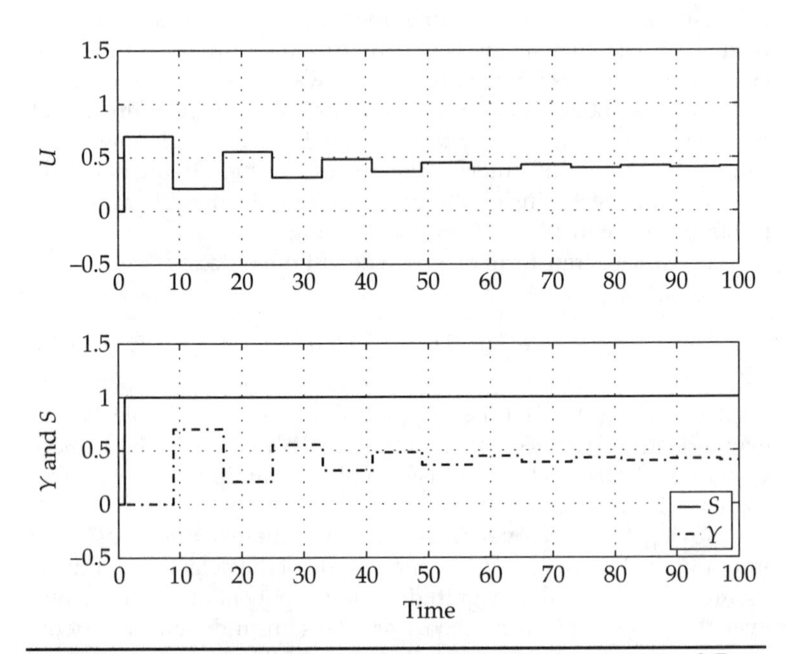

FIGURE 4-26 Proportional-only control of pure dead-time process, $k = 0.7$.

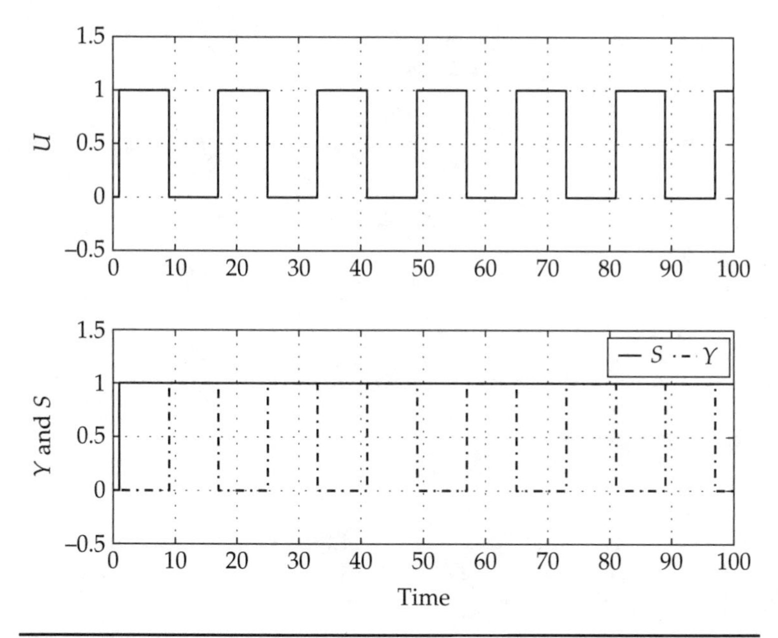

FIGURE 4-27 Proportional-only control of pure dead-time process, $k = 1.0$.

Note the step in the controller output at the time of the step in the set point. For the period of the dead time there is no action until the effect of the controller passes through the dead time. Also, note the offset between the set point and the process output. In Fig. 4-27 the control gain is set to unity and the response is on the cusp of instability.

4-4-2 Integral-Only Control of a Pure Dead-Time Process

The open-loop transfer function is

$$G_pG_c(s) = e^{-sD}\frac{I}{s}$$

After using the $s = j\omega$ trick, the open-loop transfer function is

$$G_pG_c(j\omega) = e^{-j\omega D}\frac{I}{j\omega} = |G|e^{j\theta}$$

$$|G| = \frac{I}{\omega} \tag{4-15}$$

$$\theta = -\omega D - \frac{\pi}{2}$$

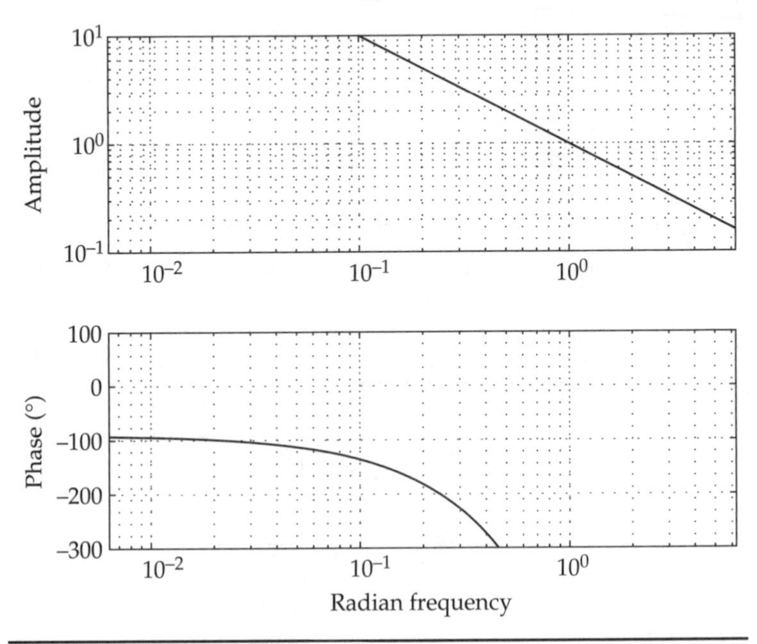

FIGURE 4-28 Bode plot for integral-only control of a pure dead-time process, $I = 1$, $D = 8$.

The presence of integral produces an additional phase lag of 90° and the dead-time contribution builds on that. The integral component also causes the amplitude ratio to decrease directly with frequency. Equation (4-15) shows that the phase is independent of the integral control gain I and is never greater than −90° so it appears as though integral-only control with this gain is not an attractive option. The Bode plot is shown in Fig. 4-28 for the case of $D = 8$. However, if we try an integral control gain of $I = 0.1$ the Bode plot is given in Fig. 4-29 and the time domain behavior is given in Fig. 4-30. Note how the control output changes continually compared to that for the proportional-only control case.

Question 4-8 Why does the control output change continually compared to the stepped behavior of the proportional-only case?

Answer During the period of the dead time after the set point has been changed, the error is constant and nonzero. As a result, the integral output ramps up.

Figure 4-31 shows the error transmission curve for the integral-only case. Disturbances with frequencies above 0.1 rad/sec are passed or even amplified.

The error transmission curve is replotted in Fig. 4-32 with linear axes. The peaks occur with a spacing of 0.125 Hz which is the reciprocal of the dead time. Notice, once again, how little of the spectrum of disturbances is attenuated.

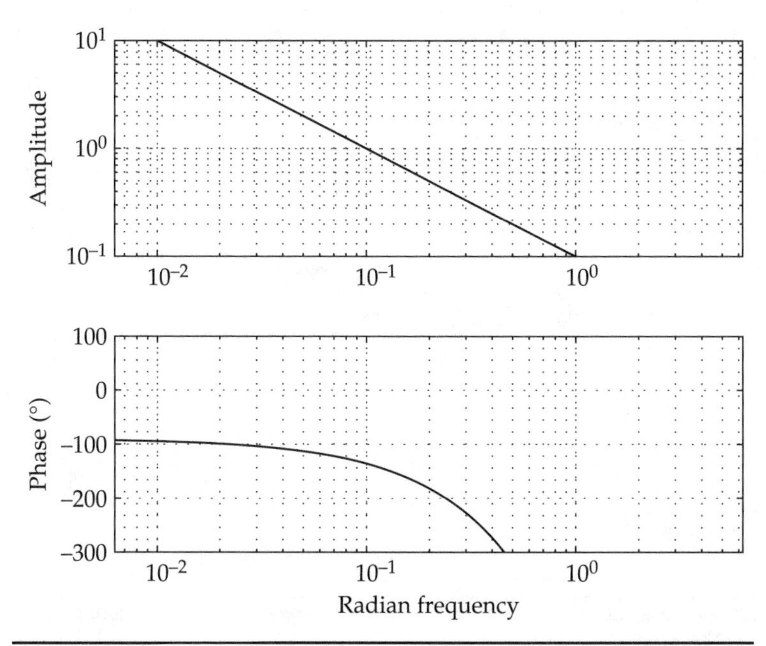

FIGURE 4-29 Bode plot for integral-only control of a pure dead-time process, $I = 0.1$, $D = 8$.

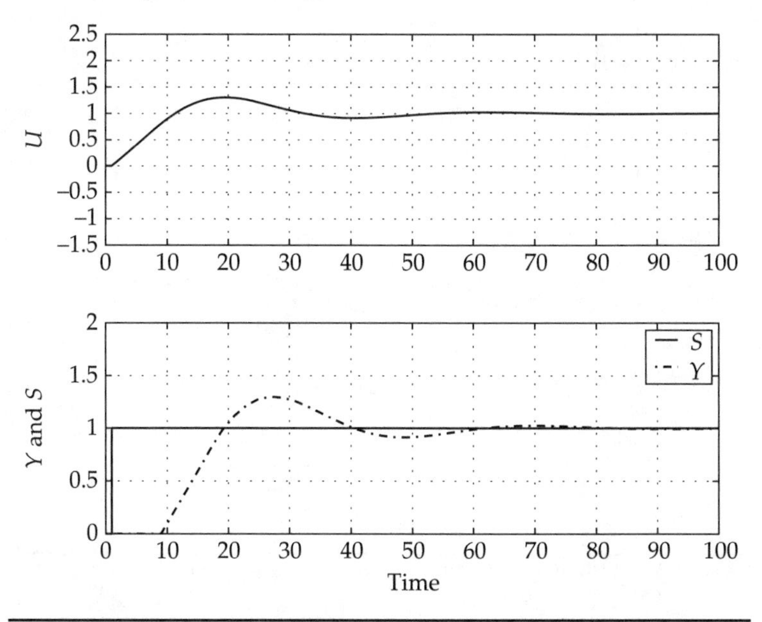

FIGURE 4-30 Bode plot for integral-only control of pure dead-time process, $I = .1$, $D = 8$ (time domain).

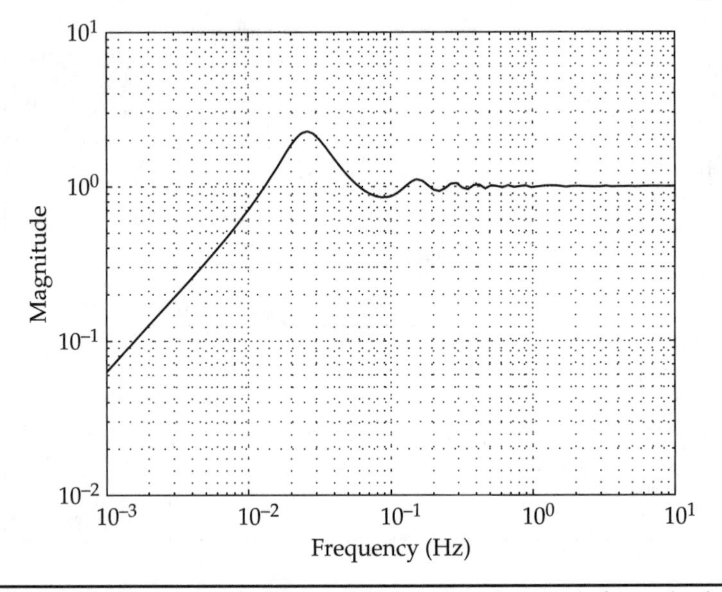

Figure 4-31 Error transmission curve for integral-only control of pure dead-time process.

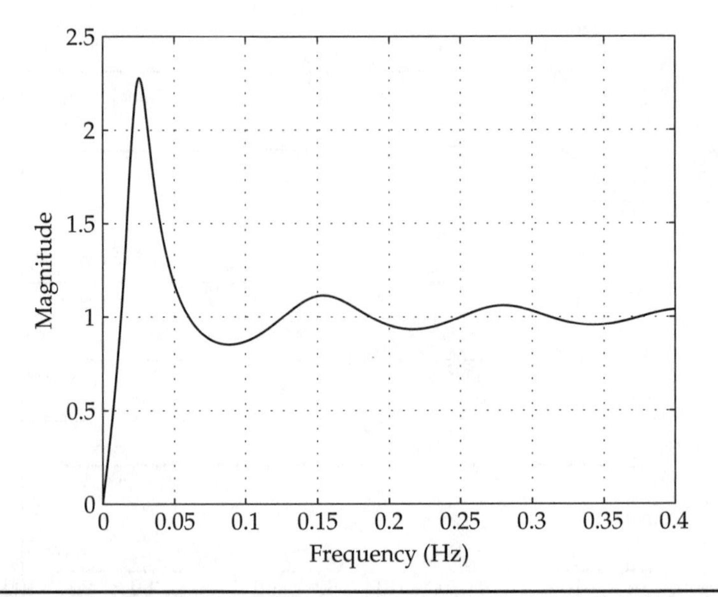

Figure 4-32 Error transmission curve for integral-only control of pure dead-time process (linear axes).

4-5 A First-Order with Dead-Time (FOWDT) Process

Consider Fig. 4-33 where the tank of liquid has been placed upstream of the pure dead-time process. The placement of the buckets-on-the-belt ahead of the tank suggests a dead time in series with a first-order process. Please do not be confused by the length of the pipe at the outlet of the tank. Let's assume that it is actually relatively short and that the pipe diameter is small so that the transit time of the liquid spent in the pipe is negligible compared to the time spent in the buckets on the belt.

Figure 4-34 shows the open-loop step-change response of the process for the case of $g = 2.5$, $\tau = 10$, $D = 8$. This is the first example process in this chapter that has had a nonunity gain.

In the continuous time domain, this model would be described by an extension of the first-order model:

$$\tau \frac{dy}{dt} + y = gU(t - D) \tag{4-16}$$

In the Laplace domain, the open-loop transfer function is

$$G_p(s) = e^{-sD} \frac{g}{\tau s + 1} \tag{4-17}$$

After applying $s = j\omega$, the magnitude and phase can be found as follows:

$$G_p(j\omega) = e^{-j\omega D} \frac{g}{j\tau\omega + 1} = |G_p| e^{j\theta}$$

$$|G_p| = \frac{g}{\sqrt{(\tau\omega)^2 + 1}} \tag{4-18}$$

$$\theta = -\tan^{-1}(\tau\omega) - \omega D$$

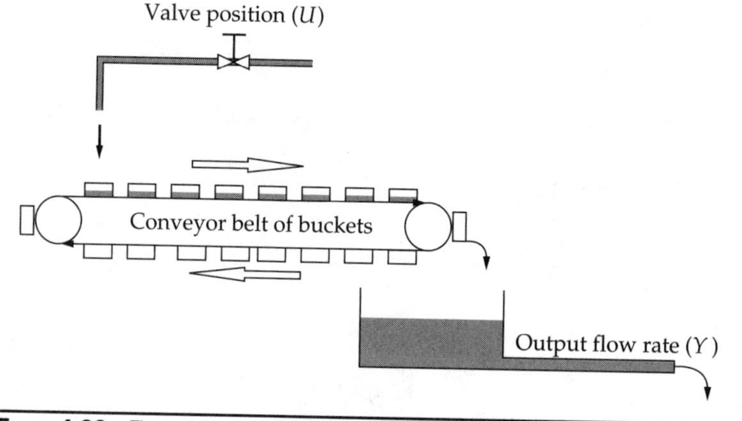

FIGURE 4-33 First-order process with dead time.

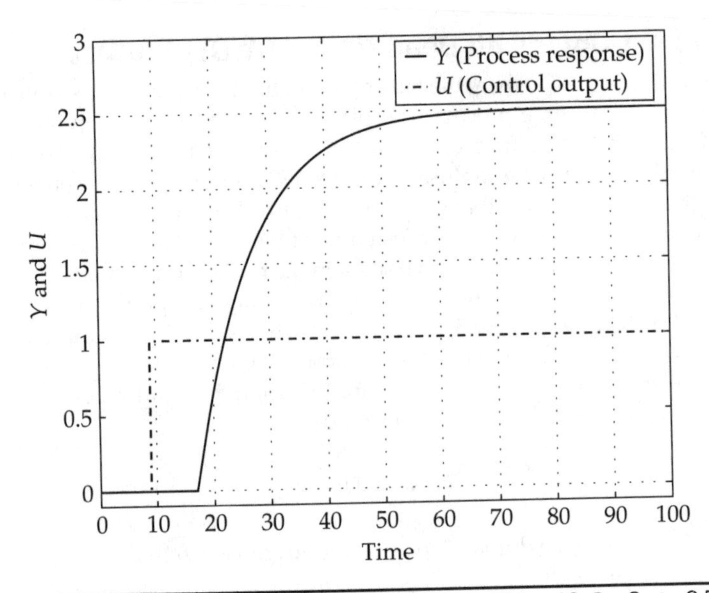

FIGURE 4-34 Open-loop step response of FOWDT process ($\tau = 10$, $D = 8$, $g = 2.5$).

The magnitude is identical to that of the dead time–less first-order model in Eq. (4-7) but the phase lag is increased by the contribution of the dead time.

The Bode plot for the open-loop transfer function is given in Fig. 4-35. Note the circles that indicate unity magnitude and –180° phase.

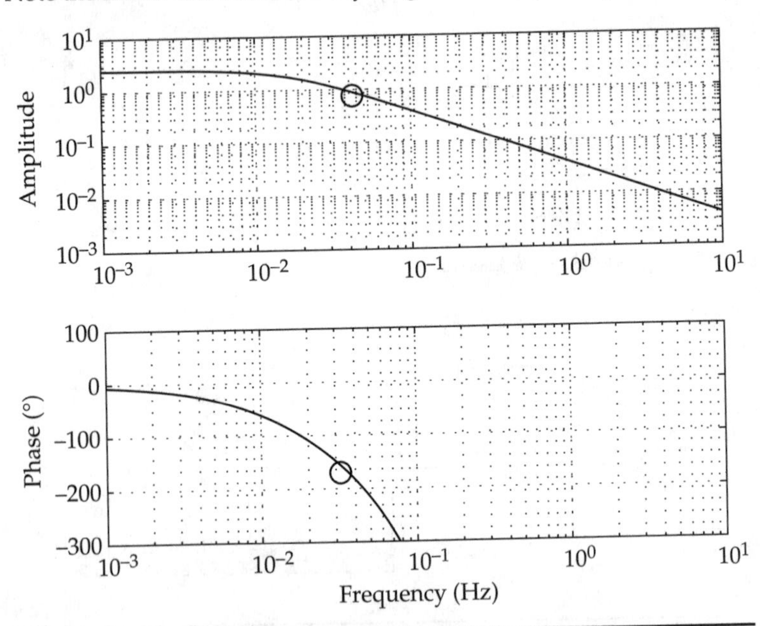

FIGURE 4-35 Bode plot for FOWDT process ($\tau = 10$, $D = 8$, $g = 2.5$).

4-5-1 The Concept of Minimum Phase

The FOWDT process is an example of a nonminimum phase model (NMP), which means there are other processes that have the same magnitude but have less phase lag. The first-order process (without dead time) is such a model. Figure 4-36 shows the open-loop Bode plot of a FOWDT process along with that of the minimum phase (MP) model. We will not use this concept but the manager may come across it and needs to be aware of it.

4-5-2 Proportional-Only Control

If the proportional-only control gain k is unity then the Bode plot for the open-loop transfer function would be identical to that in Fig. 4-35 and it would appear that this system would be unstable because at a frequency of about 0.05 Hz the phase is –180° and at that same frequency the amplitude is a little over 2.0. Likewise at a frequency of about 0.2 Hz the amplitude is about 1.0 while the phase is off-scale so it is at most –300°. This suggests that stability might be obtained by applying a control gain less than unity such that the overall gain is reduced below unity. To make it *just stable* the overall gain should equal unity or

$$gk = 1 = 2.5 \cdot k$$

$$k = 0.4$$

To obtain a little gain margin let us reduce the proportional control gain to 0.3 so that the overall gain is 0.75 instead of 2.5. Then the Bode

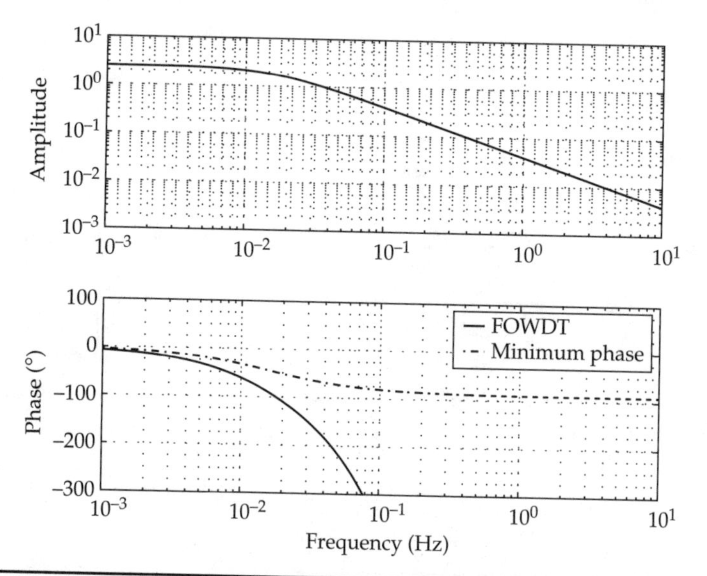

FIGURE 4-36 Bode plot for FOWDT model and the MP model ($g = 0.75$).

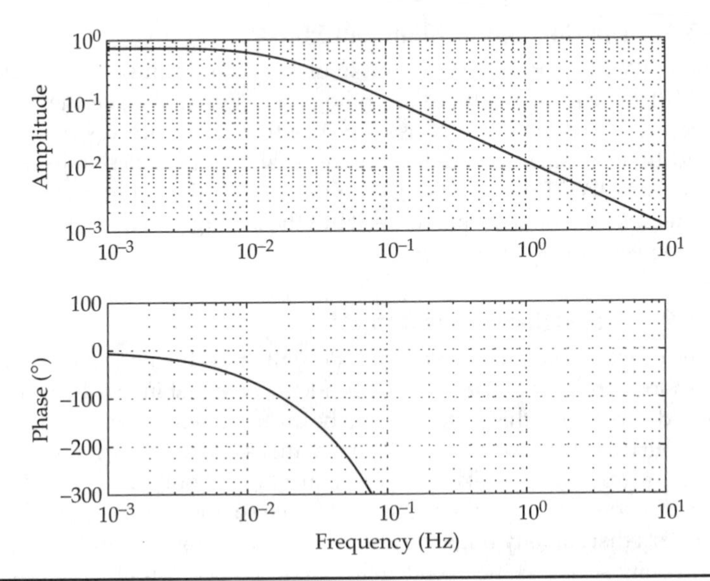

FIGURE 4-37 Bode plot for FOWDT process with P-only control ($\tau = 10$, $D = 8$, $g = 2.5*0.3$).

plot for the open-loop transfer function is shown in Fig. 4-37. Now the amplitude is always less than unity so even though the phase lag equals 180° at a frequency of about 0.05 Hz there is no concern about stability.

A simulation of proportional-only control for these conditions is shown in Fig. 4-38.

Note the offset between the set point and the process output.

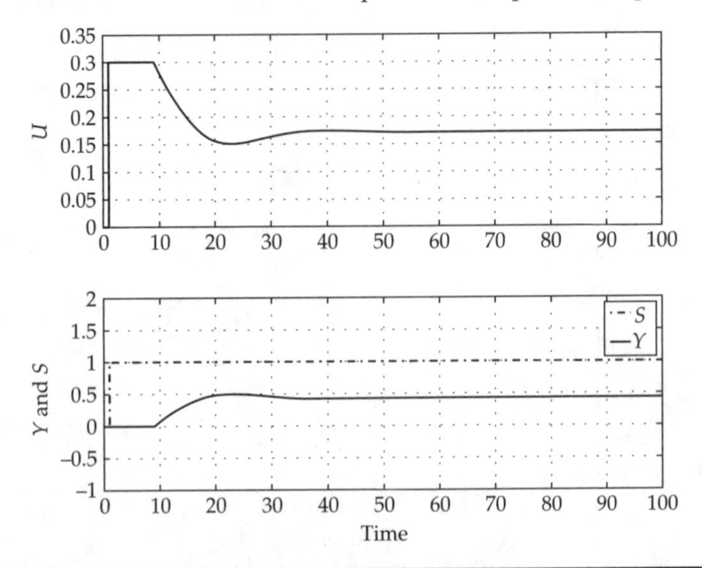

FIGURE 4-38 Bode plot for Proportional-only controlled FOWDT process to unit set-point step.

4-5-3 Proportional-Integral Control of the FOWDT Process

Adding integral control causes the open-loop transfer function to become

$$G_pG_c(s) = e^{-sD}\frac{g}{\tau s+1}\frac{ks+I}{s}$$

Applying $s = j\omega$ gives a relatively messy expression—but the reader might try wading through the following algebra—it's worth it:

$$G_pG_c(j\omega) = e^{-j\omega D}\frac{g}{\tau j\omega+1}\frac{kj\omega+I}{j\omega}$$

$$= \left(e^{-j\omega D}\right)\left(\frac{g}{\sqrt{(\tau\omega)^2+1}\,e^{j\tan^{-1}(\tau\omega)}}\right)\frac{\left(\sqrt{(k\omega)^2+1}\,e^{j\tan^{-1}\left(\frac{k\omega}{I}\right)}\right)}{\left(\omega e^{j\frac{\pi}{2}}\right)}$$

$$= \frac{g}{\sqrt{(\tau\omega)^2+1}}\frac{\sqrt{(k\omega)^2+I}}{\omega}\frac{e^{-j\omega D}e^{j\tan^{-1}\left(\frac{k\omega}{I}\right)}}{e^{j\tan^{-1}(\tau\omega)}e^{j\pi}}$$

$$= \frac{g}{\sqrt{(\tau\omega)^2+1}}\frac{\sqrt{(k\omega)^2+I}}{\omega}e^{-j\omega D+j\tan^{-1}\left(\frac{k\omega}{I}\right)-j\tan^{-1}(\tau\omega)-j\frac{\pi}{2}}$$

$$= |G|e^{j\theta}$$

$$|G| = \frac{g}{\sqrt{(\tau\omega)^2+1}}\frac{\sqrt{(k\omega)^2+I}}{\omega} \tag{4-19}$$

$$\theta = -\omega D - \tan^{-1}(\tau\omega) + \tan^{-1}\left(\frac{k\omega}{I}\right) - \frac{\pi}{2}$$

This is the most complicated transfer function yet. Note how each term was treated as a complex number z with a magnitude $|z|$ and an angle φ according to the $|z|e^{j\varphi}$ format discussed in App. B. Then the magnitudes multiplied or divided and the angles added or subtracted. The idea therefore is to break a relatively complicated structure up into its factors in the numerator and denominator, convert each factor into a complex quantity with a magnitude, and an angle and then combine the quantities according to basic algebra.

Using the Bode plot to find the correct control gains k and I is not particularly fruitful. Instead, I used time domain simulation to find

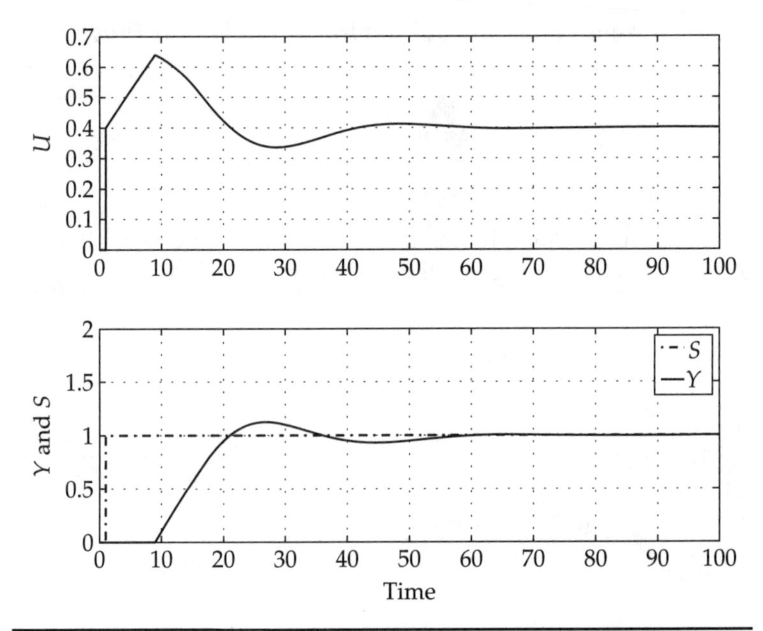

FIGURE 4-39 Response of PI controlled FOWDT process to unit set-point step.

by trial and error values that appeared to be acceptable using the results of the P-only analysis as a starting point. After a couple of trials the values of $k = 0.4$ and $I = 0.03$ resulted in the behavior shown in Fig. 4-39. The control output jumps at the moment of the set-point step and then integrates up during the dead-time period when there is no process response. Then when the process response finally gets through the dead time and starts to rise, the proportional component responds and pulls the control output back because the error is decreasing.

The error transmission curve is shown in Figs. 4-40 and 4-41. Disturbances with low frequencies are attenuated and high frequencies are passed with little or no amplification or attenuation. The ripple in the transmission curve is a consequence of resonance that occurs when a disturbing sinusoid has a frequency that is some integral factor of the reciprocal of the dead time.

The Bode plot for the FOWDT process shows that as long as there is a dead time, no matter how small, there will be a frequency for which the phase lag goes beyond 180°. Every *real* process has some dead time, no matter how small, therefore every real process can become unstable if the control gain is high enough. This is to be compared with the true first-order process without dead time which can never become unstable no matter how large the control gain is.

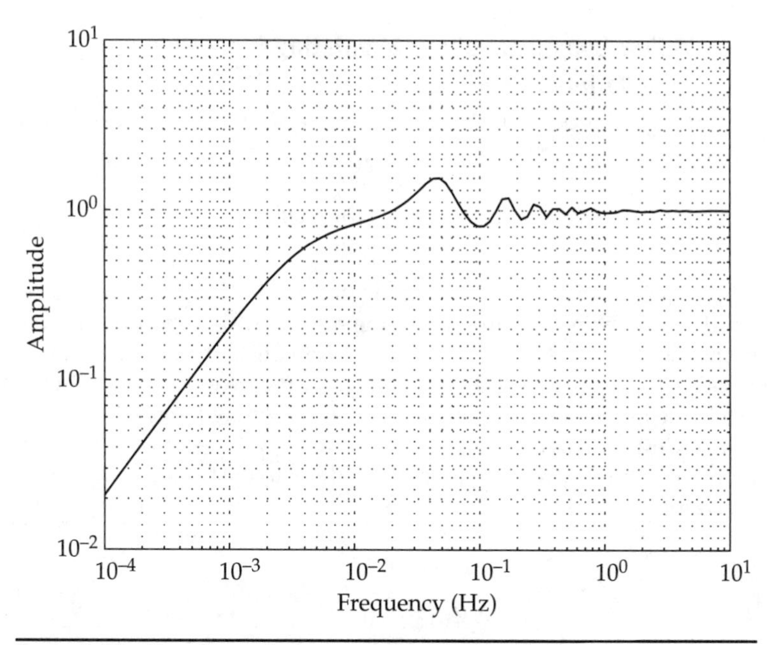

FIGURE 4-40 Error transmission curve for FOWDT with PI control.

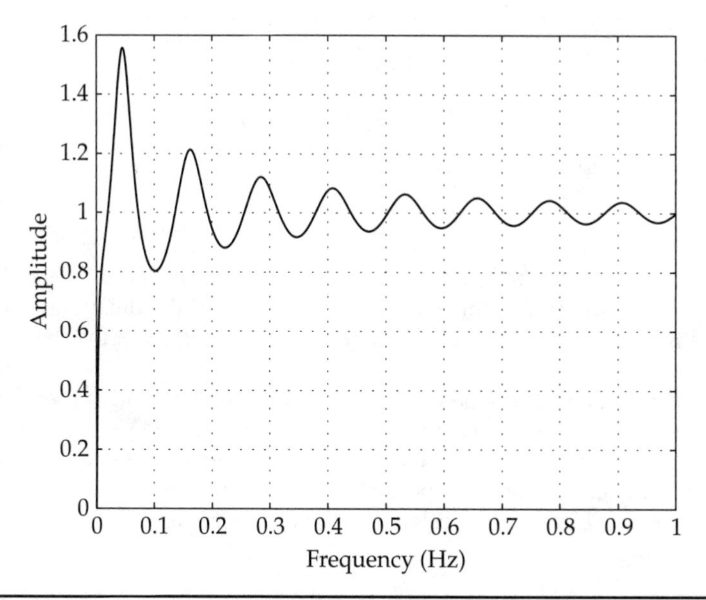

FIGURE 4-41 Error transmission curve for FOWDT with PI control, linear axes.

4-6 A Few Comments about Simulating Processes with Variable Dead Times

Consider the process idealized in Fig. 4-33 where the describing equation is

$$\tau \frac{dy}{dt} + y = gU(t - D)$$

We will discuss the discrete time domain in Chap. 9, but assume for the time being that the time domain is broken up into discrete points, $t_1, t_2, \ldots$ that are separated by a constant interval, h, as in $t_i = t_{i-1} + h$. To determine the value of y at time t we need the value of U at time $t - D$. Assume that we have an infinitely long delay vector $V(i), i = 1, 2, \ldots$ available for the storage of U. At every discrete moment of time t_i, we increment the index i, to the vector V and insert the value $U(t_i)$ as in

$$\begin{aligned} i &\leftarrow i + 1 \\ V(i) &\leftarrow U(t_i) \end{aligned} \tag{4-20}$$

Further, assume that the dead time D is an even multiple of the constant interval, as in $D = nh$.

To simulate the process we need $U(t - D)$; how do we get it? One way is to augment the simple algorithm in Eq. (4-20) as follows:

$$\begin{aligned} i &\leftarrow i + 1 \\ V(i) &\leftarrow U(t_i) \\ j &\leftarrow i - n \\ U(t - D) &\leftarrow V(j) \end{aligned} \tag{4-21}$$

In words, Eq. (4-21) says "increment the delay index, place the current value of U in the delay vector, decrement the delay index by the number of increments in the dead time D, and fetch the delayed value of U."

If the speed of the belt v is the controller output and is therefore variable, how can the correct value of U be obtained? The simplest and in my experience the most common approach uses a variable index calculated from the speed. The distance over which the belt has to carry the buckets L, is related to the dead time and the belt speed according to

$$\begin{aligned} L &= vD = vnh \\ n &= \text{int}\left(\frac{L}{vh}\right) \end{aligned} \tag{4-22}$$

where "int" means taking the integer value of $L/(vh)$.

To get a feel for how Eq. (4-22) might work, let's put in some numbers. Let $L = 100.0$, and $h = 1$. Let the nominal value of the speed be $v = 10.0$. Consequently, the nominal delay vector decrement $n = 10$. Therefore, at every instant $t_1, t_2, \ldots$ the index i is incremented, the speed is placed in the delay vector $V(i)$, the decrement n is calculated and the delayed speed $V(i-n) = V(i-10)$ is fetched. Assume that at some point in time t' the speed is decreased from 10.0 to 5.0 because of a control move. The index is incremented to i', the new speed is placed in the delay vector and a new decrement is calculated from

$$n' = \text{int}\left(\frac{100}{5}\right) = 20$$

The delayed speed is fetched from $V(i'-20) = V(i+1-20) = V(i-19)$ which is a problem. On the previous instant the delayed value was fetched from $V(i-10)$ but on the very next instant of time the delayed value is fetched from $V(i-19)$ which contains speed that is older than the one just fetched. This violates our common sense and more importantly could cause problems if the fetched value is being used in a simulation, which in turn is being used for control purposes.

To solve this problem one must realize that Eq. (4-22) represents a steady-state model and is not valid when the speed is varying. In fact, a common sense (and rigorous) definition of the dead time requires the use of the integral to take account of the varying speeds, as in

$$L = \int_{t-D(t)}^{t} v(u)du \tag{4-23}$$

which says that the integral of the conveying speed $v(t)$ over the period of the dead time $D(t)$, which is now a variable, equals the distance over which the buckets are conveyed. In the special case where the speed is constant, $v(t) = v_c$, then Eq. (4-23) gives

$$L = \int_{t-D(t)}^{t} v_c du = v_c D = v_c hn$$

which is the same as Eq. (4-22).

To actually solve Eq. (4-23) online, the integral can be approximated, as in

$$L = h[v(t_i) + v(t_i - h) + v(t_i - 2h) + \cdots + v(t_i - hn)] \tag{4-24}$$

Now, the problem is to find n such that the right-hand side of Eq. (4-24) equals L. This can be done quickly with an iterative method of the engineer's choosing.

This situation has actually occurred in my experience and this algorithm does in fact work. I suggest that this is a good example of using simple calculus to solve a problem that is often overlooked.

4-7 Partial Summary and a Slight Modification of the Rule of Thumb

Our approach has been to find a proportional control gain, called the critical gain, that makes the open-loop amplitude ratio unity or the open-loop phase lag 180°. In the former case we reduce the critical proportional control gain to make the phase lag less than 180° by about 45°. In the latter case we reduce the critical proportional control gain by a factor of 0.5.

Therefore, as a starting point we are trying the find the critical values of ω and k, namely ω_c and k_c, such that the open-loop gain has a magnitude of unity and a phase of –180°. The App. B shows that the complex number –1.0 has a magnitude of unity and a phase of –180°, so we are really trying to find values of ω_c and k_c that satisfy the following equation:

$$G_p(j\omega_c)G_c(j\omega_c) = G(j\omega_c) = -1$$

or

$$\left|G(j\omega_c)\right|e^{j\theta} = -1 = 1e^{-j\pi} \tag{4-25}$$

or

$$\left|G(j\omega_c)\right| = 1$$

$$\theta(j\omega_c) = -\pi$$

Since several of the closed-loop transfer functions have a denominator of $1+G_cG_p$, it follows that finding the poles of these transfer functions is equivalent to solving Eq. (4-25).

If the proportional gain is set equal to k_c the performance should be on a cusp between instability and stability. That is, the process with the controller should experience sustained oscillations.

The critical values for proportional-only control of the FOWDT process would be the solution of the following two equations that come directly from Eq. (4-25):

$$\frac{gk_c}{\sqrt{(\tau\omega_c)^2+1}}=1$$
$$-\omega_c D-\tan^{-1}(\tau\omega_c)=-\pi \tag{4-26}$$

Equation (4-26) gives two equations in two unknowns: k_c and ω_c. A closed form solution to this problem is not straightforward (and probably not possible, at least for me) so a numerical solution based on a two-dimensional minimization (using `fminsearch` in Matlab) yielded the following values for $g = 2.5$, $\tau = 10$, $D = 8$, $k_c = 1.06$, $f_c = 0.039$. Had we tried the same approach for PI control there would still be two equations but now there would be three unknowns.

Using a proportional gain of 1.06 in a simulation for this process shows sustained oscillations (see Fig. 4-42) suggesting that the application of the critical values does indeed provide marginal stability. In a previous simulation shown in Fig. 4-38 the proportional-only gain of 0.3 was used and the performance was acceptable. Here the gain margin was 0.3/1.06 = 0.283 or 10.9 dB which is more conservative than the rule of thumb mentioned in Sec. 4-7.

Question 4-9 Consider this algebraic approach for the *pure* dead-time process under proportional-only and integral-only control described in Eqs. (4-14) and (4-15). What kind of problems would occur?

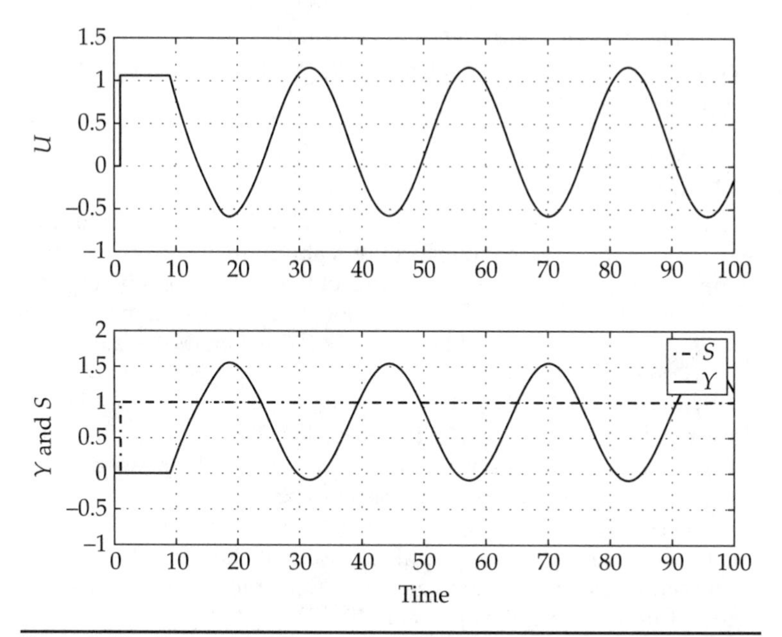

FIGURE 4-42 Response of PI controlled FOWDT process to unit set-point step indicating marginal stability.

Answer The equations used to solve for k_c and ω_c for the case of proportional-only control of a pure dead-time process are

$$G_pG_c(j\omega_c) = e^{-j\omega_c D}k = |G|e^{j\theta} = -1$$

$$|G| = k_c = 1$$

$$\theta = -\omega_c D = -\pi$$

By inspection $k_c = 1$ and $\omega_c = \pi / D$ or $f_c = \omega_c / (2\pi) = (\pi / D) / (2\pi) = 1 / (2D)$.
Since $D = 8$, it follows that $f_c = 0.125$.
For integral-only control the equations are

$$G_pG_c(j\omega_c) = e^{-j\omega_c D}\frac{I_c}{j\omega_c} = \frac{I_c}{e^{j\pi/2}\omega_c} = |G|e^{j\theta} = -1 = e^{-j\omega_c}$$

$$|G| = \frac{I}{\omega_c} = 1$$

$$\theta = -\omega_c D - \pi / 2 = -\pi$$

Therefore, $I_c = \omega_c$ and $\omega_c = \pi / (2D)$ or $f_c = 1 / (4D) = .0625$.

The reader might conclude that the Bode plot is effectively a graphical solution of Eq. (4-25). If the reader is interested in this approach, conventional textbooks on control usually contain many methods for quickly constructing Bode plots by hand. With the incredible access to computers and software like Matlab, these graphical techniques have become less attractive to some (especially *moi*) and will not be covered here.

4-8 Summary

The frequency domain was introduced by means of the substitution $s \to j\omega$ into the Laplace transform. A stability requirement for sinusoidal forcing was developed in terms of the amplitude ratio or magnitude and the phase lag of the open-loop transfer function G_cG_p.

The phase of G_cG_p should not equal –180° when the magnitude of G_cG_p is unity.

When the amplitude ratio is unity the phase margin should be on the order of 30° to 45°. That is, the phase lag should be less than 150° when the magnitude is unity. Alternatively, when the phase is –180° the gain margin should be on the order of 0.5.

The Bode plot of the open-loop amplitude ratio and phase versus frequency provided a graphical means of checking the stability of the candidate process and controller. Bode plots were constructed for the first-order process presented in Chap. 3. An auxiliary curve of the magnitude of the transfer function, $E/N = -1/(1 + G_cG_p)$, called the error transmission curve, provided insight into the ability of the controlled system to deal with disturbances or noise N of different frequencies.

The pure dead-time process and the first-order with dead-time or FOWDT process were introduced and the frequency domain analysis was applied. Because of the dead time, both processes could become unstable if the controller parameters were aggressive. Experience has shown that the FOWDT model is often a good approximation to industrial processes. It has the overall qualitative response of most processes and it is a good test for control algorithms because it can become unstable in a closed-loop configuration.

In the next chapter more processes and a new mathematical tool, matrices, will be introduced.

CHAPTER 5

Matrices and Higher-Order Process Models

In Chap. 4 the first-order with dead-time (FOWDT) process model was presented. In this chapter higher-order models will be introduced. The simplest third-order model is constructed from three cascaded first-order models which come from the water tank process. The mathematical bookkeeping required by higher-order models sometimes gets involved. To ameliorate this problem, matrices can often provide aid. Appendix G contains an elementary introduction to matrices in case the reader is a bit rusty in this area. Matrices form the backbone of the state-space approach which will make its debut in this chapter. All of the higher-order models covered in this chapter will be written as differential equations in the time domain, as transfer functions in the Laplace s-domain, as magnitudes and phases in the frequency domain, and as matrix differential equations back in the time domain.

5-1 Third-Order Process without Backflow

Figure 5-1 shows three independent tanks—independent in the sense that each downstream tank does not influence its upstream neighbor. Each tank in the series of three can be treated like the single tank we treated earlier except that the outlet flow rate of the upstream tank feeds into the next tank down the line. The single tank is described by

$$\rho A \frac{dL}{dt} + \frac{L}{R} = F$$

or (5-1)

$$\tau \frac{dL}{dt} + L = RF \qquad \tau = \rho AR$$

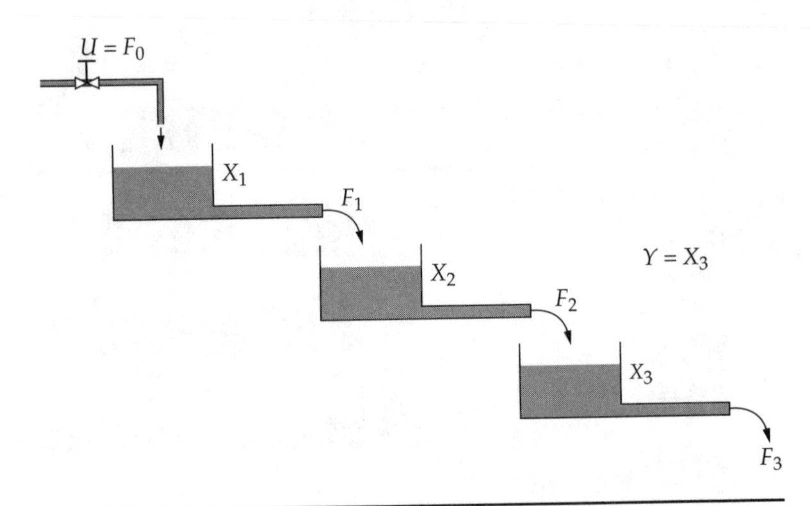

FIGURE 5-1 Three independent tanks.

For the tanks in the three-tank system let X_1, X_2, and X_3 represent the three levels in the tanks. Similarly let F_0, F_1, F_2, and F_3 represent the four flow rates shown in Fig. 5-1. The overall process input is $U = F_0$ and the overall process output is $Y = X_3$.

The equations describing their behavior can be derived from Eq. (5-1) by inspection:

$$
\begin{aligned}
&\rho A_1 \frac{dX_1}{dt} + \frac{X_1}{R_1} = F_0 \\
&\rho A_2 \frac{dX_2}{dt} + \frac{X_2}{R_2} = \frac{X_1}{R_1} \\
&\rho A_3 \frac{dX_3}{dt} + \frac{X_3}{R_3} = \frac{X_2}{R_2} \\
&Y = X_3 \qquad U = F_0
\end{aligned}
\tag{5-2}
$$

As we will see later on, X_1, X_2, and X_3 are the *states* of the system and the last line in Eq. (5-2) says that the process output Y is the third element of the state.

The Laplace Transform Version

By inspection, the reader should be able to rewrite Eq. (5-2) in the Laplace domain.

$$\left(\rho A_1 s + \frac{1}{R_1}\right)\tilde{X}_1(s) = \tilde{U}(s)$$

$$\left(\rho A_2 s + \frac{1}{R_2}\right)\tilde{X}_2(s) = \frac{1}{R_1}\tilde{X}_1(s) \qquad (5\text{-}3)$$

$$\left(\rho A_3 s + \frac{1}{R_3}\right)\tilde{X}_3(s) = \frac{1}{R_2}\tilde{X}_2(s)$$

$$\tilde{Y}(s) = \tilde{X}_3(s)$$

These equations can be combined in the Laplace domain, eliminating the $\tilde{X}_i$. First, rewrite Eq. (5-3) slightly, introducing three time constants.

$$(R_1\rho A_1 s + 1)\tilde{X}_1(s) = \tilde{U}(s) \qquad \tau_1 = R_1\rho A_1$$

$$(R_2\rho A_2 s + 1)\tilde{X}_2(s) = \frac{R_2}{R_1}\tilde{X}_1(s) \qquad \tau_2 = R_2\rho A_2$$

$$(R_3\rho A_3 s + 1)\tilde{X}_3(s) = \frac{R_3}{R_2}\tilde{X}_2(s) \qquad \tau_3 = R_3\rho A_3$$

Starting with the last equation, eliminate the three $\tilde{X}_i$ and develop an expression for the process transfer function G.

Second, eliminate $\tilde{X}_1$ and $\tilde{X}_2$ to get $\tilde{X}_3$:

$$\tilde{X}_3(s) = \frac{1}{\tau_3 s + 1}\frac{R_3}{R_2}\tilde{X}_2(s)$$

$$= \frac{1}{\tau_3 s + 1}\frac{R_3}{R_2}\frac{1}{\tau_2 s + 1}\frac{R_2}{R_1}\tilde{X}_1(s)$$

$$= \frac{1}{\tau_3 s + 1}\frac{R_3}{R_2}\frac{1}{\tau_2 s + 1}\frac{R_2}{R_1}\frac{R_1}{\tau_1 s + 1}\tilde{U}(s)$$

Next, use the fact that $\tilde{Y}(s) = \tilde{X}_3(s)$ and find the transfer function:

$$\tilde{Y}(s) = \frac{R_3}{\tau_3 s + 1}\frac{1}{\tau_2 s + 1}\frac{1}{\tau_1 s + 1}\tilde{U}(s) = G\tilde{U}(s)$$

$$G = \frac{R_3}{\tau_3 s + 1}\frac{1}{\tau_2 s + 1}\frac{1}{\tau_1 s + 1} \qquad (5\text{-}4)$$

$$= \frac{R_3}{(\tau_1\tau_2\tau_3)s^3 + (\tau_1\tau_2 + \tau_1\tau_3 + \tau_2\tau_3)s^2 + (\tau_1 + \tau_2 + \tau_3)s + 1}$$

The last line shows that the denominator of G is a third-order polynomial in s results, so the system is indeed a third-order system. Equation (5-4) also shows that the transfer function G has three poles at

$$s = -\frac{1}{\tau_1}, \quad -\frac{1}{\tau_2}, \quad -\frac{1}{\tau_3}$$

Having hopefully survived this fusillade of equations in the time and Laplace domains, let's look at the step-change response of this third-order process for the case of $\rho = 1$, all τ's being 10.0, and all R's being 10.0 (Fig. 5-2).

Note how the first tank level behaves just like the single tank studied in Chap. 3, but tanks two and three start to have inflection points (points where the rate of change of the slope of the curve changes, that is, where the second derivative changes sign). Also, note that the levels of the three tanks all line out at the same value.

Question 5-1 Why do all three tanks line out at the same value?

Answer Each tank has the same parameters and each tank is described by the same equation. At steady state the outlet flow from tank three has to equal that from tank two. Also, it must equal that from tank one and finally it must equal the inlet flow rate. Each tank is experiencing the same inlet and outlet flows and has the same dimensions. Since each tank's outlet flow rate depends only on its level, all of the levels will be the same at steady state.

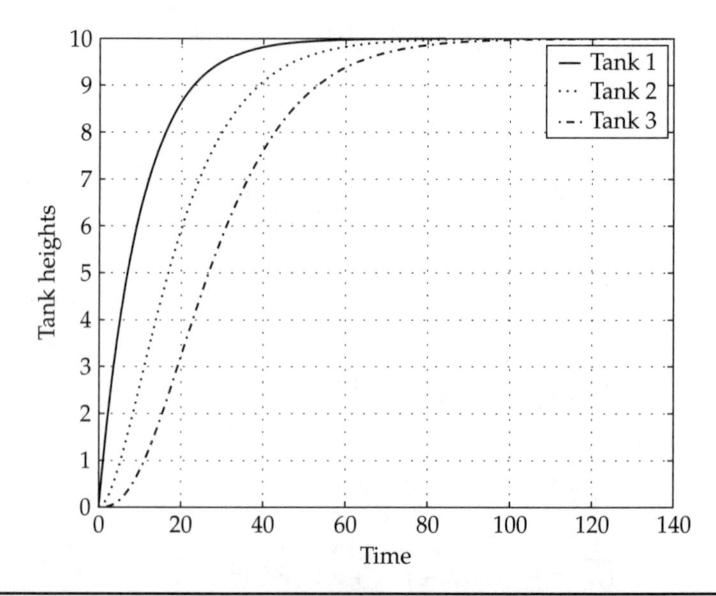

Figure 5-2 Step-change response of three independent tanks with no backflow.

The Frequency Domain Version

As we have done for all our example processes, let's move on to the frequency domain. Since we know that a single tank exhibits amplitude attenuation and phase lag with increasing frequency, what do you expect with this process? The tanks are in series. The input to the second and third tanks is the output from the first and second tanks, respectively, so, the amplitude attenuation and phase increase should be accumulative. This can be demonstrated easily by using $s = j\omega$ in Eq. (5-4).

$$G(s) = \frac{R_3}{\tau_3 s+1}\frac{1}{\tau_2 s+1}\frac{1}{\tau_1 s+1}$$

$$\begin{aligned} G(j\omega) &= \frac{R_3}{\tau_3 j\omega+1}\frac{1}{\tau_2 j\omega+1}\frac{1}{\tau_1 j\omega+1} \\ &= \frac{R_3}{\sqrt{(\tau_3\omega)^2+1}\,e^{j\theta_3}}\frac{1}{\sqrt{(\tau_2\omega)^2+1}\,e^{j\theta_2}}\frac{1}{\sqrt{(\tau_1\omega)^2+1}\,e^{j\theta_1}} \\ &= \frac{R_3 e^{-j\theta_3}e^{-j\theta_2}e^{-j\theta_1}}{\sqrt{(\tau_3\omega)^2+1}}\frac{1}{\sqrt{(\tau_2\omega)^2+1}}\frac{1}{\sqrt{(\tau_1\omega)^2+1}} \\ &= \frac{R_3 e^{-j(\theta_3+\theta_2+\theta_1)}}{\sqrt{(\tau_3\omega)^2+1}\sqrt{(\tau_2\omega)^2+1}\sqrt{(\tau_1\omega)^2+1}} \\ \theta_i &= \tan^{-1}(\tau_i\omega) \qquad i = 1,2,3 \end{aligned} \tag{5-5}$$

Equation (5-5) shows that the amplitude attenuations for each tank multiply and the phase lags for each tank add. Figure 5-3 supports this. Each tank is first order and contributes 90° of phase lag and Fig. 5-3 shows that the phase lag of the three-tank process approaches 270° at high frequencies. Figure 5-3 also shows that the magnitudes at low frequencies are the same.

Note that at high frequencies, the slopes of the magnitude plots for tanks 1, 2, and 3 are –20 dB/decade, –40 dB/decade, and –60 dB/decade, respectively.

The Matrix (State-Space) Version

Return to the time domain and rearrange Eq. (5-2) slightly

$$\begin{aligned} \frac{dX_1}{dt} &= -\frac{X_1}{\tau_1}+\frac{R_1}{\tau_1}F_0 \\ \frac{dX_2}{dt} &= \frac{R_2}{R_1}\frac{X_1}{\tau_2}-\frac{X_2}{\tau_2} \\ \frac{dX_3}{dt} &= \frac{R_3}{R_2}\frac{X_2}{\tau_3}-\frac{X_3}{\tau_3} \\ Y &= X_3 \end{aligned} \tag{5-6}$$

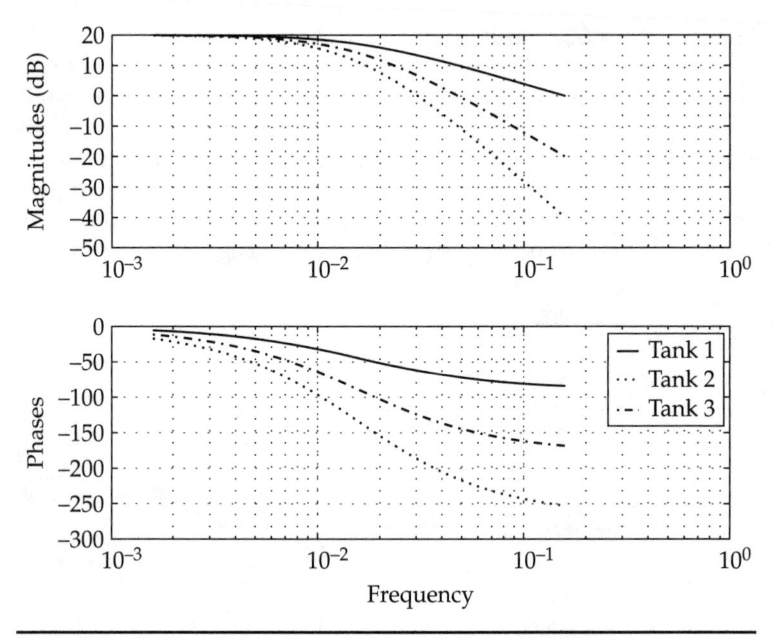

FIGURE 5-3 Bode plot of three-tank process with no backflow.

These equations can be rewritten compactly using matrices which are discussed in App. G. It is shown there that Eq. (5-6) can be written as

$$\frac{d}{dt}\begin{pmatrix} X_1 \\ X_2 \\ X_3 \end{pmatrix} = \begin{pmatrix} -\frac{1}{\tau_1} & 0 & 0 \\ \frac{R_2}{R_1}\frac{1}{\tau_2} & -\frac{1}{\tau_2} & 0 \\ 0 & \frac{R_3}{R_2}\frac{1}{\tau_3} & -\frac{1}{\tau_3} \end{pmatrix}\begin{pmatrix} X_1 \\ X_2 \\ X_3 \end{pmatrix} + \begin{pmatrix} \frac{R_1}{\tau_1} \\ 0 \\ 0 \end{pmatrix} F_0$$

$$Y = (0 \quad 0 \quad 1)\begin{pmatrix} X_1 \\ X_2 \\ X_3 \end{pmatrix} \tag{5-7}$$

In general, many linear models of processes can be written to fit the general format of

$$\frac{d}{dt}X = AX + BU$$

$$Y = CX \tag{5-8}$$

For this particular process the vectors and matrices occurring in Eq. (5-8) are

$$X=\begin{pmatrix}X_1\\X_2\\X_3\end{pmatrix} \qquad A=\begin{pmatrix}-\frac{1}{\tau_1} & 0 & 0\\ \frac{R_2}{R_1}\frac{1}{\tau_2} & -\frac{1}{\tau_2} & 0\\ 0 & \frac{R_3}{R_2}\frac{1}{\tau_3} & -\frac{1}{\tau_3}\end{pmatrix} \qquad B=\begin{pmatrix}\frac{R_1}{\tau_1}\\0\\0\end{pmatrix} \qquad U=F_0$$

$$C=(0 \quad 0 \quad 1) \qquad Y=X_3$$

Here, the process input U has one element that is presented to the model via the column vector B. The process output Y has one element that is extracted from the state vector via the row vector C. The state of the process X has three elements. This is called the *state-space* approach to presenting the process model.

To illustrate the flexibility of the state-space approach, consider the slightly modified process in Fig. 5-4.

The process input is now two-dimensional: the flows V_1 and V_2 replace the old single input flow F_0. The process output is also two-dimensional: we are interested in the tank levels of tanks two and three. What changes? The basic first-order matrix differential equation is unchanged:

$$\frac{d}{dt}X = AX + BU$$

$$Y = CX$$

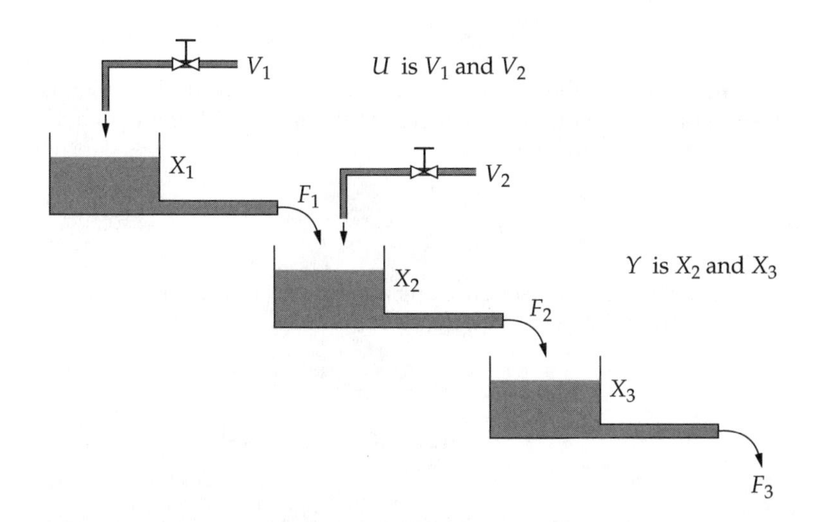

FIGURE 5-4 Three independent tanks with different inputs and outputs.

The matrices B and C change to accommodate the two-dimensional process, input and output, which have become two-dimensional column vectors. The matrix A and the state X are the same.

$$X = \begin{pmatrix} X_1 \\ X_2 \\ X_3 \end{pmatrix} \quad A = \begin{pmatrix} -\frac{1}{\tau_1} & 0 & 0 \\ \frac{R_2}{R_1}\frac{1}{\tau_2} & -\frac{1}{\tau_2} & 0 \\ 0 & \frac{R_3}{R_2}\frac{1}{\tau_3} & -\frac{1}{\tau_3} \end{pmatrix} \quad B = \begin{pmatrix} \frac{R_1}{\tau_1} & 0 \\ 0 & \frac{R_2}{\tau_2} \end{pmatrix} \quad U = \begin{pmatrix} V_1 \\ V_2 \end{pmatrix}$$

$$Y = CX$$

$$C = \begin{pmatrix} 0 & 1 & 0 \\ 0 & 0 & 1 \end{pmatrix} \quad Y = \begin{pmatrix} X_2 \\ X_3 \end{pmatrix}$$

Using the "across the row and down the column" matrix multiplication rule, the reader should check that these equations do indeed describe the process in Fig. 5-4.

In terms of the matrices and vectors, the state-space formulation appears to be first order. This suggests that there is a solution of the form

$$X_h = Ce^{\alpha t}$$

for the homogeneous form of the state-space equation which is

$$\frac{dX_h}{dt} = AX_h$$

Here, C is a column vector and α is a scalar. If this trial solution is inserted into the homogeneous part of the matrix differential equation the following results:

$$\frac{d}{dt}Ce^{\alpha t} = ACe^{\alpha t} \tag{5-9}$$

Since the rule for differentiating a matrix is simply the derivative of the elements in the matrix, Eq. (5-9) becomes

$$C\alpha e^{\alpha t} = ACe^{\alpha t}$$

$$\alpha IC = AC$$

or (5-10)

$$(A - \alpha I)C = 0$$

As is shown in App. G, solving Eq. (5-10) yields the eigenvectors and eigenvalues of the matrix A. When the eigenvectors have negative real parts, the process represented by Eq. (5-7) is stable.

Following the development in App. G, the eigenvalues of the A matrix can be determined by finding the values of λ that satisfy

$$|A-\lambda I| = \left\| \begin{pmatrix} -\frac{1}{\tau_1}-\lambda & 0 & 0 \\ \frac{R_2}{R_1}\frac{1}{\tau_2} & -\frac{1}{\tau_2}-\lambda & 0 \\ 0 & \frac{R_3}{R_2}\frac{1}{\tau_3} & -\frac{1}{\tau_3}-\lambda \end{pmatrix} \right\| = 0$$

$$= \left(-\frac{1}{\tau_1}-\lambda\right)\left(-\frac{1}{\tau_2}-\lambda\right)\left(-\frac{1}{\tau_3}-\lambda\right) = 0$$

Therefore, the eigenvalues are equal to the poles of the transfer function G.

It is important that the reader not move on until the material from this section (and App. G) has been thoroughly digested.

5-2 Third-Order Process with Backflow

Figure 5-5 shows an interconnected three-tank system with forward *and* backflow. If we treat each tank separately, the equations of Chap. 3 derived from mass balances can be applied immediately.

$$\begin{aligned} \rho A_1 \frac{dX_1}{dt} &= U - \frac{X_1 - X_2}{R_{12}} \\ \rho A_2 \frac{dX_2}{dt} &= \frac{X_1 - X_2}{R_{12}} - \frac{X_2 - X_3}{R_{23}} \\ \rho A_3 \frac{dX_3}{dt} &= \frac{X_2 - X_3}{R_{23}} - \frac{X_3}{R_3} \\ Y &= X_3 \end{aligned} \qquad (5\text{-}11)$$

The variables X_1, X_2, and X_3 represent the levels in each of the tanks. The net flow leaving the first tank is

$$\frac{X_1 - X_2}{R_{12}}$$

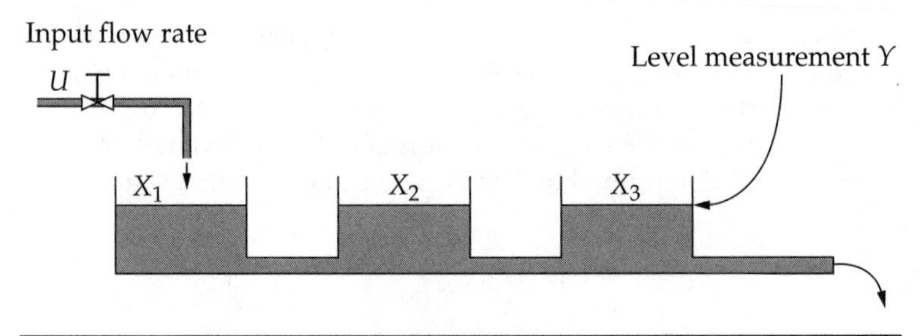

Figure 5-5 A three-tank system with backflow.

The $A_i, i = 1,2,3$, represent the respective cross-sectional areas of the tanks and the R's represent the resistances in the connections between the tanks and at the outlet. The quantities X_1, X_2, and X_3 are the *states* of the system and the last element of Eq. (5-11) says that the process output is the third element of the state. There are three equations and if all the equations were combined to eliminate the states, a third-order differential equation would be generated.

Alternatively let's move directly to the Laplace domain and do some algebra.

$$\begin{aligned}
\rho A_1 \tilde{X}_1 s &= \tilde{U} - \frac{\tilde{X}_1 - \tilde{X}_2}{R_{12}} \\
\rho A_2 \tilde{X}_2 s &= \frac{\tilde{X}_1 - \tilde{X}_2}{R_{12}} - \frac{\tilde{X}_2 - \tilde{X}_3}{R_{23}} \\
\rho A_3 \tilde{X}_3 s &= \frac{\tilde{X}_2 - \tilde{X}_3}{R_{23}} - \frac{\tilde{X}_3}{R_3} \\
\tilde{Y} &= \tilde{X}_3
\end{aligned} \tag{5-12}$$

Solving for $\tilde{Y}$ yields the following:

$$\tilde{Y} = \frac{R_3 \tilde{U}}{c_3 s^3 + c_2 s^2 + c_1 s + 1}$$

$$\begin{aligned}
c_3 &= \rho^3 A_3 \; R_{23} \, R_3 \, A_2 \, R_{12} \, A_1 \\
c_2 &= \rho^2 (R_3 \, A_2 \, R_{12} \, A_1 + A_2 \, R_{12} \, R_{23} \, A_1 + A_3 \, R_{23} \, R_3 \, A_2 \\
&\quad + A_3 \, R_{23} \, R_3 \, A_1 + A_3 \, R_3 \, A_1 \, R_{12}) \\
c_1 &= \rho (A_1 \, R_{12} + A_2 \, R_{23} + R_{23} \, A_1 + A_3 \, R_3 + R_3 \, A_2 + R_3 \, A_1)
\end{aligned} \tag{5-13}$$

Before the reader gets too impressed, I did not do this algebra by hand. Rather, I used the Matlab Symbolic toolbox–the algebra is just too tedious and the opportunity to make an algebraic bookkeeping mistake is too large. Note that all of the coefficients c_i in the denominator's polynomial are positive.

As an aside, the Matlab script that I used to develop Eq. (5-13) is

```
% interconnect.m
clear
syms A1 A2 A3 R12 R23 R3 s I2 U rho
syms x y z Z1 Z2 R33
syms R12A1 R12A2 R23A2 R23A3 R3A3
syms lam w
syms T1 T2 T3 % declare these variables as symbolic

% tanks with back flow
S=solve( rho*A1*s*x+(x-y)/R12-U , rho*A2*s*y-((x-y)/
R12-(y-z)/R23) , rho*A3*s*z-((y-z)/R23-z/R3));
Z1=S.z;
Z1=collect(Z1,s);
pretty(Z1)
```

Although I do not expect you to be adept at creating Matlab scripts, I do think you can browse the above code and get a feel for how simple it is to have the computer do the algebra.

On looking at Eq. (5-13) carefully, one sees the combination ρAR occurs frequently. This combination has units of seconds and could be considered a time constant of sorts. However, finding the poles of the transfer function is not as straightforward as for Eq. (5-4).

In any case, Eq. (5-13) shows that the highest power of the Laplace operator s is three, meaning that the equations describe a third-order system. Figure 5-6 shows the response of the three tank levels for a step in the input flow rate. The parameter values used were $A_1 = 0.1$, $A_2 = 0.1$, $A_3 = 0.1$, $R_{12} = 10$, $R_{23} = 10$, and $R_3 = 10$. Note how the steady-state levels are all different.

Question 5-2 Why are the steady-state levels different?

Answer At steady state all the flows must be the same. The net flow between tanks one and two is

$$(X_1 - X_2)/R_{12} = U \qquad \text{or} \qquad X_1 = X_2 + R_{12}U$$

Since U is nonzero and positive, X_1 must be greater than X_2. A similar argument shows that X_2 must be greater than X_3. Since the levels drive the flows, they must also be different.

The frequency domain behavior can be obtained in a manner similar to that for the three-tank process with no backflow and is

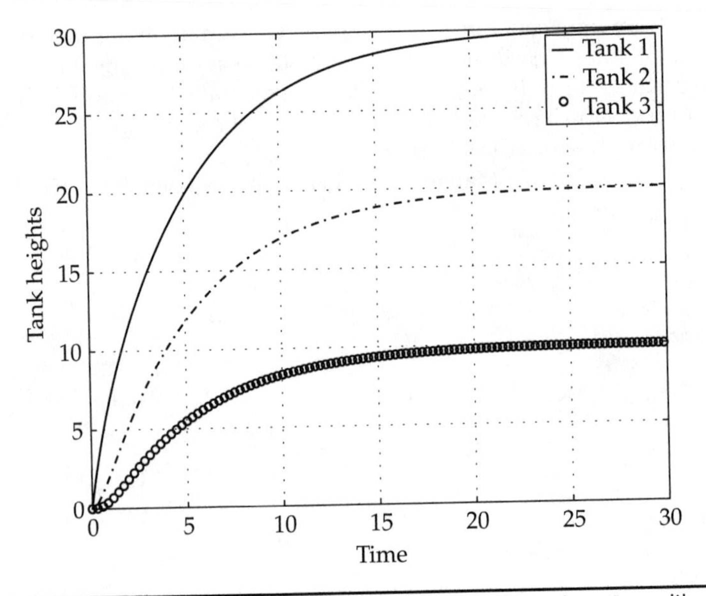

FIGURE 5-6 Step response of third-order process—three-tank system with backflow.

given in Fig. 5-7. Note that, unlike the case with no backflow, the three amplitude curves do not converge to a common asymptote as the frequency decreases because, as we mentioned in question 5-2, the steady-state levels are not the same.

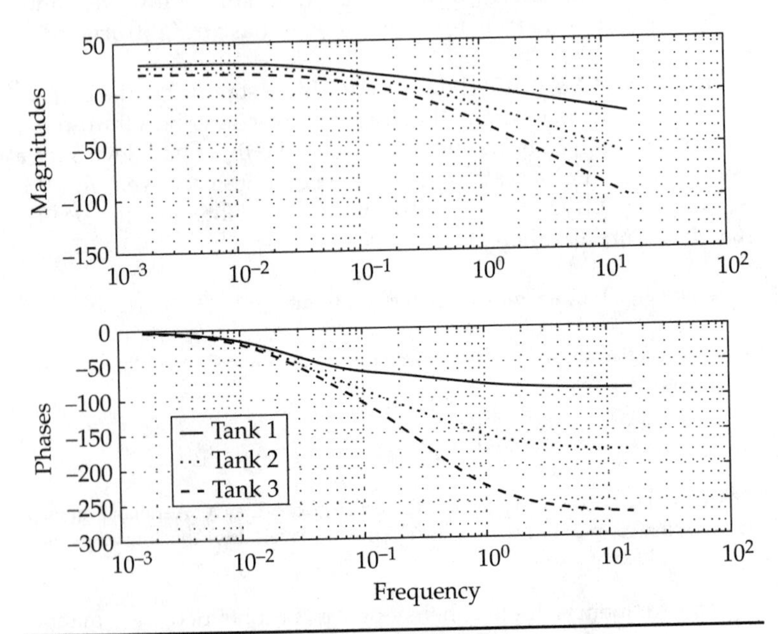

FIGURE 5-7 Bode plot for three tanks with backflow.

The State-Space Version

Based on the reader's experience with studying the no-backflow case, he should be able to rewrite Eq. (5-12) as follows. (Actually, the reader might better off to try this as an exercise.)

$$\frac{d}{dt}\begin{pmatrix} X_1 \\ X_2 \\ X_3 \end{pmatrix} =$$

$$\begin{pmatrix} -\dfrac{1}{\rho A_1 R_{12}} & \dfrac{1}{\rho A_1 R_{12}} & 0 \\ \dfrac{1}{\rho A_2 R_{12}} & -\left(\dfrac{1}{\rho A_2 R_{12}} + \dfrac{1}{\rho A_2 R_{23}}\right) & \dfrac{1}{\rho A_2 R_{23}} \\ 0 & \dfrac{1}{\rho A_3 R_{23}} & -\left(\dfrac{1}{\rho A_3 R_{23}} + \dfrac{1}{\rho A_3 R_3}\right) \end{pmatrix} \begin{pmatrix} X_1 \\ X_2 \\ X_3 \end{pmatrix} + \begin{pmatrix} \dfrac{1}{\rho A_1} \\ 0 \\ 0 \end{pmatrix} U$$

Note that only the 1, 3 and 3, 1 positions in the A matrix are zero. The full second row shows that the second tank is coupled with the other two tanks.

Question 5-3 If, for the three-tank system, the A matrix were diagonal, what would that mean physically and mathematically?

Answer The absence of off-diagonal terms in the A matrix means that there is no cross coupling and that each tank acts completely independent of the others. A diagonal A matrix also means that, instead of a set of three connected differential equations, there are three separate first-order differential equations that can be solved separately using techniques already presented in this book.

5-3 Control of Three-Tank System with No Backflow

These two example processes have a potential for control problems because at high frequencies the phase lag approaches 270°. To make it even more interesting, let's try integral-only control which we know adds an immediate 90° of phase lag to whatever is being controlled. With integral-only control the open-loop transfer function for the three-tank process with no backflow becomes

$$G(s) = G_p(s)G_c(s) = \frac{R_3}{\tau_3 s+1}\frac{1}{\tau_2 s+1}\frac{1}{\tau_1 s+1}\frac{I}{s} \qquad G_c = \frac{I}{s}$$

$$G(j\omega) = \frac{R_3}{\tau_3 j\omega+1}\frac{1}{\tau_2 j\omega+1}\frac{1}{\tau_1 j\omega+1}\frac{I}{j\omega}$$

$$= \frac{R_3}{\sqrt{(\tau_3\omega)^2+1}\,e^{j\theta_3}}\frac{1}{\sqrt{(\tau_2\omega)^2+1}\,e^{j\theta_2}}\frac{1}{\sqrt{(\tau_1\omega)^2+1}\,e^{j\theta_1}}\frac{I}{e^{j\frac{\pi}{2}}}$$

$$= \frac{R_3 e^{-j\theta_3} e^{-j\theta_2} e^{-j\theta_1} e^{-j\frac{\pi}{2}}}{\sqrt{(\tau_3\omega)^2+1}} \frac{1}{\sqrt{(\tau_2\omega)^2+1}} \frac{1}{\sqrt{(\tau_1\omega)^2+1}} \frac{I}{\omega}$$

$$= \frac{R_3 e^{-j(\theta_3+\theta_2+\theta_1+\frac{\pi}{2})}}{\sqrt{(\tau_3\omega)^2+1}\sqrt{(\tau_2\omega)^2+1}\sqrt{(\tau_1\omega)^2+1}} \frac{I}{\omega}$$

$$\theta_i = \tan^{-1}(\tau_i\omega) \quad i = 1,2,3 \tag{5-14}$$

The Bode plot of the process with (I = 1) and without integral control is shown in Fig. 5-8. Note how the addition of integral-only control raises the amplitude curve and lowers the phase curve.

The critical points where the phase equals –180° or where the amplitude or overall gain equals unity (or 0 dB) can be found graphically from Fig. 5-8. This suggests that if $I = 1$ were to be used, instability would result. (Can you convince yourself of this?) To get the magnitude plot down below unity, when the phase equals –180°, requires that we lower I significantly. Figure 5-8 suggests that we might want to lower the gain by at least as much as 50 dB or a factor of 0.003. It is a bit difficult to estimate this using the graph.

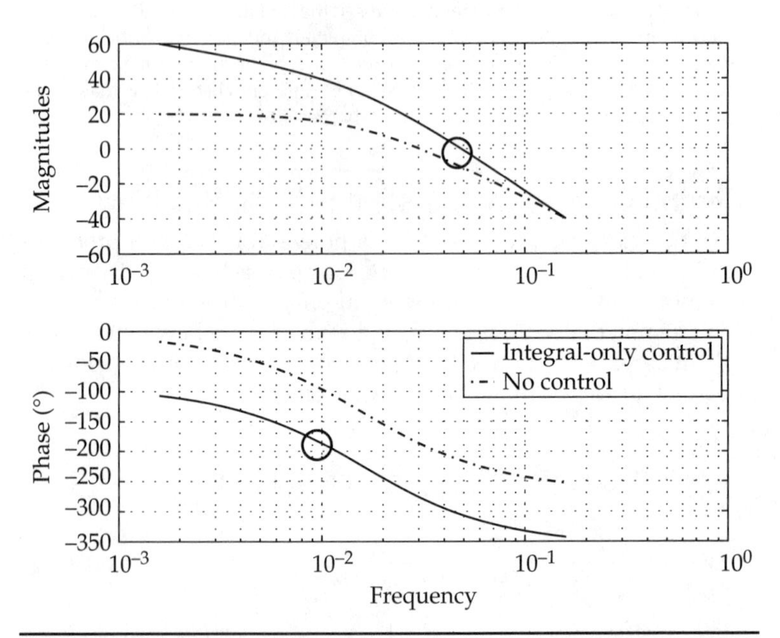

FIGURE 5-8 Bode plot for no backflow three-tank process with and without integral control.

Alternatively, a two-dimensional valley-seeking algorithm can be easily constructed to find the values of I and ω that minimize the following quantity

$$\left(\frac{R_3}{\sqrt{(\tau_3\omega)^2+1}\sqrt{(\tau_2\omega)^2+1}\sqrt{(\tau_1\omega)^2+1}}\frac{I}{\omega}-1\right)^2 + \left(\tan^{-1}(\tau_1\omega)+\tan^{-1}(\tau_2\omega)+\tan^{-1}(\tau_3\omega)+\frac{\pi}{2}+\pi\right)^2 \tag{5-15}$$

which is equivalent to solving Eq. (4-25) in Chap. 4 which is repeated here:

$$G_p(j\omega_c)G_c(j\omega_c) = G(j\omega_c) = -1$$

or

$$|G(j\omega_c)|e^{j\theta} = -1 = 1e^{-j\pi}$$

or

$$|G(j\omega_c)| = 1 \qquad |G(j\omega_c)| - 1 = 0$$

or

$$\theta(j\omega_c) = -\pi \qquad \theta(j\omega_c) + \pi = 0$$

For this case, the quantity in Eq. (5-15) is minimized by $I = 0.0089$ and $f = \omega/2\pi = 0.00929$ Hz.

A Matlab script to carry out this minimization is

```
clear
close all
x0=[1  .001];
x=fminsearch('ThirdCrit',x0);
disp('for Third Order process no back flow')
disp(['kc = ' num2str(x(1)) ' fc = ' num2str(x(2)) ])
% get mag and angle at this w and k
freq=x(2);
k=x(1);
R=10;
tau=10;
w=2*pi*freq;
y1=k*R/(sqrt( (tau*w)^2+1 ))^3; % magnitude at w and k
y2=-atan(tau*w)-atan(tau*w)-atan(tau*w);
```

```
y3=R/(sqrt( (tau*w)^2+1 ))^3; % magnitude at w and
kdisp(['at critical point mag = ' num2str(y1) ' angle
(deg) = ' num2str(y2*180/pi)])
disp(['process ampl at critical freq = ' num2str(y3) ' '
num2str(20*log10(y3)) ' dB'])
disp('________')
% now add integral-only to three tank process with no
back flow
clear
close all
x0=[1  .001];
x=fminsearch('ThirdCritI',x0);
disp('for Third Order process with integral-only')
disp(['I = ' num2str(x(1)) ' fc = ' num2str(x(2)) ])
freq=x(2);
I=x(1);
R=10;
tau=10;
w=2*pi*freq;
y1=I*R/(w*(sqrt( (tau*w)^2+1 ))^3); % magnitude at w and k
y2=-atan(tau*w)-atan(tau*w)-atan(tau*w)-pi/2;
y3=R/(w*(sqrt( (tau*w)^2+1 ))^3); % magnitude at w and k
disp(['at critical point mag = ' num2str(y1) ' angle
(deg) = ' num2str(y2*180/pi)])
disp(['process ampl at critical freq = ' num2str(y3) ' '
num2str(20*log10(y3)) ' dB'])
%-------------------
function y=ThirdCrit(x)
% called by critpars.m
k=x(1);
w=x(2)*(2*pi);
R=10;
tau=10;
y1=k*R/(sqrt( (tau*w)^2+1 )^3) - 1;
y2=-atan(tau*w)-atan(tau*w)-atan(tau*w)+pi;
y=y1^2+y2^2;
%--------------------------------
function y=ThirdCritI(x)
% called by critpars.m
k=x(1);
w=x(2)*(2*pi); % convert to radian freq
R=10;
tau=10;
y1=k*R/(w*(sqrt( (tau*w)^2+1 ))^3) - 1;
y2=-atan(tau*w)-atan(tau*w)-atan(tau*w)-pi*.5+pi;
y=y1^2+y2^2;
```

Note that the above script calls two functions, `ThirdCrit` and `ThirdCritI`. It also uses a built-in Matlab function `fminsearch`.

Using the integral gain from these calculations for a simulation would result in marginal stability, namely sustained but not-growing oscillations. Decreasing the integral gain by a factor of 0.5 to 0.004 provides a little better gain margin and produces a new Bode plot shown in Fig. 5-9. Note that the circles in this

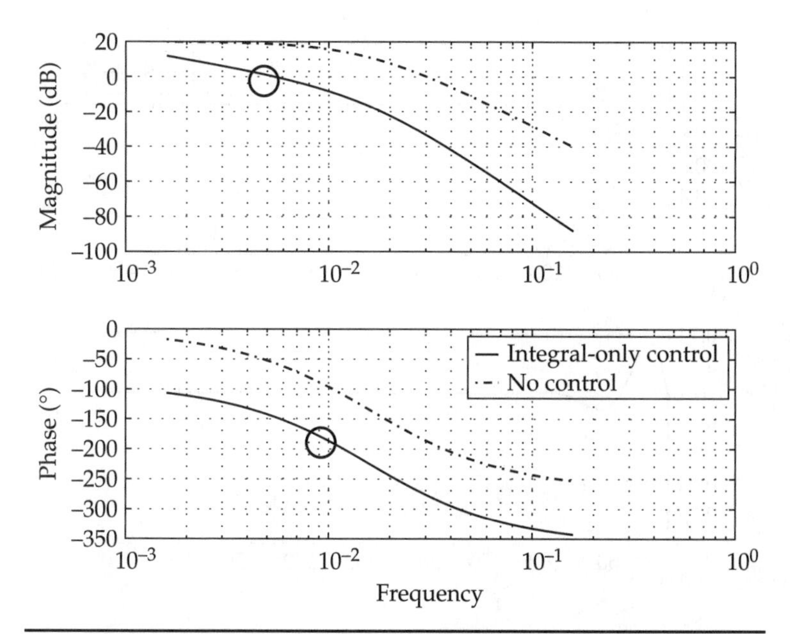

FIGURE 5-9 Bode plot for no backflow three-tank process, integral-only control with $i = 0.004$.

figure show that when the phase is –180° the gain is less than unity (or 0 dB). Likewise, when the gain is unity the phase is greater than –180°. The time domain performance for this gain is shown in Fig. 5-10.

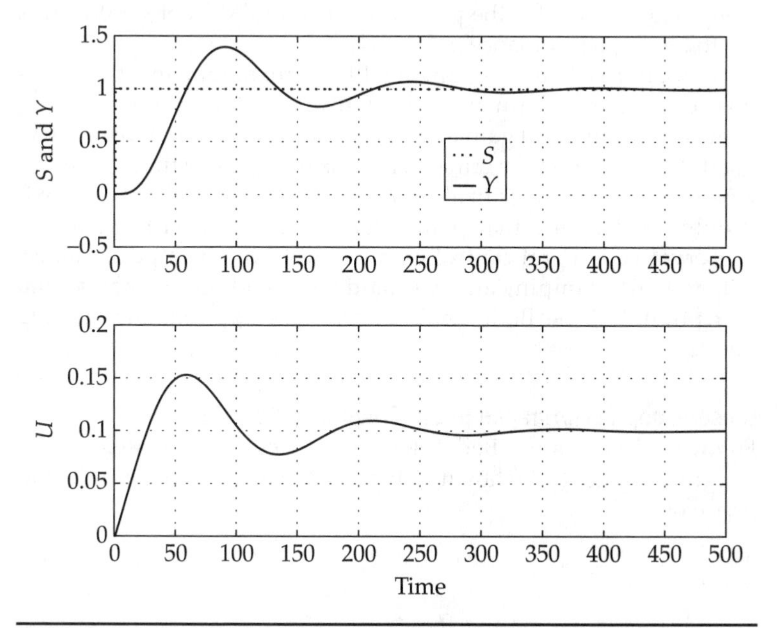

FIGURE 5-10 Integral-only control of three-tank process without backflow.

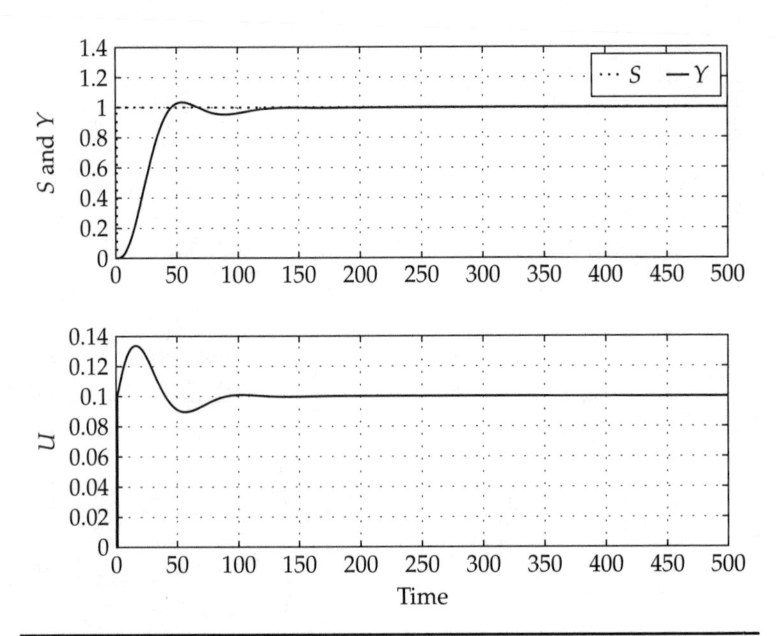

FIGURE 5-11 Three-tank process without backflow (0.004), adding proportional control ($P = 0.1$).

The closed-loop performance can be improved by adding proportional control. Using the above conservative integral gain of 0.004 along with a guess for the proportional control gain of $k = 0.1$ gives the following performance in Fig. 5-11.

Our approach to designing the PI control parameters that gives this acceptable performance is partially trial and error. We arrived at the critical integral gain of 8.9×10^{-3} and cut it in half before trying it. Had we done the critical analysis with proportional-only we would have found a critical proportional control gain of approximately 0.8. Halving that parameter and using it with the above conservative integral control gain would have given performance with too little damping and we could have made adjustments from that point. So, the Bode analysis can give approximate starting points.

Closed-Loop Performance in the Frequency Domain

Figure 5-12 shows the Bode plot for the *closed*-loop system under integral-only control (shown in Fig. 5-10). Here the magnitude and phase of

$$\frac{\tilde{Y}}{\tilde{S}} = \frac{G_c G_p}{1 + G_c G_p}$$

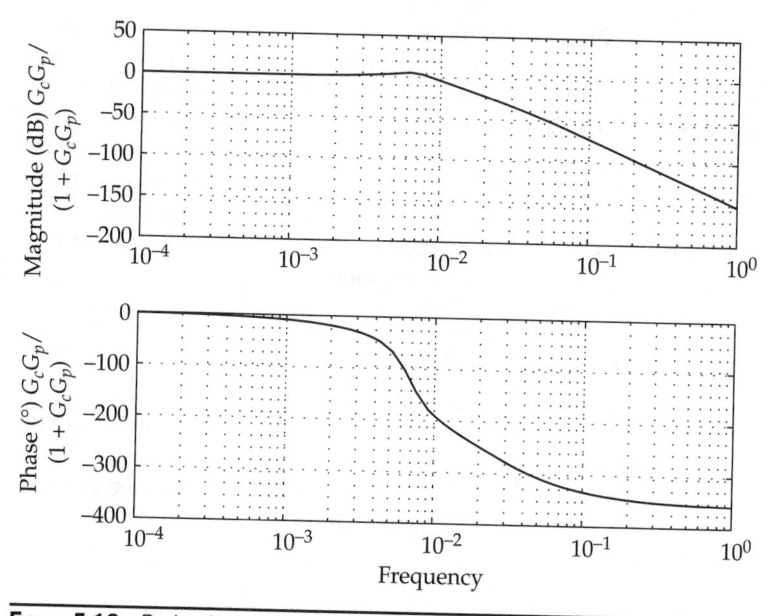

FIGURE 5-12 Bode plot of no backflow, integral-only control, three-tank, closed-loop.

is plotted against frequency. This figure is to be compared with Fig. 5-9 which dealt with the *open*-loop Bode plot for G_cG_p. Figure 5-12 shows that for low frequencies the magnitude of the Y/S is near unity and the phase lag is relatively low.

In other words, the process output Y follows the set point S fairly well at low frequencies, say below 5×10^{-4} Hz. As the frequency of S increases, the process output Y does not follow the set point as well—there is an attenuation of Y relative S and there is phase lag. Physically, this is to be expected. At low frequencies, the inertia of the process is not enough to keep the process output from effectively following the set point. However, at high frequencies, the inertia is such to cause the process output not to follow the set point.

5-4 Critical Values and Finding the Poles

In Chap. 3, the poles of the appropriate transfer function for the controlled system could be found by solving Eq. (3-50) or

$$G_cG_p = -1 \tag{5-16}$$

In these last couple of examples, we have shown that, after replacing s with $j\omega$, Eq. (5-16) essentially becomes a complex equation

$$G_c(j\omega_c)G_p(j\omega_c) = e^{j\pi} \tag{5-17}$$

where the dependence on the critical radian frequency ω_c is shown. (Note that Eq. (5-17) is a consequence of the expression $e^{j\pi} = -1$ that we mentioned in App. B). Actually, Eq. (5-17) depends on both ω_c and the critical control gain k_c (if the control is proportional-only). If the control is integral-only then the critical control gain would be I_c. Since Eq. (5-17) is now a complex equation, there are real and imaginary parts. Therefore, there are two equations in the two unknowns, ω_c and k_c. This argument suggests that the pole-finding approach and the Bode plot approach are basically the same.

5-5 Multitank Processes

Expand the concept presented in Fig. 5-1 to N tanks, each with no backflow, and specify that all N tanks have the same volume and that the interconnecting piping is the same. Therefore, all tanks will have the same time constant, say 1.0 after scaling, and the same resistance to flow. The ith tank will be described by

$$\tau \frac{dx_i}{dt} + x_i = x_{i-1} \qquad i = 1, \ldots, N$$

where x_0 will be the inlet flow rate and x_i will be the flow rate leaving the ith tank. The step-change response of the tanks is shown in Fig. 5-13.

Look at the first curve to the left of the graph for $N = 1$. The response reaches a value of 0.63 at $t = 1$ which is the time constant of

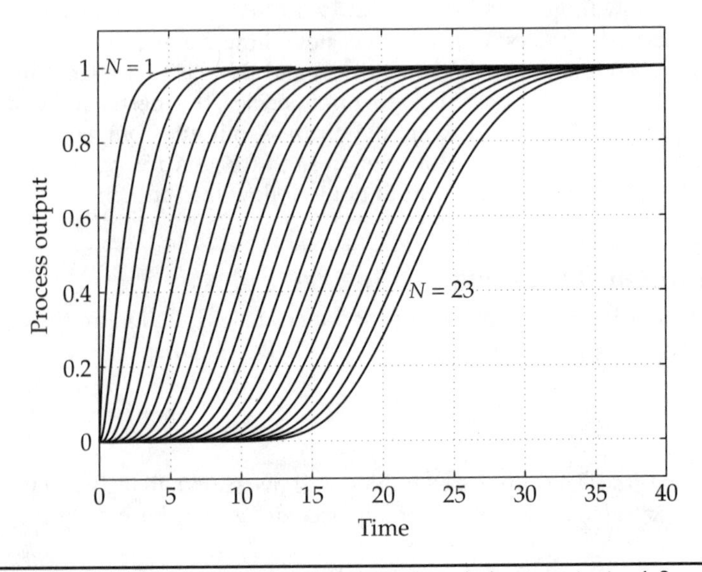

Figure 5-13 Step response of N tanks—each tank time constant = 1.0.

the single-tank process. The next curve on the right is for $N = 2$ and the response reaches a value of 0.63 at about $t = 2$. Note that as N increases the response curves take on a sharper inflection and the time at which the process output passes the value of 0.63 increases. For large N, the process appears to have a significant dead time even though there is no explicit dead time in the model.

Now, repeat this thought experiment except make the time constant of each tank decrease as N increases such that the total effective time constant is held constant at 1.0, for each N.

$$\tau_i \frac{dx_i}{dt} + x_i = x_{i-1} \qquad i = 1,\ldots,N$$

$$\tau_i = \frac{1}{N}$$

The step-change response of this system is shown in Fig. 5-14.

Note that the process output of all the tanks tends to pass 0.63 at a time of 1.0 and as N increases the inflection point becomes sharper. In the limit of an infinite number of tanks, the step response will become a sharp step at $t = 1.0$ identical to that of the pure dead-time process with a dead time of 1.0.

The Bode plot given in Fig. 5-15 is an extension of that in Fig. 5-3.

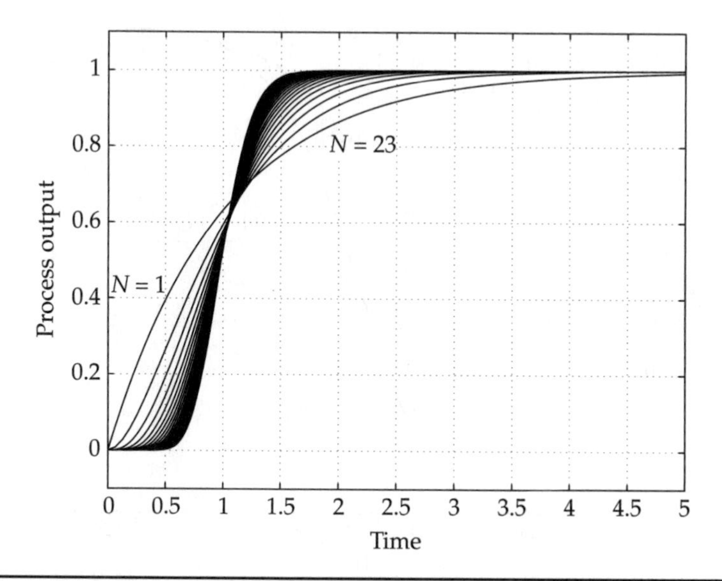

FIGURE 5-14 Step response of N tanks—total time constant = 1.0.

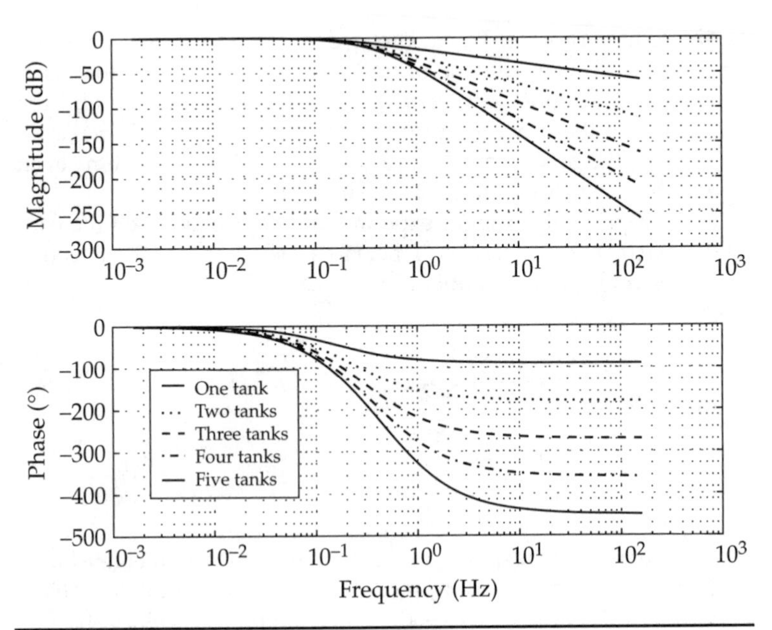

Figure 5-15 Bode plot of multitank system.

Matching the *N*-Tank Model with a FOWDT Model

The FOWDT model introduced in Chap. 4 has quite a bit of flexibility and is simple to use. Figure 5-16 shows how a 20-tank model with each tank having a time constant of 1/20 can be approximated by a FOWDT having a time constant of 0.2 and a dead time of 0.732.

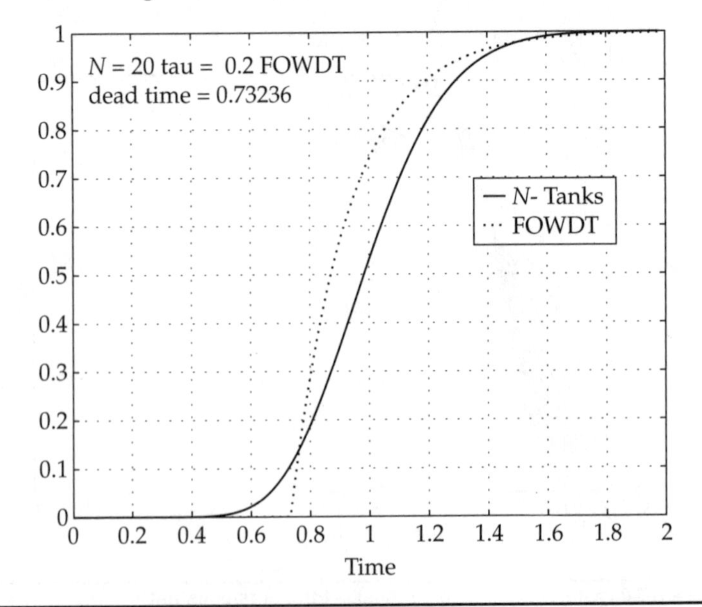

Figure 5-16 Matching a FOWDT model with *N*-tanks in the time domain.

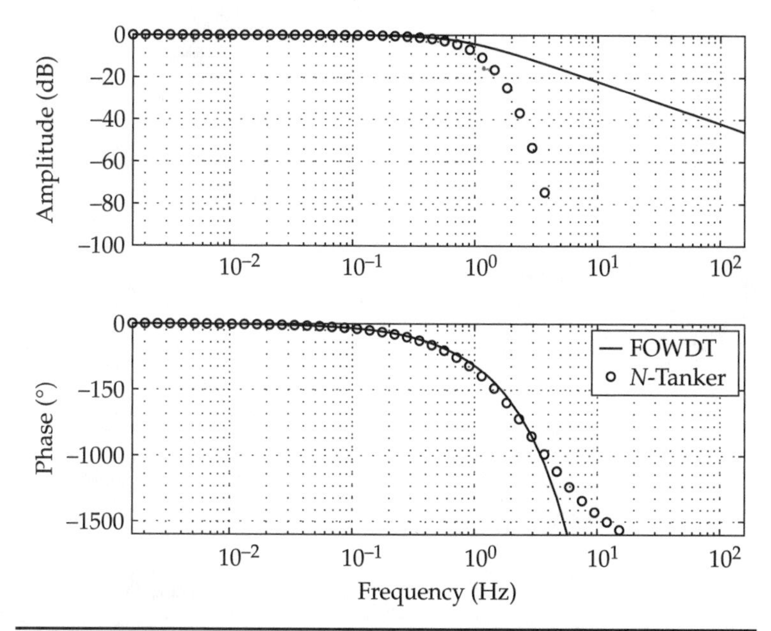

FIGURE 5-17 Matching a FOWDT model with multitank system in the frequency domain.

For a complementary point of view, the two models are compared in the frequency domain in Fig. 5-17. The corner frequencies of the two models are approximately the same but for higher frequencies the amplitude drops off much more quickly for the 20-tank model. This makes sense because the FOWDT model is basically first order so the magnitude drops off at 20 dB/decade while the other model's attenuation rate is 20 times that. On the other hand, the phase plot shows that the FOWDT model with the true dead time has a much higher drop-off rate. Furthermore, we know that each of the tanks in the 20-tank model contributes 90° of phase so the maximum phase lag at infinite frequency will be 1800°.

For simulation purposes and control algorithm testing, the FOWDT model might provide a simple computational approximation to the multiple tank process.

5-6 Summary

The toolkit of model processes has been expanded to include multiple-stage systems. The mathematical toolkit has been expanded to include matrix methods to simplify and enhance the mathematical bookkeeping and to pave the road to more methods using state-space concepts. In the next chapter our process toolkit will be enlarged to include yet another model process—one that rings.

CHAPTER 6

An Underdamped Process

6-1 The Dynamics of the Mass/Spring/Dashpot Process

All of the example processes mentioned so far have been "overdamped" in that the open-loop step response does not generate overshoot or oscillations of any kind. The first-order process really has no choice—its behavior is dictated by its gain and time constant. The three tank third-order process has an inflection point in the step response but it will never oscillate or "ring" when subjected to a step change in the process input with no feedback control. These overdamped processes are typical of most of the real-live industrial processes that I faced for most of my career. However, near the end I got involved in some new photonics processes that were underdamped and posed many new challenges.

When we close the control loop on the overdamped processes we could get underdamped and even unstable behavior when the feedback was aggressive but the processes by themselves could not exhibit this kind of performance.

Not so with the so-called mass/spring/dashpot process shown in Fig. 6-1. To derive an equation that describes its behavior one needs to apply Newton's second law of motion:

$$\sum F = m\frac{d^2y}{dt^2} \tag{6-1}$$

The sum of the forces acting on the mass causes the mass to accelerate. The displacement of the mass is given by y. The first component of the forces is due to the spring that applies a force proportional to the extension of the mass's position y, the process output, from equilibrium. The spring constant is k. The direction of this force, $-ky$, is opposite to the direction of the mass's movement. The second force is the friction of the dashpot. It acts in proportion to the speed of the mass and is also in a direction opposite to the motion of the mass, as

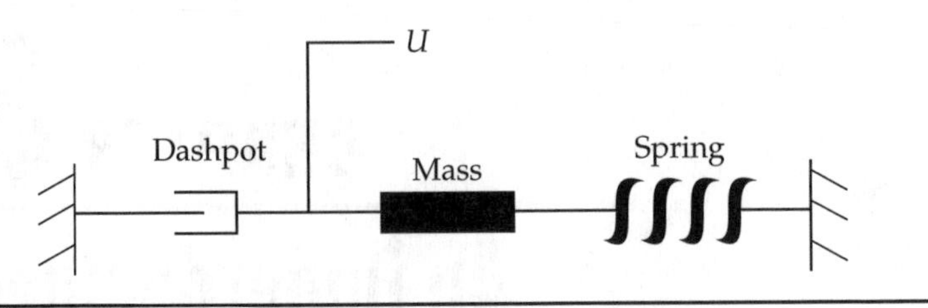

FIGURE 6-1 Mass/spring/dashpot process.

in $-B(dy/dx)$. The coefficient of friction is B. Finally, the third component is the applied external force U, which is also the process input.

With this in mind, Eq. (6-1) becomes

$$m\frac{d^2y}{dt^2} = -B\frac{dy}{dt} - ky + gU \tag{6-2}$$

By convention, Eq. (6-2) is rewritten as

$$\frac{d^2y}{dt^2} + 2\zeta\omega_n\frac{dy}{dt} + \omega_n^2 y = g\omega_n^2 U$$
$$\omega_n = \sqrt{\frac{k}{m}} \qquad \zeta = \frac{B}{2\sqrt{km}} \tag{6-3}$$

where the damping factor ζ and the natural frequency ω_n appear as functions of the mass, spring constant, and coefficient of friction. When the damping factor ζ varies between 0 and 1 the behavior is underdamped. When $\zeta = 1$ the behavior is critically damped and when $\zeta > 1$ the behavior is overdamped. The natural frequency is effectively the frequency of the "ringing" that the mass experiences after a disturbance. A higher natural frequency means a faster response and higher frequency ringing. The natural frequency has units of radians/sec and is related to f_n, the frequency in cycles/sec, as follows:

$$\omega_n = 2\pi f_n$$

Alternatively, Eq. (6-3) can be written as

$$\tau^2\zeta^2\frac{d^2y}{dt^2} + 2\tau\zeta^2\frac{dy}{dt} + y = gU$$
$$\tau = \frac{2m}{B} \qquad \zeta = \frac{B}{2\sqrt{km}} \tag{6-4}$$

Figure 6-2 shows the step-change response of the mass/spring/dashpot process for various values of the damping coefficient.

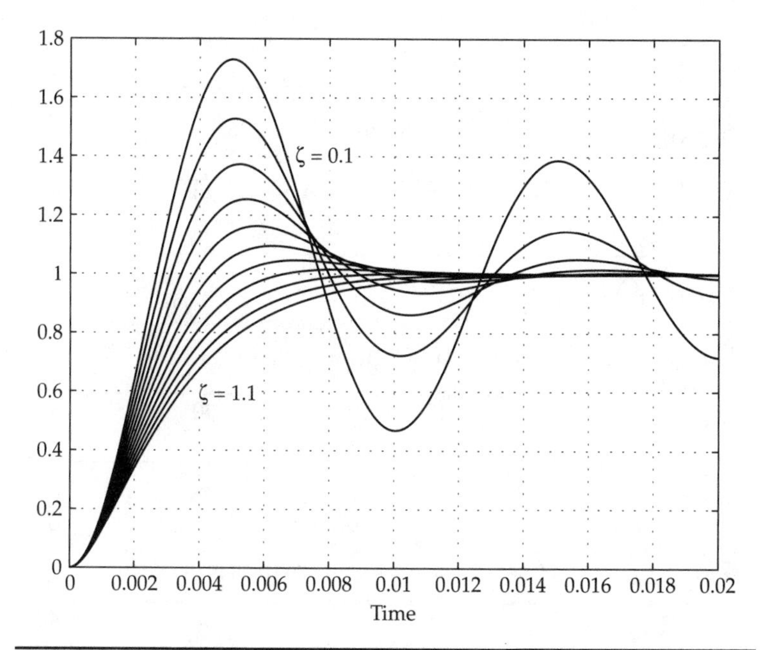

Figure 6-2 Step response of mass/spring/dashpot process—typical second-order Bode plot; $g = 1$, $\omega_n = 100$.

Question 6-1 How do the physical parameters of the mass/spring/dashpot process relate to your intuition?

Answer The damping factor is proportional to the coefficient of friction B. Thus, as the dashpot provides more resistance, the damping factor increases and the response becomes less oscillatory. However, the time constant will decrease.

The damping factor is inversely proportional to the spring constant k. But, the natural frequency is directly proportional to the spring constant. Therefore, as the spring gets stiffer (as k increases) the damping factor decreases and the natural frequency increases. This means there will be more underdamped behavior and the frequency of the oscillations will be higher. The time constant is unaffected by the spring constant.

Finally, both the damping factor and the natural frequency are inversely proportional to the mass but the time constant is directly proportional to the mass. Thus, as the mass increases, the natural frequency decreases and the damping decreases. So, with more mass the process will exhibit more underdamped behavior, the frequency of the oscillations will decrease and the time constant will decrease. That the damping will increase with less mass may be a little counter-intuitive.

To help get a feel for this consider Fig. 6-3 where the spring constant and the dashpot friction are kept constant but the mass is varied. Here the spring constant k is 5 and the coefficient of friction B is 1. As the mass increases, the natural frequency and the damping coefficient both decrease giving a more drawn-out underdamped behavior shown in Fig. 6-3. Figure 6-4 shows how the dynamics change with the spring constant and Fig. 6-5 shows the same thing with the coefficient of friction.

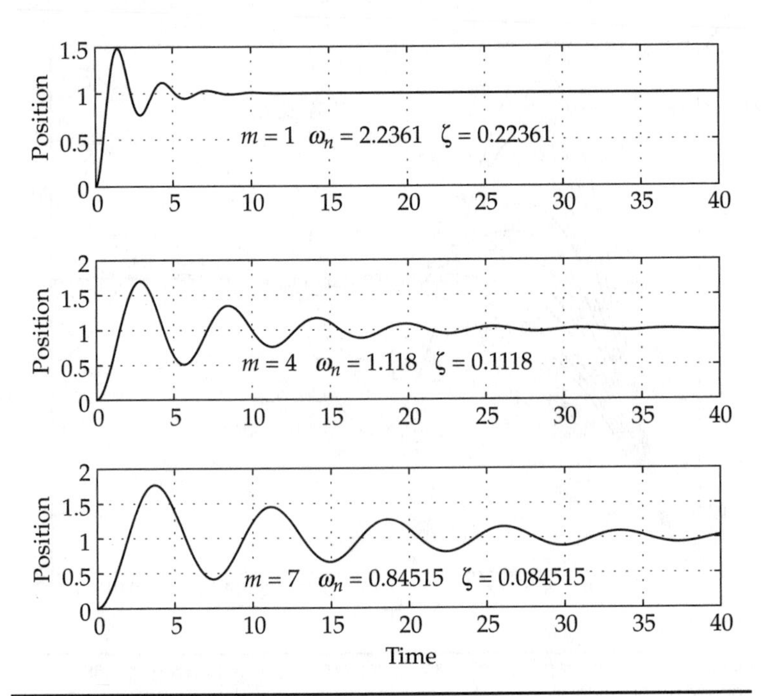

FIGURE 6-3 Effect of mass on dashpot dynamics.

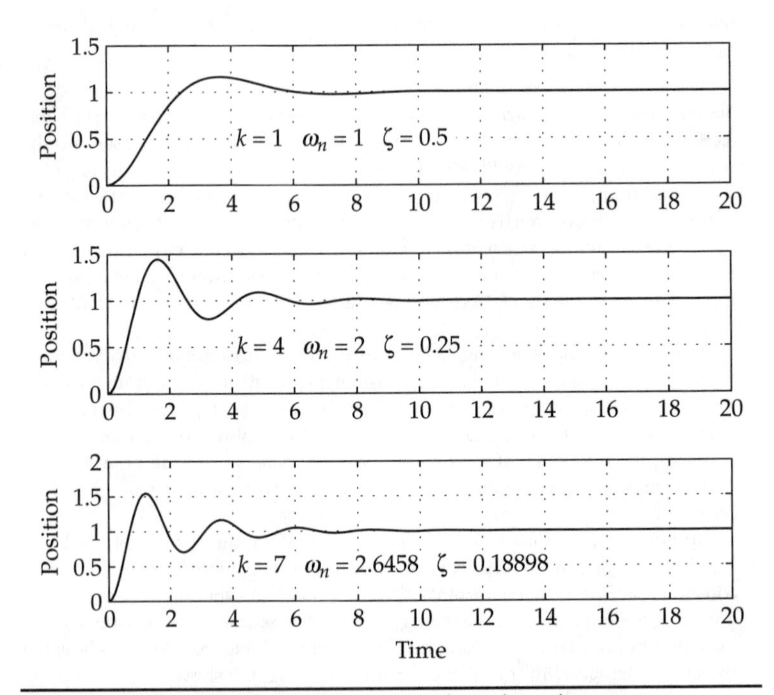

FIGURE 6-4 Effect of spring constant on dashpot dynamics.

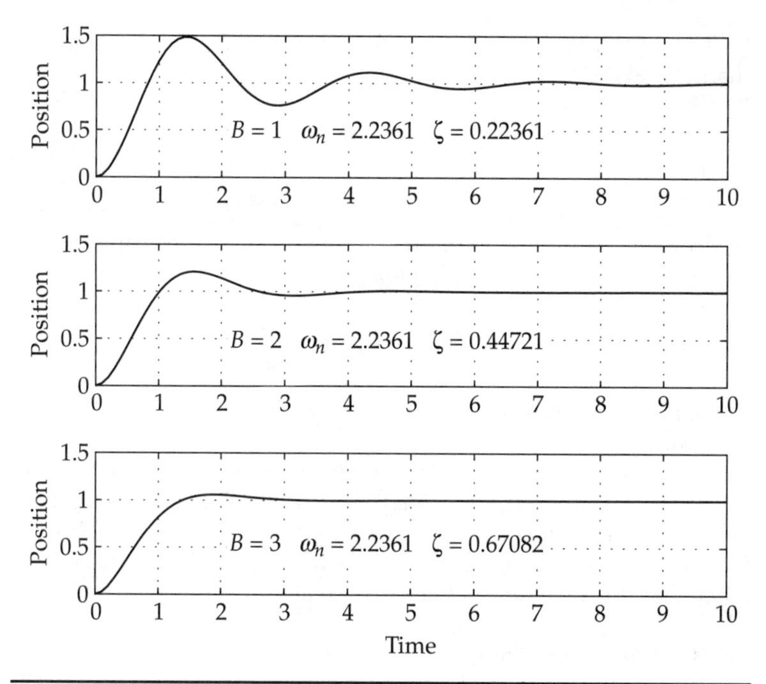

FIGURE 6-5 Effect of friction on dashpot dynamics.

6-2 Solutions in Four Domains

6-2-1 Time Domain

As is our usual approach, we could attempt to solve Eq. (6-3) by trying a solution for the homogeneous part of the form:

$$y_h = Ce^{at}$$

This would generate a quadratic equation for a, which would lead to a homogeneous solution that had two exponential terms. For the time being this approach will be sidestepped.

6-2-2 Laplace Domain Solution

Alternatively, let's go directly to the Laplace domain and take the Laplace transform of Eq. (6-3), as in

$$s^2\tilde{y}(s) + 2\zeta\omega_n s\tilde{y}(s) + \omega_n^2\tilde{y}(s) = g\omega_n^2\tilde{U}(s)$$

$$\frac{\tilde{y}(s)}{\tilde{U}(s)} = G_p(s) = \frac{g\omega_n^2}{s^2 + 2\zeta\omega_n s + \omega_n^2} = \frac{g\omega_n^2}{(s-p_1)(s-p_2)} \tag{6-5}$$

The poles of the transfer function are located at the roots of the quadratic in the denominator:

$$p_1, p_2 = -\zeta\omega_n \pm \omega_n\sqrt{\zeta^2 - 1} = a \pm jb$$

If the damping factor ζ is less than unity, these poles become complex conjugates and the solution will contain sinusoidal components suggesting underdamped behavior, as in

$$y_h(t) \sim C_1 e^{p_1 t} + C_2 e^{p_2 t} = C_1 e^{(a+jb)t} + C_2 e^{(a-jb)t}$$

where Euler's formula $e^{a+jb} = e^a[\cos(b) + j\sin(b)]$ can be used to bring in the sinusoids.

Figure 6-6 shows how the roots (or poles) move in the s-plane as the damping factor changes from 0.1 to 1.1. For this case, the natural frequency was kept constant at 100 Hz. When $\zeta = 1.1$, the poles are both real but when $\zeta = 0.1$ both poles nearly lie on the imaginary axis. When $\zeta = 1$ the poles are the same and real.

6-2-3 Frequency Domain

Letting $s = j\omega$ in Eq. (6-5), which gives

$$\frac{\tilde{y}(j\omega)}{\tilde{U}(j\omega)} = \frac{g\omega_n^2}{(j\omega)^2 + 2\zeta\omega_n(j\omega) + \omega_n^2} \tag{6-6}$$

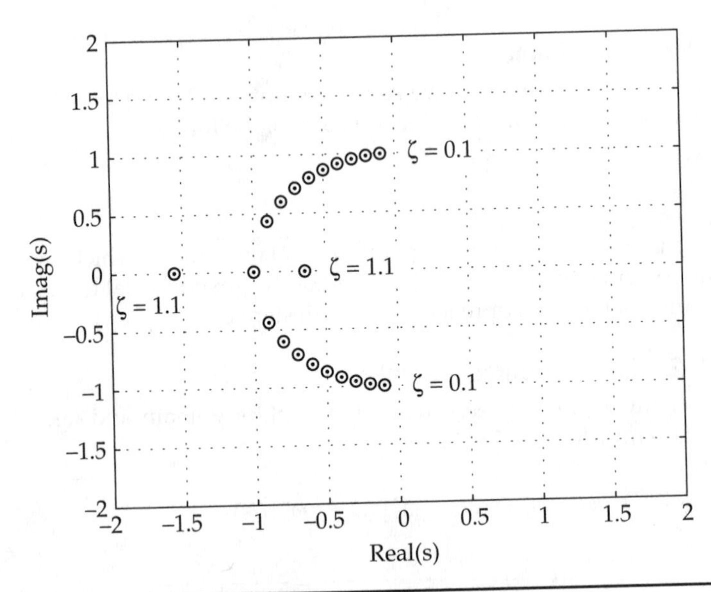

Figure 6-6 Poles of second-order model; $\zeta = 0.1$ to 1.1.

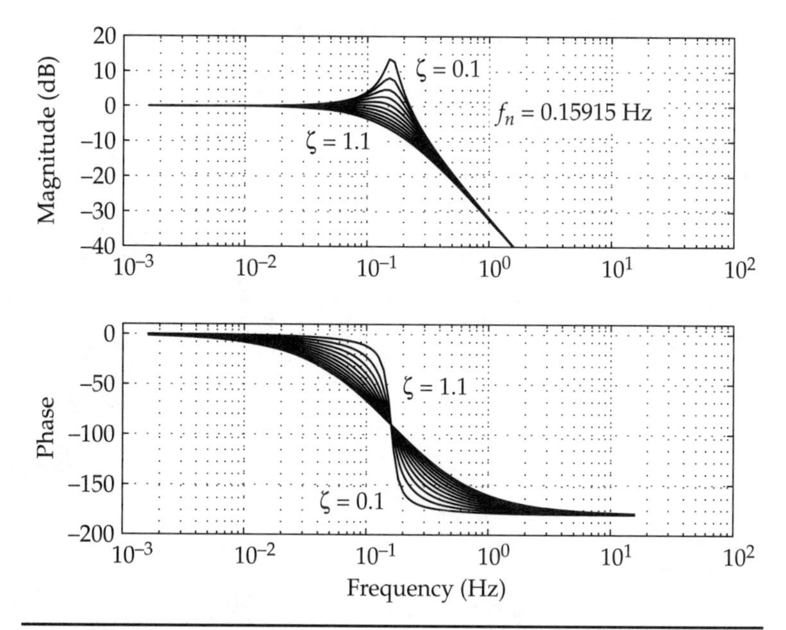

FIGURE 6-7 Typical second-order Bode diagram showing effect of damping.

Figure 6-7 shows the Bode plot constructed for the magnitude and phase from Eq. (6-6). Note that as the damping decreases a peak develops in the amplitude plot suggesting the start of a resonance at the natural frequency, which for this example is at 1.0 rad/sec or 0.159 Hz. Therefore, lightly damped systems will have oscillations or "ringing" at the natural frequency which will die off in time. In the phase diagram, as the damping decreases, the slope of the phase curve increases sharply at the natural frequency. Note that the maximum phase lag is 180°.

6-2-4 State-Space Representation

Let's start with the time domain representation:

$$\frac{d^2y}{dt^2} + 2\zeta\omega_n \frac{dy}{dt} + \omega_n^2 y = g\omega_n^2 U \tag{6-7}$$

we construct two elements of the state as

$$x_1 = y$$

$$x_2 = \frac{dy}{dt}$$

Substituting these in Eq. (6-7) generates two first-order equations

$$\frac{dx_1}{dt} = x_2$$

$$\frac{dx_2}{dt} + 2\zeta\omega_n x_2 + \omega_n^2 x_1 = g\omega_n^2 U$$

These two equations can be written in matrix form as follows:

$$\frac{d}{dt}\begin{pmatrix} x_1 \\ x_2 \end{pmatrix} = \begin{pmatrix} 0 & 1 \\ -\omega_n^2 & -2\zeta\omega_n \end{pmatrix}\begin{pmatrix} x_1 \\ x_2 \end{pmatrix} + \begin{pmatrix} 0 \\ g\omega_n^2 \end{pmatrix} U$$

$$\frac{dx}{dt} = Ax + BU$$

$$A = \begin{pmatrix} 0 & 1 \\ -\omega_n^2 & -2\zeta\omega_n \end{pmatrix} \tag{6-8}$$

$$B = \begin{pmatrix} 0 \\ g\omega_n^2 \end{pmatrix}$$

The eigenvalues of the A matrix are the values of λ that satisfy the following equation:

$$\left| \begin{pmatrix} 0 & 1 \\ -\omega_n^2 & -2\zeta\omega_n \end{pmatrix} - \lambda \begin{pmatrix} 1 & 0 \\ 0 & 1 \end{pmatrix} \right| = 0$$

$$\left| \begin{pmatrix} 0-\lambda & 1 \\ -\omega_n^2 & -2\zeta\omega_n - \lambda \end{pmatrix} \right| = 0$$

$$(-\lambda)(-2\zeta\omega_n - \lambda) - (1)(-\omega_n^2) = 0$$

$$\lambda^2 + 2\zeta\omega_n\lambda + \omega_n^2 = 0$$

Therefore, the poles of the transfer function in Eq. (6-5) are the eigenvalues of the A matrix in Eq. (6-8).

6-2-5 Scaling and Round-Off Error

The quadratic equation whose roots yield the eigenvalues contain terms that have widely varying numerical values and this can provide round-off errors in the computation of simulations and Bode plots. To address this, the time and the dependent variables can be scaled. To make the bookkeeping less messy we start with primes to indicate the time t' and the dependent variable y'. Our starting point is

$$\frac{d^2y'}{dt'^2} + 2\zeta\omega_n \frac{dy'}{dt'} + \omega_n^2 y' = g\omega_n^2 U$$

Introduce a new independent variable $t' = t/\omega_n$ and a new dependent variable $y' = gy$. Applying these substitutions gives

$$\omega_n^2 \frac{d^2(gy)}{dt^2} + 2\zeta\omega_n^2 \frac{d(gy)}{dt} + \omega_n^2 gy = g\omega_n^2 U$$

$$\frac{d^2y}{dt^2} + 2\zeta\frac{dy}{dt} + y = U$$

The scaled process has a natural frequency and a gain of unity. Note that if the natural frequency is 1.0 rad/sec then the natural frequency will also be

$$f = \frac{1.0 \text{ rad/sec}}{2\pi \text{ rad/cycle}} = 0.159 \text{ cycles/sec}$$

The equation yielding the eigenvalues becomes

$$\lambda^2 + 2\zeta\lambda + 1 = 0$$

and the eigenvalues or poles become

$$\lambda_1, \lambda_2 = -\zeta \pm \sqrt{\zeta^2 - 1} = a \pm jb$$

The next batch of computations will deal with the scaled process.

6-3 PI Control of the Mass/Spring/Dashpot Process

You have been exposed to attempts to control the first-order process (the single water tank) and the third-order process (the three-tank process). The approach has been to feed the process output back and subtract it from the set point and generate an error signal. Then an adjustment to the process input was developed based on signals proportional to the error and proportional to the integral of the error. The time domain has been used to demonstrate the effectiveness (or lack thereof) of these methods. The frequency domain has been used to get an estimate for one of the control gains by making sure that the open-loop combination of the process and controller, represented by G_pG_c, had sufficient gain margin when the phase was −180° (or sufficient phase margin when the gain was unity).

For the mass/spring/dashpot process with proportional-integral control G_pG_c looks like

$$G_pG_c = \frac{1}{s^2 + 2\zeta s + 1}\frac{ks + I}{s}$$

The real negative zero at $s=-I/k$ from G_c will not cancel either of the complex conjugate poles at $-\zeta \pm \sqrt{\zeta^2-1}$ from G_p. This suggests that proportional plus integral control may not have much impact on the complex poles that produce the underdamped behavior. The presence of these complex conjugate poles poses a different control challenge relative to that posed by the single water tank and the three-tank process.

A more detailed approach will be given to tuning PI control algorithms in Chap. 11 but for control of the dashpot process with $\zeta=0.1$ a crude trial and error approach will suffice here. First, the proportional gain k was chosen to be unity because the process has unity gain and a good rule of thumb suggests that the process gain and the control gain be reciprocals. The integral gain was another matter. We started with conservative values of I and increased them until a semblance of acceptable behavior was arrived at with $I=0.3$.

Figure 6-8 shows the response to a step in the set point. Note that the average value tends to be the set point but the oscillations take a while to damp out. Figure 6-9 shows that open-loop Bode plot for G_pG_c. Note that the low frequencies are amplified and that the phase changes dramatically at the natural frequency of 0.159 Hz. For frequencies beyond the natural frequency, the phase lag stays relatively constant in the neighborhood of 180°.

Figure 6-10 shows the closed-loop Bode plot for $(G_pG_c)/[1+G_pG_c]$. Note that again there is a dramatic change near the natural frequency.

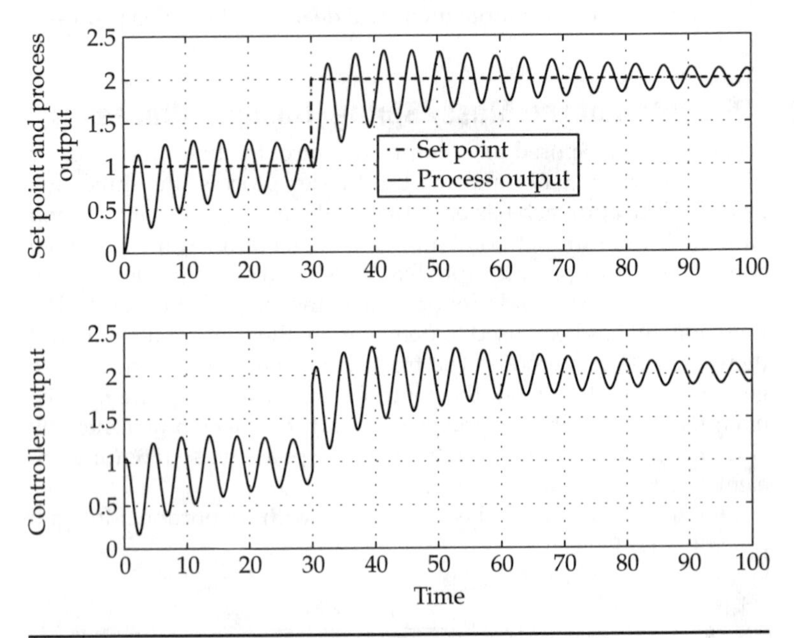

FIGURE 6-8 Set-point step-change response under PI control.

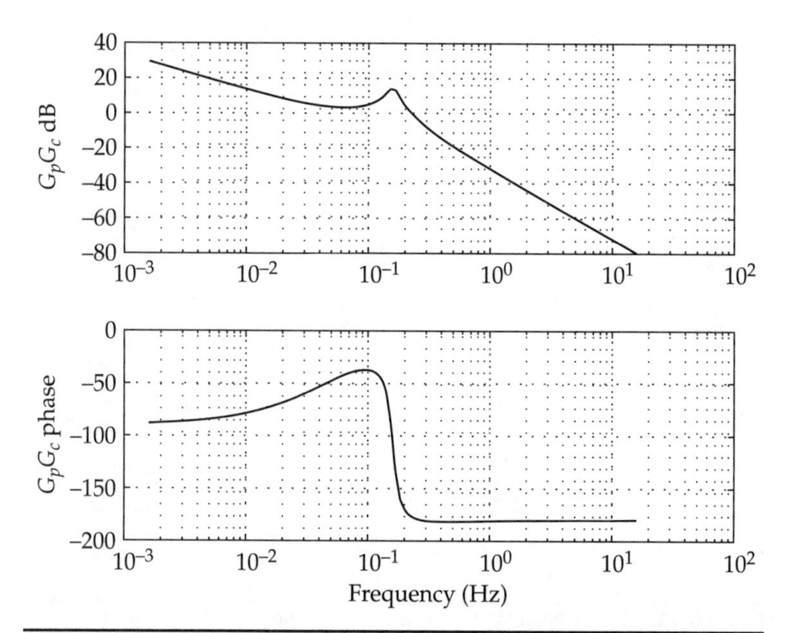

Figure 6-9 Open-loop Bode plot with PI control—mass/spring/dashpot process.

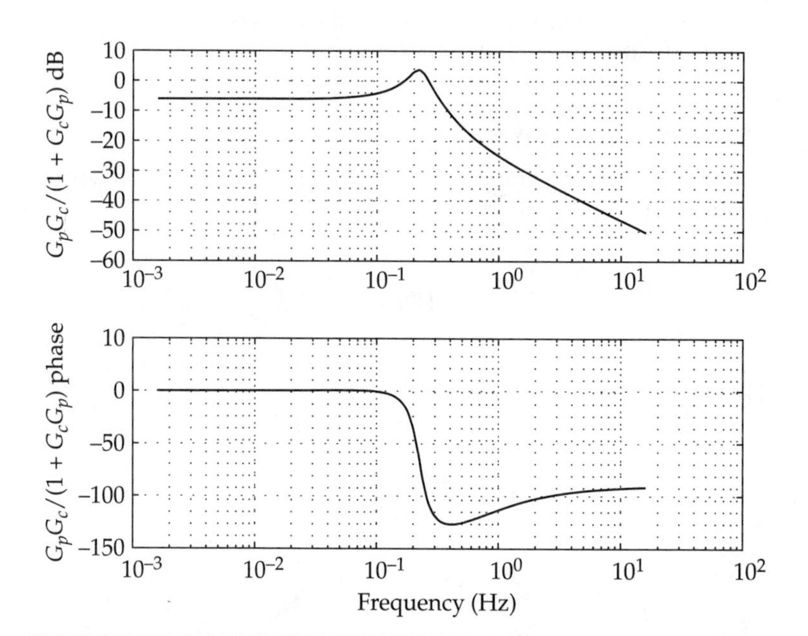

Figure 6-10 Closed-loop Bode plot with PI control—mass/spring/dashpot process.

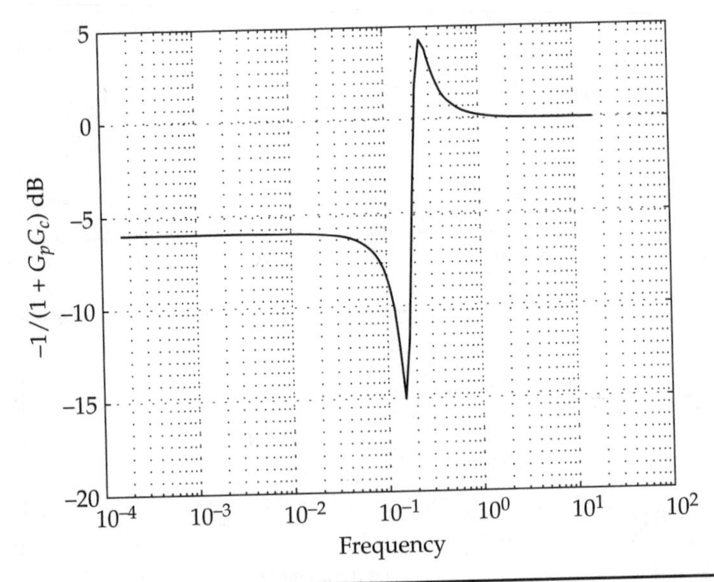

Figure 6-11 Error transmission curve for mass/spring/dashpot with PI control.

The error transmission curve for this case is shown in Fig. 6-11. Note how disturbances with frequencies below the natural frequency are all attenuated by about 6 dB while those with frequencies above the natural frequency are passed with neither attenuation nor amplification.

6-4 Derivative Control (PID)

The proportional-integral-derivative (PID) control algorithm has an additional term proportional to the derivative of the error:

$$U(t) = K_c\left(e(t) + I\int_0^t du\,e(u) + D\frac{de(t)}{dt}\right)$$

$$\tilde{U}(s) = K_c\left(\tilde{e}(s) + \frac{I}{s}\tilde{e}(s) + Ds\tilde{e}(s)\right)$$

$$= K_c\left(1 + \frac{I}{s} + Ds\right)\tilde{e}(s)$$

$$\frac{\tilde{U}(s)}{\tilde{e}(s)} = G_c = K_c\left(1 + \frac{I}{s} + Ds\right) = K_c\frac{s + I + Ds^2}{s}$$

We have ignored the initial value of $U(t)$ and the presence of the error term in the derivative—both problems will be dealt with in

Chaps. 9 and 11. Furthermore, an overall control gain K_c has been introduced to be consistent with wide usage among control engineers.

Unlike the PI control algorithm, PID has two zeros in the numerator of G_c

$$\frac{\tilde{U}(s)}{\tilde{e}(s)} = G_c = K_c \frac{s + I + Ds^2}{s} = K_c \frac{(s - s_2)(s - s_2)}{s}$$

$$s_1,\ s_2 = \frac{-1 \pm \sqrt{1 - 4DI}}{2D}$$

which can be complex conjugates if $4DI > 1$. Therefore, these potentially complex zeroes in G_c *might* ameliorate the presence of the complex poles in G_p:

$$G_p G_c = \frac{1}{s^2 + 2\zeta s + 1} K_c \frac{s + I + Ds^2}{s}$$

Tuning the PID algorithm for the dashpot process was done by trial and error. We kept the proportional and integral gains of the previous simulation for PI and started with a conservative value for D and increased it until satisfactory control was obtained with $D = 4.0$. Figure 6-12 shows the poles of G_p and the zeroes of G_c for the PID controller and for the PI controller used in Sec. 6-3. Figure 6-13 shows the poles of closed-loop transfer function $(G_p G_c)/[1 + G_p G_c]$.

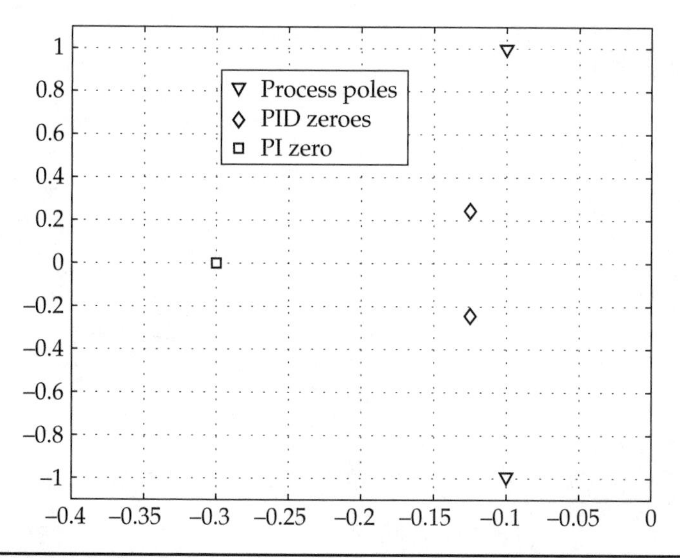

FIGURE 6-12 Poles of process and zeroes of PI and PID controller.

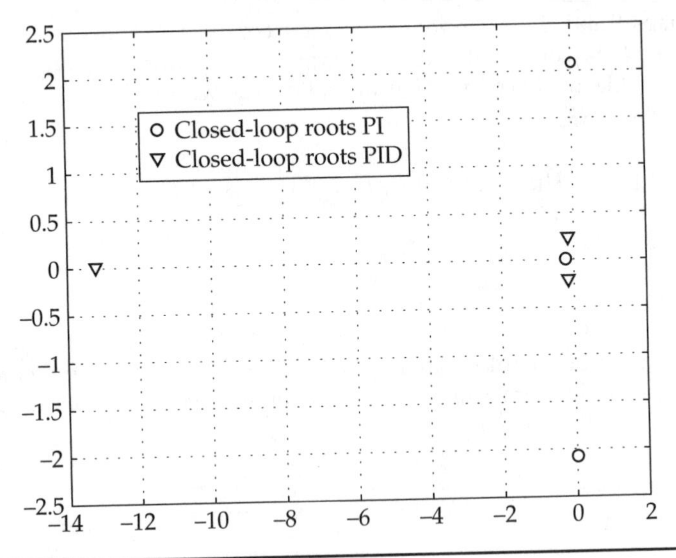

FIGURE 6-13 Closed-loop poles for PI and PID.

Note how the addition of the derivative component brought the closed-loop poles down and away from the imaginary axis.

Figure 6-14 shows the response to a step in the set point. Comparing Figs. 6-14 and 6-8, shows that the addition of derivative

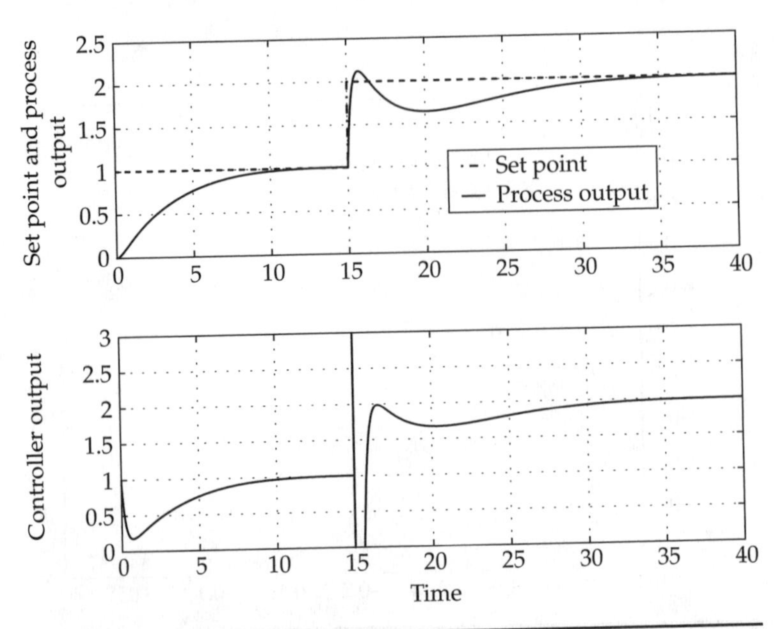

FIGURE 6-14 Set-point step-change response with PID control.

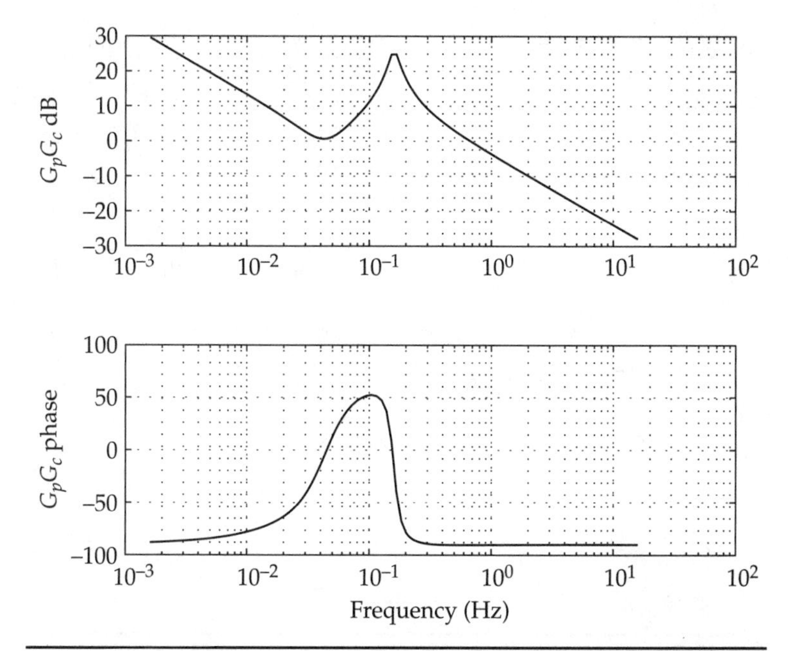

FIGURE 6-15 Open-loop Bode plot under PID control—mass/spring/dashpot process with derivatives.

appears to have solved the problem of the oscillations. But at what cost? There are two set-point changes in Fig. 6-14. The first takes place at time $t = 0$ and the Matlab `Simulink` simulation does not detect the full impact of that change. The second step at time $t = 15$ shows the effect of the derivative of a step: the control output goes off scale in both directions. In reality this output would be clamped at 0% and 100% of full scale but the extreme movement should give the reader pause on two counts. First, the extreme activity of the controller output might cause ancillary problems and second, one must be a little careful when carrying out simulations.

The Bode plot for the open loop shows how the presence of the derivative radically changes the shape of the phase curve such that the phase margin is quite large. The closed-loop Bode plot is shown in Fig. 6-16. Compare this plot with Fig. 6-11 for PI control. Figure 6-17 shows the error transmission curve. Compared to the error transmission curve for PI in Fig. 6-12, the addition of derivative changes the ability of the controller to attenuate low-frequency disturbances. The high-frequency disturbances are passed without attenuation or amplification.

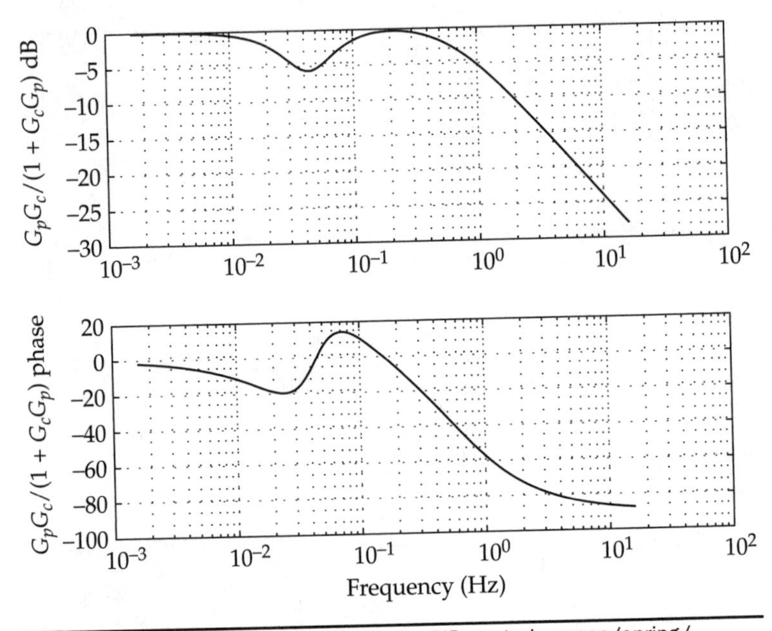

FIGURE 6-16 Closed-loop Bode plot under PID control—mass/spring/dashpot process with derivatives.

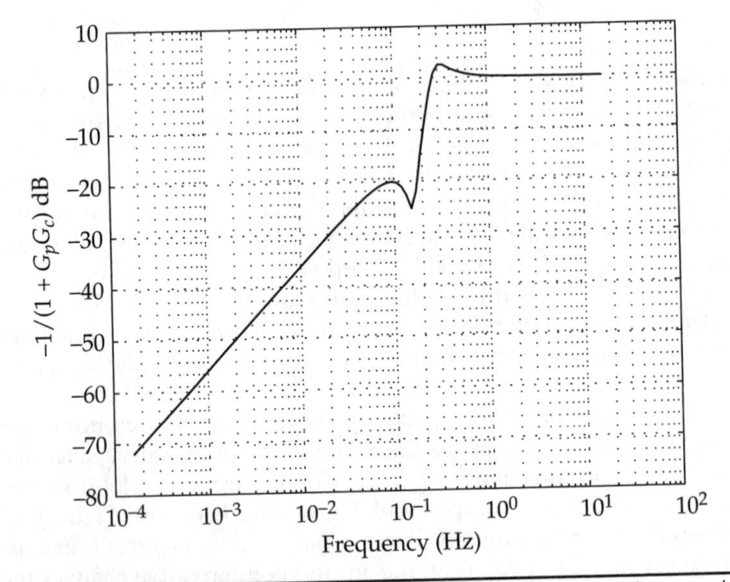

FIGURE 6-17 Error transmission curve for dashpot with PID control—mass/spring/dashpot process with derivatives.

6-4-1 Complete Cancellation

Perhaps the reader is wondering: what would happen if the zeros of the PID controller were chosen to exactly match those of the process? That is, what if:

$$\frac{-1 \pm \sqrt{1-4DI}}{2D} = -\zeta \pm \sqrt{\zeta^2 - 1}$$

$$D = \frac{1}{2\varsigma} \qquad I = \frac{1}{2\varsigma}$$

This would cause the open-loop transfer function to become

$$G_p G_c = \frac{1}{s^2 + 2\zeta s + 1} K_c \frac{s + I + Ds^2}{s} = \frac{K_c}{s}$$

and the closed-loop transfer function would be

$$\frac{Y}{S} = \frac{G_c G_p}{1 + G_c G_p} = \frac{\frac{K_c}{s}}{1 + \frac{K_c}{s}} = \frac{1}{\frac{s}{K_c} + 1}$$

which means that the response to a step in the set point would look like a unity-gain first-order process with a time constant of $1/K_c$.

In general, using controller zeros to cancel process poles can be dangerous. If a zero in the controller is used to cancel an unstable process pole, problems could occur if the cancellation is not exact. For this case, the perfect cancellation values for D and I are much larger than those used in the simulation. As an exercise you might want to use the Matlab script and `Simulink` model that I used to generate Fig. 6-14 to see what happens when these "perfect cancellation" values are applied.

6-4-2 Adding Sensor Noise

At this point, as a manager, you might be impressed to the point where you would conclude that the addition of derivative was the best thing since sliced bread (aside from the preceding comments about the extreme response to set-point steps). However, when the process output is noisy, troubles arise. For the purposes of this simulation exercise, we will add just a little white sensor noise (to be defined later) to the PI and the PID simulations. Figures 6-18 and 6-19 show the impact of adding a small amount of sensor noise on the process output signal for PI and PID. The added noise is barely discernible when PI control is used but when the same amount of

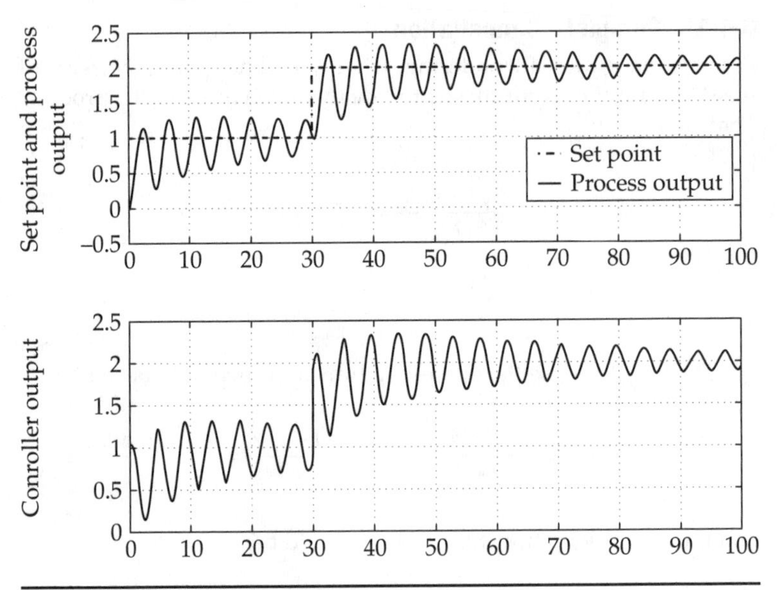

Figure 6-18 Response of PI control with added noise.

process noise is added to PID control there is quite a change in the control output. The addition of the derivative component to the control algorithm still drives the process output to set point without oscillation, but there is a tremendous price to pay in the activity of the control output. Also, note the spike in the output at $t = 30$ when the

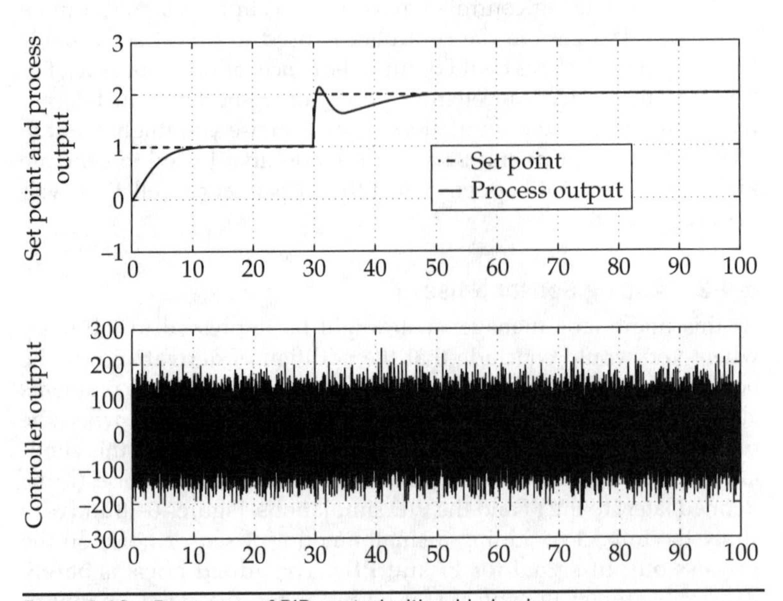

Figure 6-19 Response of PID control with added noise.

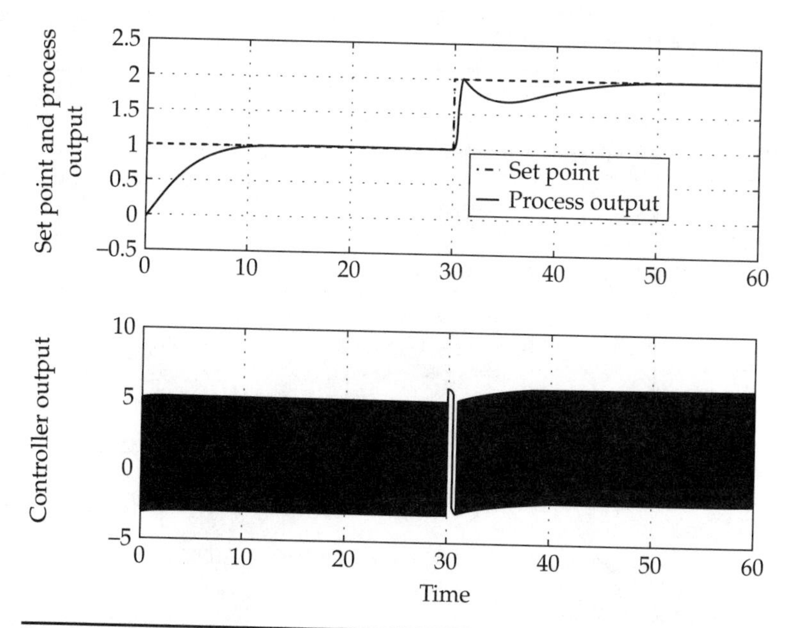

FIGURE 6-20 PID control of a set-point step in the face of process output noise.

set point is stepped. This excessive activity might wear out the control output actuator quickly and/or it might generate nonlinear responses in the real process that in turn might lead to unacceptable performance. Furthermore, the range of the control output is ±200, which is to be compared to [0, 2] for PI control.

Figure 6-20 shows that the simulated reaction to both the noise and the step in the set point for the case where the slew rate of the input to the continuous differentiator was limited to 1.0 unit per unit time. In real life there would be physical limitations depending on the hardware involved but in any case one must be careful using the derivative component.

6-4-3 Filtering the Derivative

The moral of this short story is to be careful about adding derivative because it greatly amplifies noise and sudden steps. Adding a first-order filter (with a time constant of 1.0) to the derivative partially addresses the problem as shown in Fig. 6-21. The outrageous control output activity has been ameliorated but there is still ringing.

Using the Laplace transform is the easiest way to present the filtered derivative:

$$\frac{\tilde{U}(s)}{\tilde{e}(s)} = G_c = K_c\left(1 + \frac{I}{s} + D\frac{s}{\tau_D s + 1}\right)$$

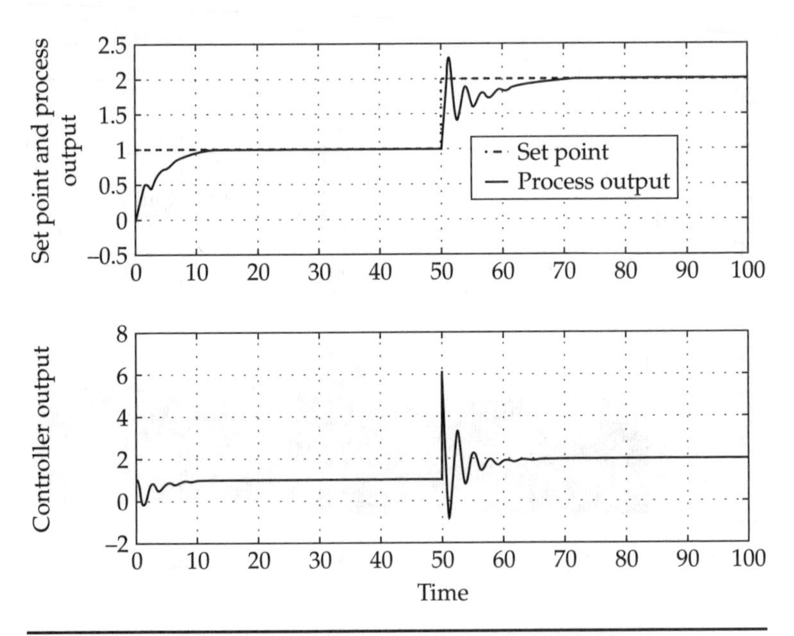

FIGURE 6-21 PIfD set-point step-change response with added noise.

Note the presence of the $1/(\tau_D s + 1)$ factor in the derivative term. For our simulation, the filter time constant τ_D was chosen to be equal to $1/\omega_n = 1$. Since this algorithm will most likely be implemented as a digital filter, its detailed discussion will be deferred until the discrete time domain is introduced in Chap. 9. However, why do you suppose that modifying the derivative term by the factor:

$$\frac{1}{\tau_D s + 1}$$

has the necessary beneficial effect? This factor (or transfer function) is the same as that for a first-order process with unity gain. We know from Chap. 4 that it will pass extremely low-frequency signals almost unaffected while attenuating high-frequency signals. The performance of the PID algorithm with a filter (or PIfD) is anchored by this ability to attenuate the higher frequency part of the sensor noise.

Some insight into this problem may be gained by studying the Bode plots of G_c for the PI, PID, and PIfD algorithms in Fig. 6-22. All three algorithms deal with low-frequency disturbances similarly. PI does nothing with disturbances having frequencies above the natural frequency as indicated by the magnitude gain of 0 dB in Fig. 6-22. PID aggressively addresses higher-frequency disturbances—in fact, the higher the frequency, the more aggressive the action. PIfD applies a

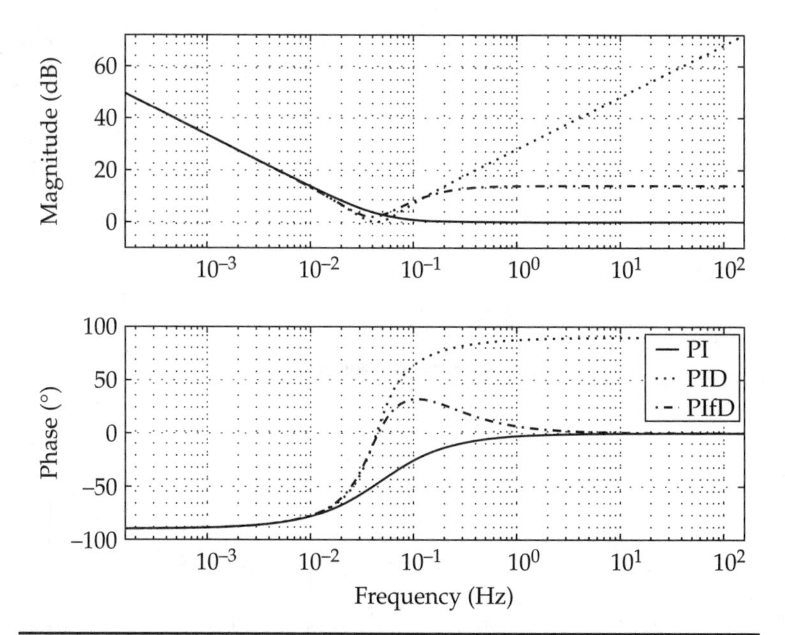

Figure 6-22 Bode plot for PI, PID, and PIfD controllers; $K_c = 1$, $I = 0.3$, $D = 4$, $\tau = 1$.

relatively constant gain of about 14 dB to disturbances greater than the natural frequency.

As an aside, Fig. 6-23 shows the Matlab `Simulink` block diagram that I used along with a Matlab script to generate the graphs for this section. It is not my goal to show you how to use Matlab and `Simulink` but you, as a manager, should be aware that these tools are somewhat de facto accessories to any control engineer that has to do computations. You might want to study the block diagram. First, there is one block for the dashpot process. Second, the PIfD algorithm is composed of several blocks, all of which should be fairly straightforward.

6-5 Compensation before Control—The Transfer Function Approach

Since the dashpot process has given us so much trouble, another approach will be taken in this section. We are going to modify the process by feeding the process variable and its first derivative back with appropriate gains. The gains will be chosen to make the modified process behave in a way more conducive to control.

Without compensation, the dashpot Laplace transform from Eq. (6-5) is

$$s^2\tilde{y}(s) + 2\zeta\omega_n s\tilde{y}(s) + \omega_n^2\tilde{y}(s) = g\omega_n^2\tilde{U}(s) \tag{6-9}$$

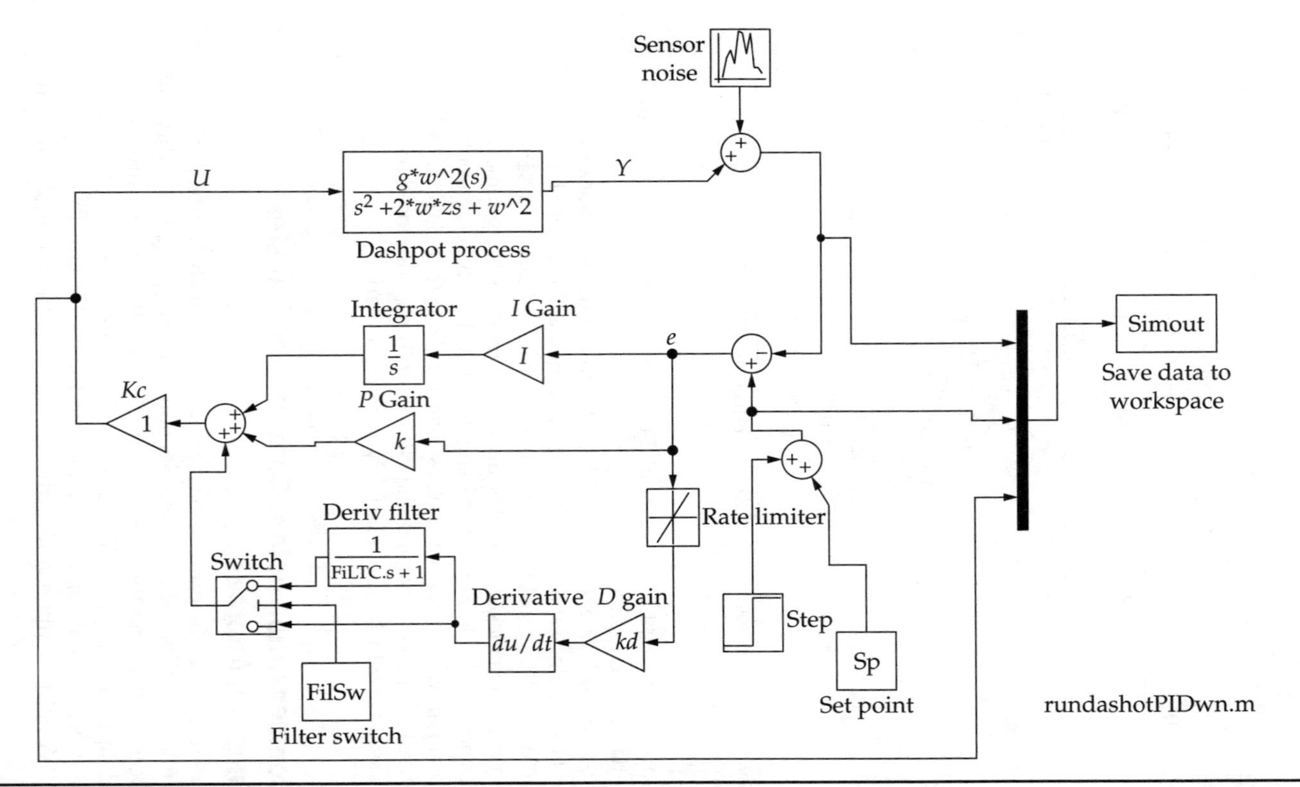

FIGURE 6-23 `Simulink` block diagram for dashpot control.

Remember that the s operator takes the derivative of what follows it. Solving Eq. (6-9) for the second derivative of y, namely $s^2\tilde{y}(s)$ or d^2y/dt^2 or y'', gives

$$s^2\tilde{y}(s) = -2\zeta\omega_n s\tilde{y}(s) - \omega_n^2\tilde{y}(s) + g\omega_n^2\tilde{U}(s) \tag{6-10}$$

In the time domain, Eq. (6-10) would look like

$$\frac{d^2y}{dt^2} = -2\zeta\omega_n\frac{dy}{dt} - \omega_n^2 y + g\omega_n^2 U$$

Note that ω_n and g have reappeared but remember that t and y can always be scaled to make both quantities unity.

The block diagram of Eq. (6-10) is given in Fig. 6-24. This block diagram is a little more complicated than that given in Fig. 3-10 in Chap. 3 for the first-order model. The reader should make sure she understands how Fig. 6-24 works before proceeding. Start where you see $s^2\tilde{y}(s)$ in the diagram. This signal passes through one integrator represented by the #1 block containing $1/s$. As a result, $s\tilde{y}(s)$ or dy/dx or y' is generated. This signal is then passed through another integrator (block #2) and y is generated. Each of these signals is fed back to summing points where they add up to form $s^2\tilde{y}(s)$, which is consistent with Eq. (6-10). Therefore the dashpot process can be considered as having internal feedback loops even when no feedback controller is present, just as with the first-order process back in Chap. 3.

Why use this block diagram form, with all the internal details exposed, rather than the simpler version where just one block represents the process and the overall transfer function? To feed y' back, we need to gain access to it. The overall transfer function block diagram does not provide a port for this signal so the "bowels" of the process have to be revealed.

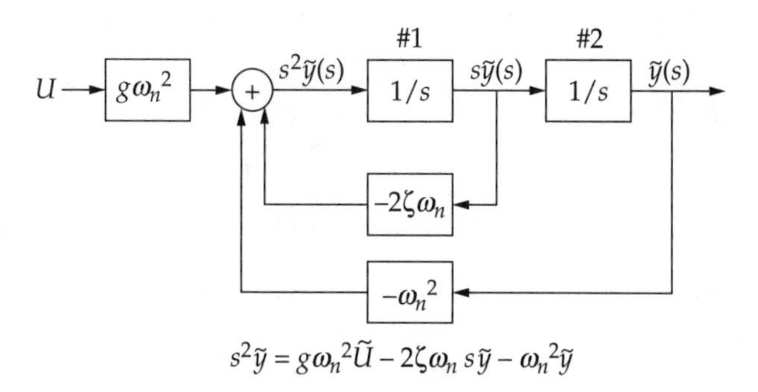

FIGURE 6-24 The dashpot model before compensation.

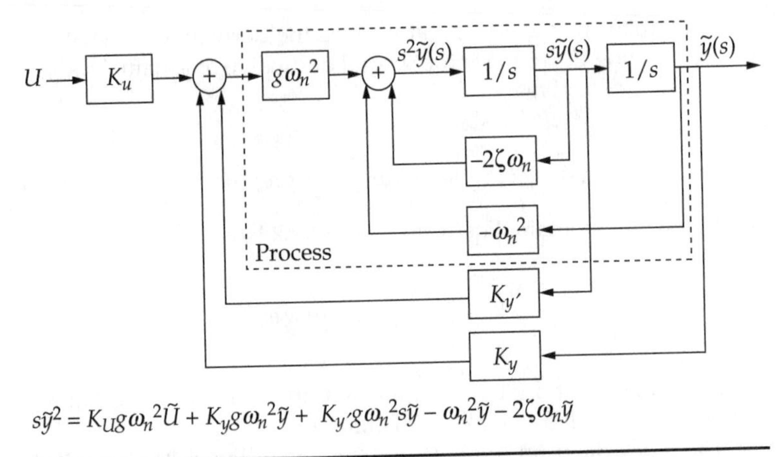

FIGURE 6-25 The dashpot model with states fed back.

Figure 6-25 shows a modified block diagram where y and dy/dx are fed back *again*, this time with gains K_y and $K_{y'}$. Note that no control is being attempted yet. We are feeding these signals back to create a new *modified* process that will have more desirable properties. Everything inside the dotted line box represents the structure of the original process. All the lines and blocks outside the box represent the added compensation. The Laplace transform of the modified system is

$$s^2\tilde{y} = K_U g\omega_n^2\tilde{U} + K_y g\omega_n^2\tilde{y} + K_{y'} g\omega_n^2 s\tilde{y} - \omega_n^2\tilde{y} - 2\zeta\omega_n\tilde{y} \tag{6-11}$$

The logic behind the structure of this block diagram is the same as that for the unmodified process shown in Fig. 6-24. Three gains, K_y, $K_{y'}$, and K_U have been introduced.

The values for these gains will be chosen so that the modified process looks like a *desired* process shown in Fig. 6-26. The Laplace transform for the desired system is

$$s^2\tilde{y}(s) = -2\zeta_D\omega_D\, s\,\tilde{y}(s) - \omega_D^2\,\tilde{y}(s) + g_D\omega_D^2\,\tilde{U}(s) \tag{6-12}$$

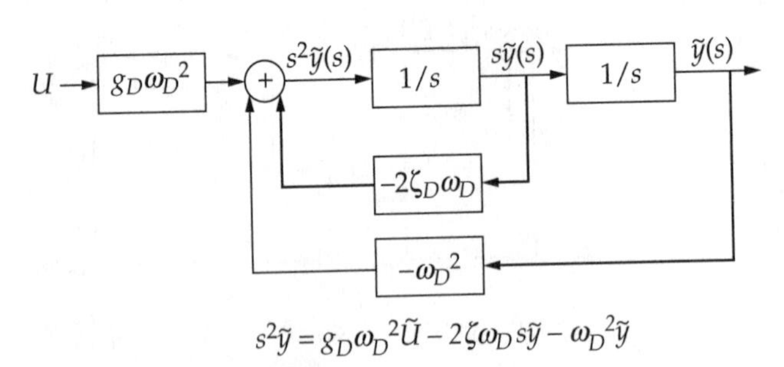

FIGURE 6-26 The desired dashpot model.

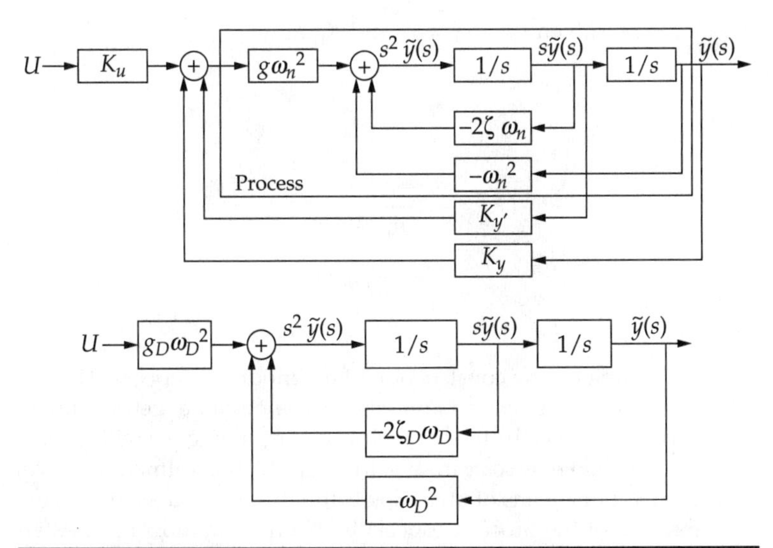

FIGURE 6-27 Choose K_u, K_y, $K_{y'}$ to make the compensated and desired models identical.

Note that this desired process has the same structure as the original process but the parameters, g_D, ζ_D, and ω_D are yet to be specified. We will specify the values and then find the values of K_y, $K_{y'}$, and K_U that will make them happen (Fig. 6-27). As you might expect, we would want the damping parameter ζ_D to be greater than that of the original process so that there is less ringing. Likewise, we might want to make the natural frequency ω_D greater than ω_n so that the response would be quicker. To make life simple, g_D is chosen to be unity.

To find the values of K_y, $K_{y'}$, and K_U after having picked values for ζ_D, ω_D, and g_D, one compares Eqs. (6-11) and (6-12), as in

$$
\begin{aligned}
s^2\tilde{y} &= K_U g\omega_n^2\tilde{U} + \left(K_y g\omega_n^2 - \omega_n^2\right)\tilde{y} + \left(K_{y'} g\omega_n^2 - 2\zeta\omega_n\right)s\tilde{y} \\
s^2\tilde{y}(s) &= g_D\omega_D^2\tilde{U} - \omega_D^2\tilde{y}(s) - 2\zeta_D\omega_D\, s\tilde{y}(s)
\end{aligned}
\tag{6-13}
$$

Comparing the coefficients of $\tilde{U}$, $\tilde{y}$, and $s\tilde{y}$ gives the following expressions.

$$
\begin{aligned}
K_U g\omega_n^2 &= g_D\omega_D^2 \\
K_y g\omega_n^2 - \omega_n^2 &= -\omega_D^2 \\
K_{y'} g\omega_n^2 - 2\zeta\omega_n &= -2\zeta_D\omega_D
\end{aligned}
$$

which can be solved for K_y, $K_{y'}$, and K_U as in

$$
\begin{aligned}
K_U &= \frac{g_D \omega_D^2}{g \omega_n^2} \\
K_y &= \frac{\omega_n^2 - \omega_D^2}{g \omega_n^2} \\
K_{y'} &= \frac{2\zeta\omega_n - 2\zeta_D \omega_D}{g \omega_n^2}
\end{aligned}
\tag{6-14}
$$

This completes the construction of the modified process. There is one nontrivial problem remaining—how does one get values of dy/dx or y' so it can be fed back? For the time being we will assume that y' is available by *some* means. In Chap. 10, the Kalman filter will be shown as one means of obtaining estimates of y and y' or, in general, the *state* of the process, especially in a noisy atmosphere. Alternatively, we might try generating dy/dt or y' by using a filtered differentiator in a manner similar to what was used in generating the PID single in the Sec. 6-4.

Figure 6-28 shows how a process with $g = 1.0$, $\omega_n = 1.0$, and $\zeta = 0.1$ can be compensated such that $g_D = 1.0$, $\omega_D = 1.5$, and $\zeta_D = 0.7$. Before one gets too excited by these results, remember that the compensation algorithm makes use of complete knowledge of the state, that is, y and y'. The estimate of the state is assumed to be perfect. How one actually estimates the state will be deferred until Chap. 10.

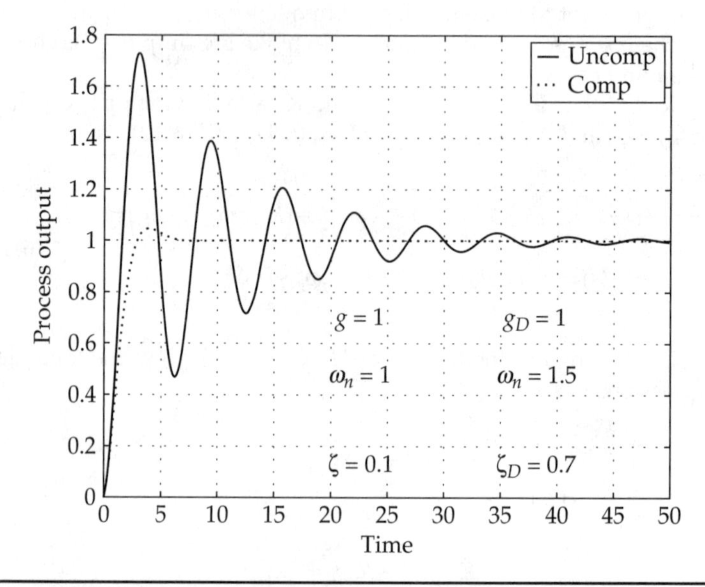

FIGURE 6-28 Effect of compensation.

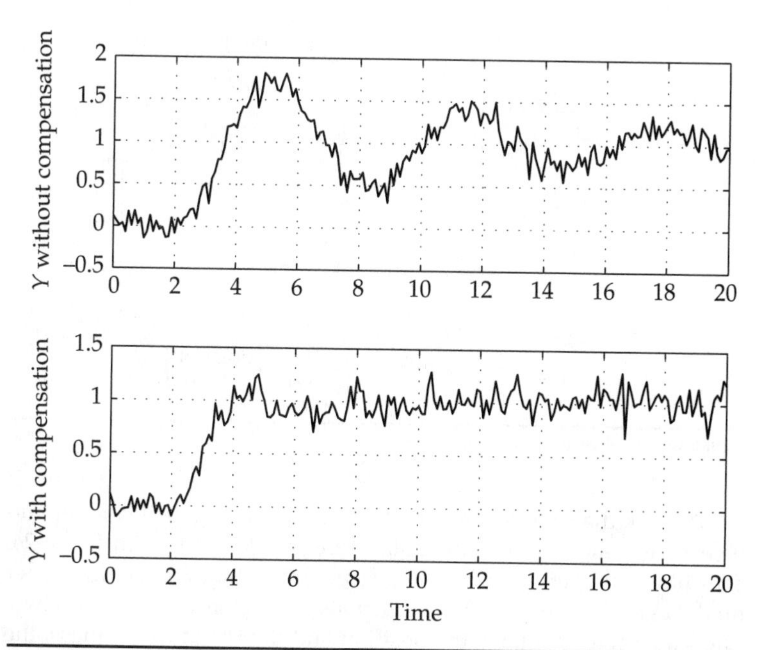

FIGURE 6-29 Effect of compensation in the face of noise using a filtered derivative for y'.

If sensor noise is added to the process output and a filtered derivative is used to estimate y' then the response to a step in the set point at $t = 2$ is shown in Fig. 6-29. The filter has a time constant of 1.0 (just as it did for the PIfD case in Sec. 6-4). The ringing appears to be removed but the "hash" riding on the process output appears to be amplified slightly.

6-6 Compensation before Control—The State-Space Approach

The state-space model for the dashpot process is

$$\frac{d}{dt}\begin{pmatrix} x_1 \\ x_2 \end{pmatrix} = \begin{pmatrix} 0 & 1 \\ -\omega_n^2 & -2\zeta\omega_n \end{pmatrix}\begin{pmatrix} x_1 \\ x_2 \end{pmatrix} + \begin{pmatrix} 0 \\ g\omega_n^2 \end{pmatrix} U$$

$$\frac{dx}{dt} = Ax + BU$$

$$s\tilde{x} = A\tilde{x} + B\tilde{U} \tag{6-15}$$

$$A = \begin{pmatrix} 0 & 1 \\ -\omega_n^2 & -2\zeta\omega_n \end{pmatrix}$$

$$B = \begin{pmatrix} 0 \\ g\omega_n^2 \end{pmatrix}$$

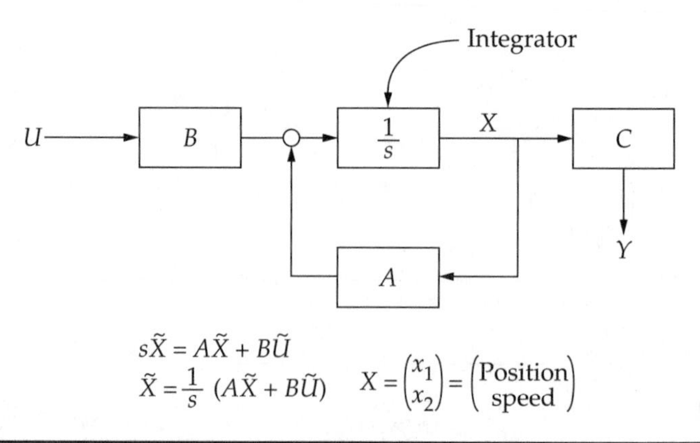

FIGURE 6-30 A state-space block diagram.

A block diagram for this model is shown in Fig. 6-30 and should be easy to follow if you understand the block diagram in Fig. 6-24. This block diagram has the same structure as that for the first-order model in Fig. 3-10 except that the signals are vectors of dimension two. The state vector contains the position and the speed of the mass, the same signals that we referred to as y and y' in the previous section.

As in the previous section, the state will be fed back such that a modified process is constructed, as in Fig. 6-31. This block diagram is general in the sense that it applies to any process model that can be described by the state-space equations, not just the dashpot model. There is only one integrator but it acts on the vector x rather than a scalar as was the case in Sec. 6-5. The gain, $K_x = \lfloor k_{x1} \quad k_{x2} \rfloor$, a row vector, has two components while the gain K_u is a scalar. The equation describing the behavior of the modified process is

$$
\begin{aligned}
s\tilde{x} &= K_u B\tilde{U} + K_x B\tilde{x} + A\tilde{x} \\
&= (A + K_x B)\tilde{x} + K_u B\tilde{U}
\end{aligned}
$$

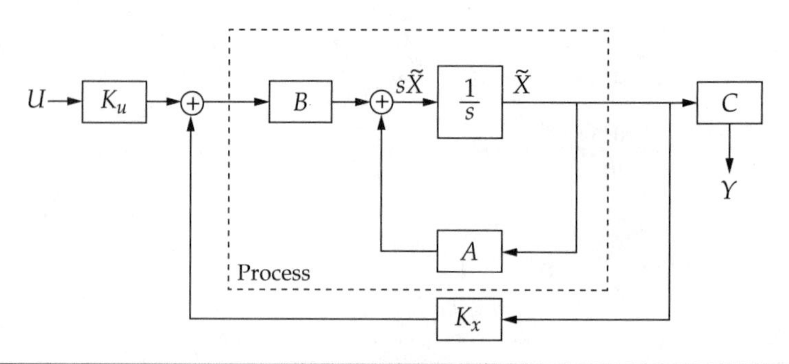

FIGURE 6-31 Compensation in state space.

or

$$s\begin{pmatrix}\tilde{x}_1\\ \tilde{x}_2\end{pmatrix}=\left[\begin{pmatrix}0 & 1\\ -\omega_n^2 & -2\zeta\omega_n\end{pmatrix}+\begin{pmatrix}0\\ g\omega_n^2\end{pmatrix}\lfloor k_{x1} \quad k_{x2}\rfloor\right]\begin{pmatrix}\tilde{x}_1\\ \tilde{x}_2\end{pmatrix}+K_u\begin{pmatrix}0\\ g\omega_n^2\end{pmatrix}\tilde{U} \quad (6\text{-}16)$$

These three gains will be chosen to make the modified process behave as a desired process defined by

$$\frac{d}{dt}\begin{pmatrix}x_1\\ x_2\end{pmatrix}=\begin{pmatrix}0 & 1\\ -\omega_D^2 & -2\zeta_D\omega_D\end{pmatrix}\begin{pmatrix}x_1\\ x_2\end{pmatrix}+\begin{pmatrix}0\\ g\omega_D^2\end{pmatrix}U$$

$$s\begin{pmatrix}\tilde{x}_1\\ \tilde{x}_2\end{pmatrix}=\begin{pmatrix}0 & 1\\ -\omega_D^2 & -2\zeta_D\omega_D\end{pmatrix}\begin{pmatrix}\tilde{x}_1\\ \tilde{x}_2\end{pmatrix}+\begin{pmatrix}0\\ g\omega_D^2\end{pmatrix}\tilde{U}$$

$$\frac{dx}{dt}=A_Dx+B_DU$$

$$s\tilde{x}=A_D\tilde{x}+B_D\tilde{U}$$

$$A_D=\begin{pmatrix}0 & 1\\ -\omega_D^2 & -2\zeta_D\omega_D\end{pmatrix}$$

$$B_D=\begin{pmatrix}0\\ g\omega_D^2\end{pmatrix} \quad (6\text{-}17)$$

To make Eqs. (6-16) and (6-17) match, the following must be true:

$$\begin{pmatrix}0 & 1\\ -\omega_n^2 & -2\zeta\omega_n\end{pmatrix}+\begin{pmatrix}0\\ g\omega_n^2\end{pmatrix}\lfloor k_{x1} \quad k_{x2}\rfloor=\begin{pmatrix}0 & 1\\ -\omega_D^2 & -2\zeta_D\omega_D\end{pmatrix}$$

or

$$\begin{pmatrix}0 & 1\\ -\omega_n^2 & -2\zeta\omega_n\end{pmatrix}+\begin{pmatrix}0 & 0\\ g\omega_n^2k_{x1} & g\omega_n^2k_{x2}\end{pmatrix}=\begin{pmatrix}0 & 1\\ -\omega_D^2 & -2\zeta_D\omega_D\end{pmatrix} \quad (6\text{-}18)$$

This single matrix equation yield two scalar equations:

$$-\omega_n^2+g\omega_n^2k_{x1}=-\omega_D^2 \qquad -2\zeta\omega_n+g\omega_n^2k_{x2}=-2\zeta_D\omega_D$$

The following must also be true:

$$K_u \begin{pmatrix} 0 \\ g\omega_n^2 \end{pmatrix} = \begin{pmatrix} 0 \\ g\omega_D^2 \end{pmatrix}$$

which yields (6-19)

$$K_u g\omega_n^2 = g\omega_D^2$$

Equations (6-18) and (6-19) are similar to Eq. (6-14).

Depending on your comfort level with the different mathematical tools, you might agree that the state-space approach is a little less cluttered and more general than the transfer-function approach. Later on, when methods of estimating the state are presented in Chap. 10, the strength of the state-space approach will become even more apparent.

6-7 An Electrical Analog to the Mass/Dashpot/Spring Process

Consider the *RLC* circuit in Fig. 6-32 where *R* refers to resistance of the resistor, *L* the inductance of the coil, and *C* the capacitance of the capacitor. The applied voltage is *V* and it will also be the process input *U*. The voltage over the resistor is *iR* where *i* is the process output *Y*. The voltage over the capacitor is

$$\frac{1}{C}\int_0^t i(u)du$$

and the voltage over the inductor is

$$L\frac{di}{dt}$$

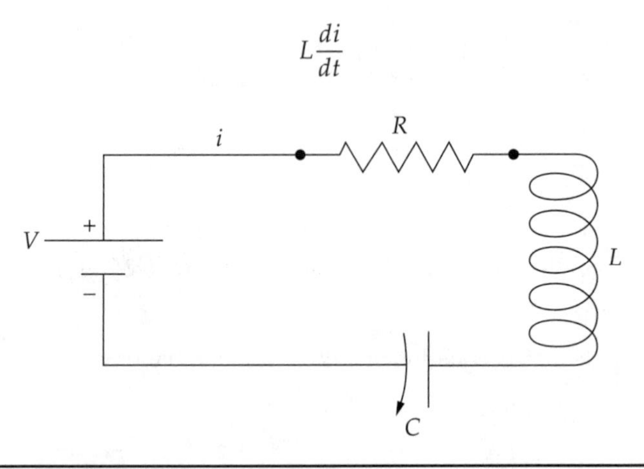

Figure 6-32 RLC circuit.

These three voltages have to add up to match the applied voltage.

$$V = iR + \frac{1}{C}\int_0^t i(u)du + L\frac{di}{dt} \qquad (6\text{-}20)$$

Eq. (6-20) could be differentiated to get rid of the integral. Alternatively, the equation could be transformed to the Laplace domain yielding

$$\tilde{V} = \tilde{i}R + \frac{\tilde{i}}{Cs} + Ls\tilde{i}$$

The output/input transfer function is

$$\frac{\tilde{i}}{\tilde{V}} = \frac{\tilde{Y}}{\tilde{U}} = G = \frac{1}{R + \frac{1}{Cs} + Ls} = \frac{Cs}{LCs^2 + RCs + 1} = \frac{\frac{1}{L}s}{s^2 + \frac{R}{L}s + \frac{1}{LC}}$$

This expression looks similar to Eq. (6-5), which is repeated here as

$$\frac{\tilde{y}(s)}{\tilde{U}(s)} = G_p(s) = \frac{g\omega_n^2}{s^2 + 2\zeta\omega_n s + \omega_n^2}$$

This suggests that

$$\omega_n^2 \Leftrightarrow \frac{1}{LC} \qquad 2\zeta\omega_n \Leftrightarrow \frac{R}{L}$$

which further suggests

$$\omega_n = \sqrt{\frac{1}{LC}} \qquad \zeta_n = \frac{R}{2}\sqrt{\frac{C}{L}}$$

Therefore, the *RLC* process has the potential of behaving in an underdamped manner similar to that of the mass/dashpot/spring process. For example, with *R*, *C*, and *L* chosen such that $\varsigma < 1$, the step response will exhibit damped oscillations with a frequency of ω_n.

Question 6-2 Can you conceive of an electrical circuit that behaves similarly to the first-order process introduced in Chap. 3?

Answer Construct a simple RC circuit by shorting the inductor in Fig. 6-32. The resulting circuit is described by

$$V = iR + \frac{1}{C}\int_0^t i(u)du \qquad \tilde{V} = \tilde{i}R + \frac{\tilde{i}}{Cs}$$

$$\frac{\tilde{i}}{\tilde{V}} = \frac{1}{R + \frac{1}{Cs}} = \frac{Cs}{RCs + 1} = \frac{Cs}{\tau s + 1}$$

Alternatively, construct a simple *RL* circuit by shorting the capacitor in Fig. 6-32. The resulting circuit is described by

$$V = iR + L\frac{di}{dt} \qquad \tilde{V} = \tilde{i}R + Ls\tilde{i}$$

$$\frac{\tilde{i}}{\tilde{V}} = \frac{1}{R + Ls} = \frac{1/R}{\frac{L}{R}s + 1} = \frac{1/R}{\tau s + 1}$$

6-8 Summary

This chapter has been devoted to just one process, the mass/spring/dashpot process, because it has the unique characteristic of ringing in response to an open-loop step input. In trying to control this process the derivative component was added to the PI algorithm. After showing that this modification could deal with the ringing but was susceptible to sensor noise, a derivative filter was added and the PIfD algorithm was conceived.

In an alternative approach, the process was modified by state feedback, after which it presented attractive nonringing dynamic behavior. The compensation was developed using transfer functions and state-space equations. To feed the state back, the mass's speed had to be estimated. A filtered derivative was used to provide that estimate. In general, the state may consist of other signals needing something other than filtered differentiation. In this case Chap. 10 will show that the state-space approach along with the Kalman filter will be needed.

CHAPTER 7

Distributed Processes

Most of the example processes presented so far have been *lumped*. That is, the example processes have been described by one or more ordinary differential equations, each representing a process element that was relatively self-contained. Furthermore, each ordinary differential equation described a "lump." A process with dead time does not yield to this "lumping" approach and can in some ways be considered a distributed process which is the subject of this chapter.

7-1 The Tubular Energy Exchanger—Steady State

Consider Fig. 7-1 which shows a jacketed tube of length L. A liquid flows through the inside tube. The jacket contains a fluid, say steam, from which energy can be transferred to the liquid in the tube. To describe how this process behaves in steady state, a simple energy balance can be made, not over the whole tube but over a small but finite section of the tube. Several assumptions (and idealizations) must be made about this new process.

1. The steam temperature T_s in the jacket is constant along the whole length of the tube. The tube length is L. The steam temperature can vary with time but not space.
2. The tube is cylindrical and has a cross-sectional area of $A = \pi D^2 / 4$ where D is the diameter of the inner tube.
3. The liquid flows in the tube as a plug at a speed v. That is, there is no radial variation in the liquid temperature. There is axial temperature variation of the liquid due to the heating effect of the steam in the jacket but there is no axial transfer of energy by conduction within the fluid. This is equivalent to saying the radial diffusion of energy is infinite compared to axial diffusion. The temperature of the flowing liquid therefore is a function of the axial displacement z, as in $T(z)$.

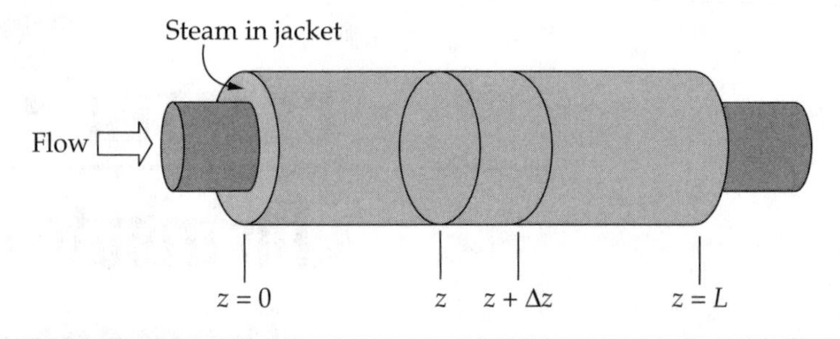

FIGURE 7-1 A jacketed tube.

4. There is a small disc placed at some arbitrary location z along the tube that has cross-sectional area A and thickness Δz. This disc will be used to derive the model describing equation.
5. The liquid properties of density ρ, heat capacity C_p, thermal conductivity k are constant (independent of position and of temperature).
6. The flux of energy between the steam in the jacket and the flowing liquid is characterized by an overall heat transfer coefficient U.

A thermal energy balance over the disc of thickness Δz at location z will describe the steady-state behavior of the tube exchanger. The result is given in Eq. (7-1) which is boxed below. You might want to skip to that location if derivations are not your bag. Otherwise, the derivation proceeds as follows.

Energy rate in at z due to convection: $vA\rho C_p T(z)$

Energy rate out at $z+\Delta z$ due to convection: $vA\rho C_p T(z+\Delta z)$

Energy rate in from jacket: $U(\pi D\Delta z)\left[T_s - T\left(\frac{z+z+\Delta z}{2}\right)\right]$

In this last term the energy rate is proportional to the difference between the jacket temperature T_s and the liquid temperature in the middle of the disc, at the point

$$(z+z+\Delta z)/2$$

The energy balance then becomes

$$vA\rho C_p T(z) + U(\pi D\Delta z)\left[T_s - T\left(\frac{z+z+\Delta z}{2}\right)\right] = vA\rho C_p T(z+\Delta z)$$

After a slight rearrangement and after dividing all terms by Δz one gets

$$\frac{vA\rho C_p T(z+\Delta z) - vA\rho C_p T(z)}{\Delta z} = U\pi D\left[T_s - T\left(\frac{z+z+\Delta z}{2}\right)\right]$$

The thickness of the disc is decreased to differential or infinitesimal size as in

$$\lim_{\Delta z \to 0} \frac{vA\rho C_p T(z+\Delta z) - vA\rho C_p T(z)}{\Delta z} = \lim_{\Delta z \to 0} U\pi D \times \left[T_s - T\left(\frac{z+z+\Delta z}{2}\right)\right]$$

From App. A one sees that the above equation contains the definition of the derivative of T with respect to z, as in

$$vA\rho C_p \frac{dT}{dz} = U\pi D[T_s - T(z)] \tag{7-1}$$

This ordinary differential equation describes the steady-state behavior of the idealized jacketed tube energy exchanger. From Chap. 3 we already know how to solve this equation if we know an inlet temperature, as in $T(0) = T_0$.

If Eq. (7-1) is rearranged slightly, the reader can see the similarity to the equation for the liquid tank presented in Chap. 3.

$$\frac{vA\rho C_p}{U(\pi D)} \frac{dT}{dz} + T = T_s$$

$$\psi \frac{dT}{dz} + T = T_s \tag{7-2}$$

$$\psi = \frac{vA\rho C_p}{U\pi D} = \frac{vD\rho C_p}{4U}$$

The reader has seen Eq. (7-2) before, at least structurally. By inspection, the reader can arrive at a solution to Eq. (7-2) as

$$T(z) = T_0 e^{-\frac{z}{\psi}} + \left(1 - e^{-\frac{z}{\psi}}\right) T_s \tag{7-3}$$

The parameter ψ can be considered as a kind of "space constant," somewhat analogous to the time constant used in transient analysis. In fact, ψ is the tube length needed for $T(z)$ to reach 63% of the jacket temperature T_s.

The reader should spend a few moments looking at Eq. (7-3) to see how $T(z)$ changes as various parameters change. First, it shows that as one travels down the tube axially the temperature $T(z)$ approaches T_s. Second, as the overall heat transfer coefficient U increases, the parameter ψ decreases. Thus, $T(z)$ approaches T_s more

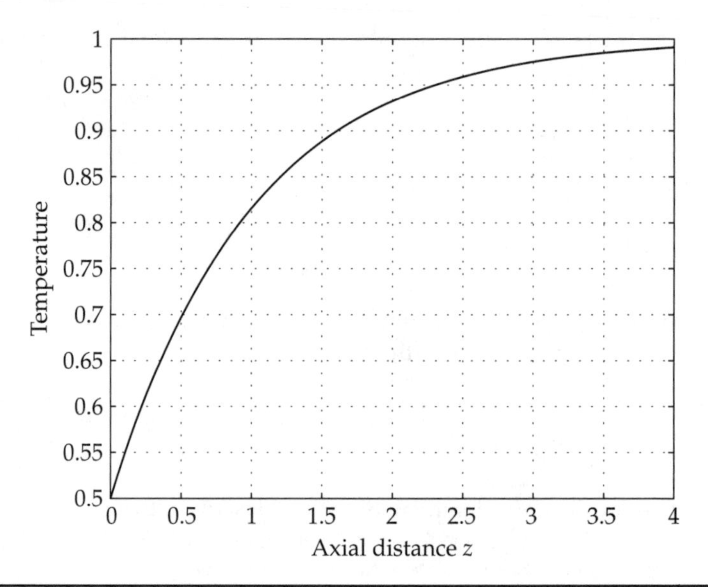

FIGURE 7-2 Steady-state temperature profile $t(0) = 0.5$.

rapidly with respect to axial distance (not time—remember this is a steady-state solution). Conversely, as the flow rate v of the liquid increases, ψ increases and it will require more tube length for $T(z)$ to reach the same temperature value than it would for a lower flow rate. Figure 7-2 shows a steady-state temperature profile for a tubular heat exchanger of length 2.0 with $\psi = 1$, $T_s = 1$, $T(0) = 0.5$.

Note that the liquid temperature reaches 63% of the ultimate value of T_s when $z = \psi$, so the reader sees that ψ plays the same role in the spatial domain that τ played in the time domain for the first-order process. This process is called a *distributed* process because the variation of the process variable T is distributed over the length of the tube as is the effect of the steam in the jacket.

7-2 The Tubular Energy Exchanger—Transient Behavior

The dynamic behavior can be described by a partial differential equation that also evolves from a thermal energy balance over a small disc of length Δz located somewhere in the interior of the tube and over a moment in time of length Δt. The balance proceeds as in Sec. 7-1 but with one more term—the temporal accumulation of thermal energy in the disc. The temperature now depends on both the axial distance z and the time t, as in $T(z,t)$. Furthermore, the jacket temperature T_s may now depend on time but, as specified above, it is not a function of axial position. A second balance could be written for the steam in the jacket; however, for the time being, the dynamics of the steam are assumed to be much quicker than those of the liquid flowing through the tube.

As with the steady-state derivation in Sec. 7-1, the reader can skip to the result in Eq. (7-6) which is boxed. For the adventurous, the balance proceeds as follows.

Energy rate at z due to convection at time t during the interval Δt:

$$vA\rho C_p T(z,t)\Delta t$$

Energy rate out at $z+\Delta z$ due to convection at time t during the interval Δt:

$$vA\rho C_p T(z+\Delta z,t)\Delta t$$

Energy rate in from jacket at time t during the interval Δt:

$$U(\pi D\Delta z)\left[T_s - T\left(\frac{z+z+\Delta z}{2},t\right)\right]\Delta t$$

Accumulation of energy in the disc between time t and time $t+\Delta t$ in the volume $A\Delta z$:

$$A\Delta z\rho C_p T\left(\frac{z+z+\Delta z}{2},t+\Delta t\right) - A\Delta z\rho C_p T\left(\frac{z+z+\Delta z}{2},t\right)$$

Entering the various elements into the balance equation

$$\text{In} - \text{out} = \text{accumulation}$$

gives

$$vA\rho C_p T(z,t)\Delta t + U(\pi D\Delta z)\left[T_s - T\left(\frac{z+z+\Delta z}{2},t\right)\right]\Delta t - vA\rho C_p T(z+\Delta z,t)\Delta t$$

$$= A\Delta z\rho C_p T\left(\frac{z+z+\Delta z}{2},t+\Delta t\right) - A\Delta z\rho C_p T\left(\frac{z+z+\Delta z}{2},t\right)$$

Dividing by $\Delta t\Delta z$ and doing a little rearranging gives

$$vA\rho C_p \frac{T(z,t)-T(z+\Delta z,t)}{\Delta z} + U(\pi D)\left[T_s - T\left(\frac{z+z+\Delta z}{2},t\right)\right]$$

$$= A\rho C_p \frac{T\left(\frac{z+z+\Delta z}{2},t+\Delta t\right) - T\left(\frac{z+z+\Delta z}{2},t\right)}{\Delta t}$$

If the space element Δz and the time element Δt are both decreased to an infinitesimally small size, then the following partial differential equation results.

$$-vA\rho C_p \frac{\partial T}{\partial z} + U(\pi D)[T_s - T(z,t)] = A\rho C_p \frac{\partial T}{\partial t} \tag{7-4}$$

Dividing all terms by $A\rho C_p$ and remembering that $A = \pi D^2/4$, gives

$$\frac{\partial T}{\partial t} + v\frac{\partial T}{\partial z} = \frac{4U}{D\rho C_p}[T_s - T(z,t)] \tag{7-5}$$

The quantity $D\rho C_p / 4U$ has units of sec, so Eq. (7-5) could be written as

$$\begin{aligned} \frac{\partial T}{\partial t} + v\frac{\partial T}{\partial z} &= \frac{1}{\tau_T}[T_s - T(z,t)] \\ \tau_T &= \frac{D\rho C_p}{4U} = \frac{\psi}{v} \end{aligned} \tag{7-6}$$

where τ_T has units of time and is a time constant. Equation (7-6) is a *partial* differential equation describing the time–space behavior of the temperature in the tube. It is subject to initial conditions, such as $T(z,0) = T_0$, $0 \le z \le L$, and a boundary condition on the inlet, such as $T(0,t) = T_i$, $t > 0$. Since we have added the dependence on time, this process model can be used in simulations to test control algorithms.

As an aside, the quantity

$$\frac{\partial T}{\partial t} + v\frac{\partial T}{\partial z}$$

is often called the *total derivative* or the *convective derivative* of temperature and is sometimes given the symbol DT / Dt.

7-2-1 Transfer by Diffusion

The model in Eq. (7-6) describes the transfer of energy along the tube by convection. Energy can also be transported axially by molecular diffusion where the rate is proportional to the axial gradient of temperature, as in $-k(\partial T / \partial z)$ where k is the thermal conductivity. If one modifies the above energy balance on an element of length Δz by adding the contribution of diffusion, the result is

$$\frac{\partial T}{\partial t} + v\frac{\partial T}{\partial z} = \frac{4U}{D\rho C_p}[T_s - T(z,t)] + \frac{k}{\rho C_p}\frac{\partial^2 T}{\partial z^2} \tag{7-7}$$

where the added mechanism of transport is described by the term $(k / \rho C_p) / (\partial^2 T / \partial z^2)$. The presence of this term makes the solution procedure significantly more difficult and we will not refer to Eq. (7-7) until later in this chapter when lumping is discussed.

7-3 Solution of the Tubular Heat Exchanger Equation

There are a variety of approaches to solve Eq. (7-6) but we will pick the one using the tools already developed in this book and the one that will lend itself to using the frequency domain to gain insight. This means transforming the time dependence out of Eq. (7-6) using the Laplace transform. This will leave us with a first-order ordinary differential equation in the spatial dimension z which we can solve using standard techniques. The details are given in App. F.

The result of applying the Laplace transform to Eq. (7-6) is

$$s\tilde{T} + v\frac{d\tilde{T}}{dz} = \frac{1}{\tau_T}(\tilde{T}_s - \tilde{T}) \tag{7-8}$$

You should convince yourself that Eq. (7-8) is indeed the result of multiplying Eq. (7-6) by $\exp(-st)$ and integrating over $[0,\infty]$ with respect to time. In any case, after the dust has settled, Eq. (7-8) is a first-order ordinary differential equation of the form

$$v\frac{d\tilde{T}}{dz} + \left(s + \frac{1}{\tau_T}\right)\tilde{T} = \frac{\tilde{T}_s}{\tau_T} \tag{7-9}$$

where the Laplace variable s is just a parameter. Remember that $\tilde{T}_s$ is the Laplace transform of the jacket temperature which we specified could be a function of time but not of axial position, that is, T_s or $\tilde{T}_s$ is not a function of z.

Now, how do we solve Eq. (7-9)? We could apply the Laplace transform again with a different variable, say p, instead of s and remove the spatial dimension or we could solve the ordinary differential equation by trying a solution of the form Ce^{cz}. Both of these approaches have been used elsewhere in this book. The details of the solution are presented in App. F and the result is

$$\tilde{T}(z,s) = \tilde{T}_0 e^{az} + \tilde{T}_s \frac{1 - e^{az}}{\tau_T s + 1} \tag{7-10}$$

where

$$a = -\frac{1}{v}\left(s + \frac{1}{\tau_T}\right)$$

So, we have solved Eq. (7-9) for the spatial dependence of the temperature with the Laplace transform s as a parameter. Alternatively, we could look at Eq. (7-10) as the Laplace transform of $T(z,t)$ with the spatial dimension z as a parameter.

7-3-1 Inlet Temperature Transfer Function

Equation (7-10) contains two transfer functions of interest. The first transfer function shows how the inlet temperature affects the outlet temperature (at $z = L$):

$$\frac{\tilde{T}(L,s)}{\tilde{T}_0(s)} = e^{-\frac{L}{v}\left(s+\frac{1}{\tau_T}\right)} = e^{-s\frac{L}{v}} e^{-\frac{L}{v\tau_T}} = e^{-st_D} e^{-\frac{t_D}{\tau_T}} \tag{7-11}$$

where $t_D = L/v$ is the average residence time or delay time for the tube. Equation (7-11) ignores the impact of T_s and shows that $T(L,t)$ lags $T(0,t)$ by t_D and is attenuated by a constant factor of e^{-t_D/τ_T}. This makes physical sense based on the assumptions of plug flow for the liquid. Thus, when T_0 is the input, Eq. (7-11) suggests that the response of $T(L,t)$ behaves as dead-time process with an attenuation factor.

Question 7-1 What does a time plot of this response look like and is it physically realistic?

Answer A sharp step in the inlet propagates through the reactor as a sharp step in the liquid temperature. Thus plug flow is idealistic because there is bound to be some axial mixing either from turbulence or diffusion. If Eq. (7-7) were solved, the propagation would be more realistic with less sharpness. Later when lumping is discussed this issue of idealistic sharpness will be revisited.

7-3-2 Steam Jacket Temperature Transfer Function

Equation (7-10) yields a second transfer function relating the steam jacket temperature to the outlet temperature.

$$\frac{\tilde{T}(L,s)}{\tilde{T}_s(s)} = \frac{1 - e^{-st_D} e^{-\frac{t_D}{\tau_T}}}{\tau_T s + 1} \tag{7-12}$$

We will use this transfer function later on when assessing the feasibility of controlling the outlet temperature by manipulating the steam temperature. The denominator of Eq. (7-12) has appeared before so we can expect τ_T to act as a time constant in a way similar to previous transient analyses.

7-4 Response of Tubular Heat Exchanger to Step in Jacket Temperature

Let the jacket temperature be a step of size U_c at time zero. Equation (7-12) becomes

$$\tilde{T}(L,s) = \frac{1 - e^{-st_D} e^{-\frac{t_D}{\tau_T}}}{\tau_T s + 1} \frac{U_c}{s} \tag{7-13}$$

Appendix F shows that the inversion of Eq. (7-13) gives

$$T(L,t) = U_c\left(1 - e^{-\frac{t}{\tau_T}}\right) - U_c e^{-\frac{t_D}{\tau_T}}\left(1 - e^{-\frac{t-t_D}{\tau_T}}\right)\hat{U}(t - t_D) \tag{7-14}$$

Appendix F also explains the nature of the unit step function $\hat{U}(t)$.

7-4-1 The Large-Diameter Case

Figure 7-3 shows the behavior of $T(L,t)$ for the case where $U_c = 1$, $L = 1$, $\tau_T = 1$, $t_D = 1$, and $v = 1$. Note for $t > t_D$, the outlet temperature is constant at

$$T(L,\infty) = U_c\left(1 - e^{-\frac{t_D}{\tau_T}}\right) = 0.6321$$

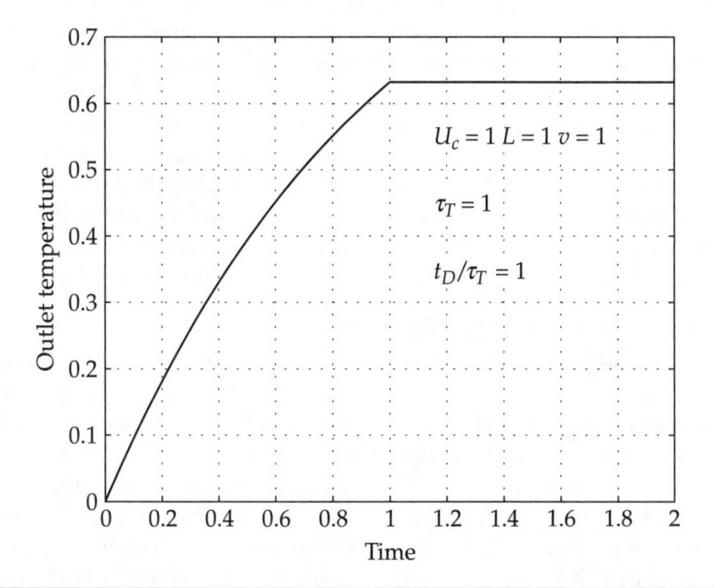

FIGURE 7-3 Response of large-diameter tubular heat exchanger to step in jacket temperature.

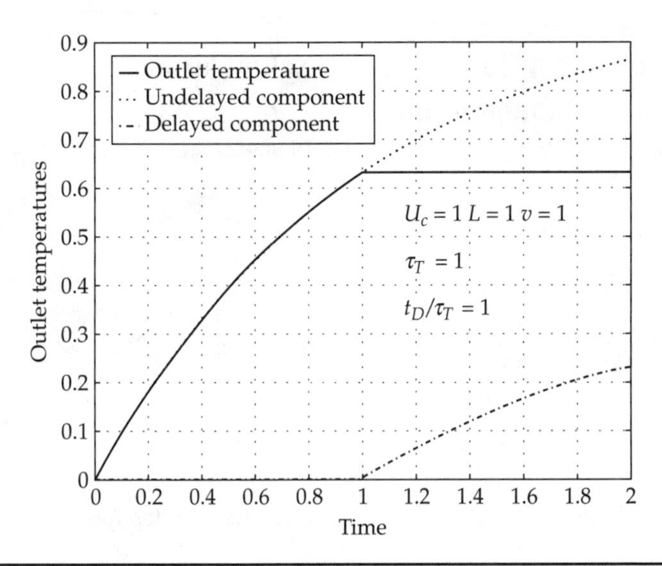

FIGURE 7-4 Components of the outlet temperature for large-diameter tube.

This makes physical sense because t_D seconds are required for the liquid entering the tube to pass completely through. Because $\tau_T = 1$, the liquid temperature only reaches 63% of the steady-state value before it exits the tube. After that time it does not increase because it no longer sees the jacket temperature. Remembering that $\tau_T = D\rho C_p / 4U$ suggests that the time constant could be decreased if the tube had a smaller diameter. This makes sense because a smaller-diameter tube would allow the energy to be transferred from the steam to the liquid more quickly. Let us agree to have this current collection of parameters describe the *large-diameter* tube exchanger. This large-diameter tube exchanger might pose control problems if we try to adjust the jacket temperature to drive the liquid outlet temperature to set point.

Figure 7-4 shows the same outlet temperature along with the two components in Eq. (7-14): the undelayed first-order response and the delayed first-order response which has the attenuation factor of e^{-t_D/τ_T}.

7-4-2 The Small-Diameter Case

For comparison, consider the case where $U_c = 2$, $L = 1$, $\tau_T = 0.1$, and $v = 1$, shown in Fig. 7-5. The time constant τ_T is now a tenth of its value in the previous simulation. We will refer to this piece of equipment as the *small-diameter* tube exchanger.

The residence time $t_D = L / v$ is still 1.0 but because the time constant τ_T is so much smaller, the liquid flowing through the tube has time (10 time constants) to almost completely reach the jacket temperature before it exits. The liquid reaches 63% of the steady-state value after $t = \tau_T$ or 0.1 sec but the liquid spends $t_D = 1.0$ sec in the tube. Figure 7-6 shows the components of Eq. (7-14). Since τ_T

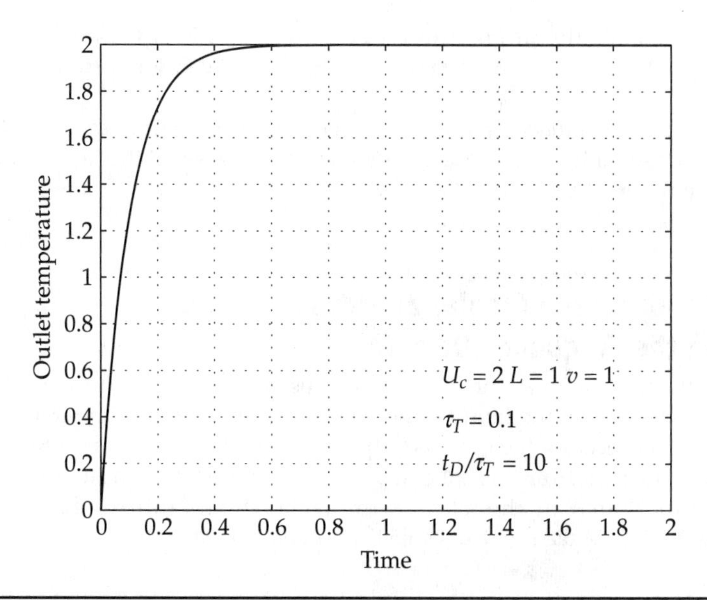

Figure 7-5 Response of small-diameter tubular heat exchanger to step in jacket temperature.

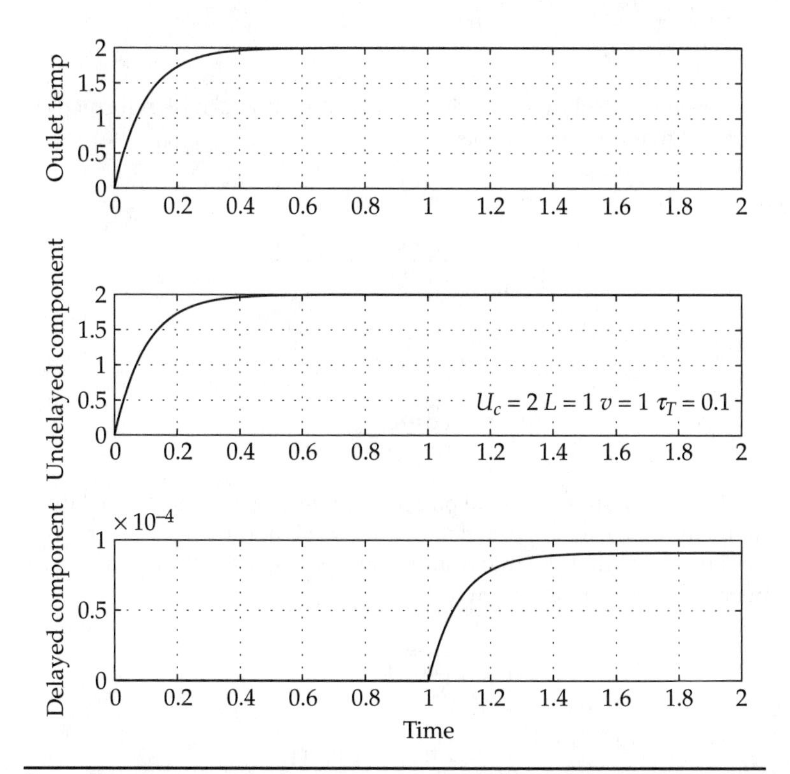

Figure 7-6 Components of outlet temperature for small-diameter tube.

is so small, the attenuation factor $e^{-t_D/\tau_T} = 4.5 \times 10^{-5}$ snuffs out the delayed component leaving only the conventional first-order component.

This *small-diameter* tube exchanger might be more amenable to the idea of adjusting the jacket temperature to control the liquid outlet temperature.

7-5 Studying the Tubular Energy Exchanger in the Frequency Domain

We wish to analyze the effect of a sinusoidal variation in the jacket temperature on the liquid outlet temperature. Start with the transfer function between the process output, which is the liquid temperature as it emerges from the tube at $z = L$, and the process input/output control which in this case is the jacket temperature, given in Eq. (7-12). Make the usual substitution of $s \to j\omega$:

$$\frac{\tilde{T}(L,j\omega)}{\tilde{T}_s(j\omega)} = \frac{1 - e^{-\frac{t_D}{\tau_T}} e^{-j\omega t_D}}{\tau_T j\omega + 1}$$

Appendix F shows that this transfer function can be reformed in terms of magnitude and phase as

$$\left|\frac{\tilde{T}(L,j\omega)}{\tilde{T}_s(j\omega)}\right| = \sqrt{\frac{1 - 2e^{-\frac{t_D}{\tau_T}} \cos^2(\omega t_D) + e^{-2\frac{t_D}{\tau_T}}}{(\tau\omega)^2 + 1}}$$

$$\theta = \tan^{-1}\left(\frac{e^{-\frac{t_D}{\tau_T}} \sin(\omega t_D)}{1 - e^{-\frac{t_D}{\tau_T}} \cos(\omega t_D)}\right) - \tan^{-1}(\tau_T j\omega) \tag{7-15}$$

Figure 7-7 shows a Bode plot for this process model for the large-diameter tube exchanger where $L = 1$, $\tau_T = 1$, and $v = 1$.

First, note that the magnitude and phase curves start to decrease near the corner frequency which is

$$\omega_{cor} = \frac{1}{\tau_T} = 1$$

$$f_{cor} = \frac{\omega_{cor}}{2\pi} = 0.159 \text{ Hz}$$

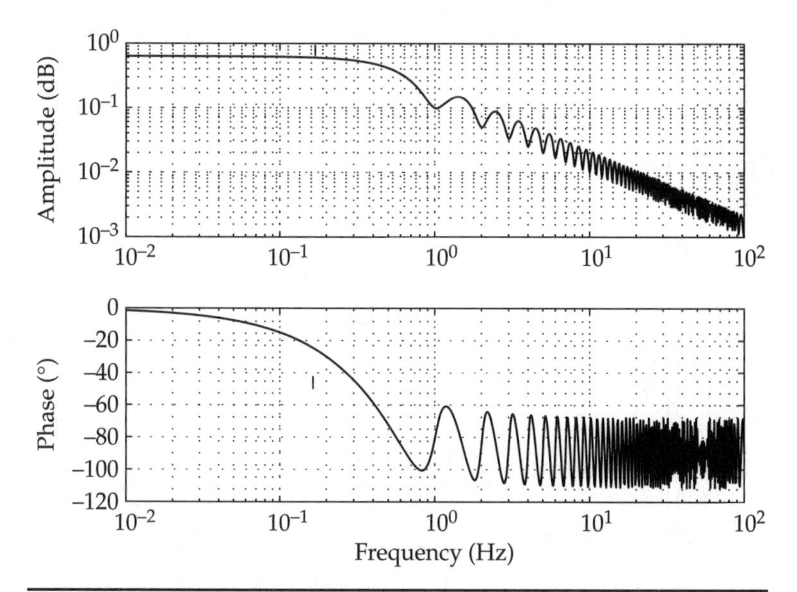

FIGURE 7-7 Bode plot for large-diameter tube exchanger.

Second, note the "resonances" or "ripples." The large-diameter tube exchanger Bode plot is replotted in Fig. 7-8. There appears to be a ripple that has peaks at multiples of 1 Hz. This makes sense because the residence time t_D is 1.0 sec.

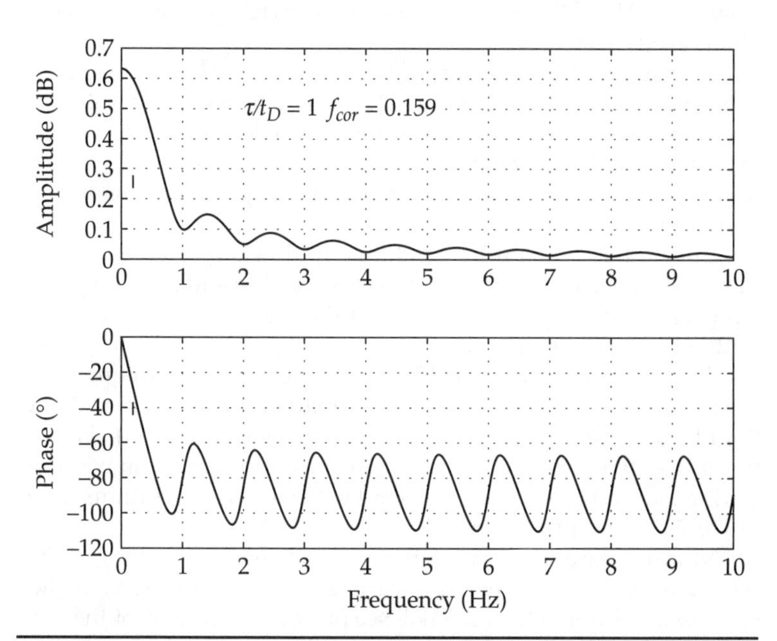

FIGURE 7-8 Linear Bode plot for small tube exchanger.

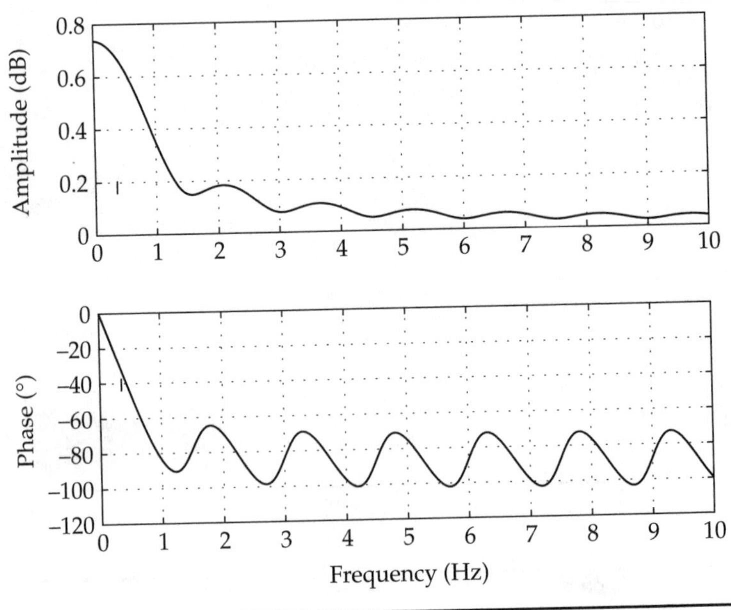

FIGURE 7-9 Bode plot for special tubular energy exchanger with $v = 1.5$, $\tau_t = 0.5$, $f_{cor} = 0.31831$, $t_D/\tau = 1.3333$.

If the flow rate v is increased from 1.0 to 1.5 and τ_T is decreased from 1.0 to 0.5 for this large-diameter tube exchanger, the linear Bode plot is given in Fig. 7-9.

The ratio of the delay time t_D to τ_T is now 1.333 and the spacing between peaks in the frequency is 1.333 Hz which is equal to t_D / τ_T. Therefore, the ripple or resonance appears to depend on the number of time constants the liquid spends in the tube. I have used terms like resonance and ripple, but keep in mind that this is taking place in the frequency domain, not the time domain.

One perhaps can understand this rippling shape in the magnitude and phase curves if one starts with a sinusoidal steam jacket temperature having a period equal to the residence time, namely, t_D. For the case shown in Fig. 7-8 where $v = 1.0$, $L = 1.0$, $\tau = 1.0$, the residence time is $t_D = 1.0$ sec. If the jacket temperature frequency is 1.0 Hz then the period of the jacket temperature will be 1.0 and liquid traveling through the tube will see one sinusoidal cycle of the jacket temperature variation, from minimum to maximum, during its residence. There is a small valley in the magnitude curve in Fig. 7-8 at $f = 1.0$.

Now, increase the jacket temperature frequency to 1.5 Hz. The period of the jacket temperature variation will be 0.67 sec so the liquid passing through the tube will see one cycle plus part of the next cycle which will contain the maximum jacket temperature amplitude.

Consequently, the outlet liquid will be slightly higher in temperature, as indicated by the local peak in the magnitude curve at about $f = 1.5$ Hz in Fig. 7-8. This argument can be continued to explain the other peaks and valleys as the frequency changes.

Also, note that the maximum amplitude at low frequencies is not unity but 0.6321 which was the maximum outlet temperature for the step-change response. This is because the tube is too short for the outlet temperature to reach full value no matter how low the steam jacket frequency. Even with these relatively strange phenomena, the phase lag never exceeds 110° and bounces around 90°.

The small-diameter tube exchanger, where $L = 1$, $\tau_T = 0.1$, and $v = 1$, is a different story in Fig. 7-10.

There are no resonances and the maximum phase lag is 90°. The magnitude and phase curves behave like a first-order system with a corner frequency of

$$\omega_{cor} = \frac{1}{\tau_T} = 10$$

$$f_{cor} = \frac{\omega_{cor}}{2\pi} = 1.59 \text{ Hz}$$

The ratio $t_D / \tau_T = 10$ indicates that the liquid spends 10 time constants in the tube. For jacket temperature frequencies below 0.1 Hz,

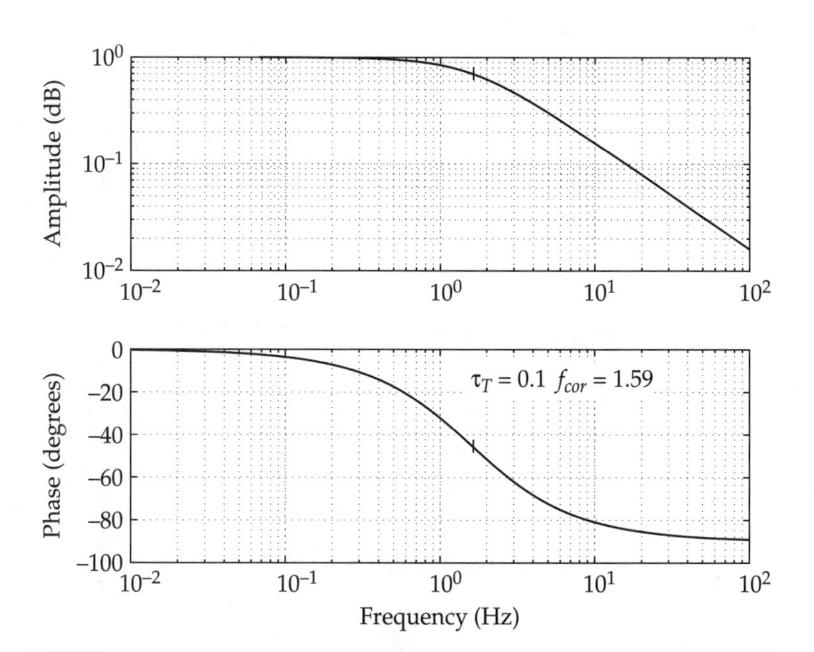

FIGURE 7-10 Bode plot for small tube exchanger.

which would have periods of 10.0 sec or more, there appears to be no resonance because the amplitude is nearly unity. This suggests that for these low frequencies, the jacket temperature variations are passed through the tube unattenuated. Thus, for low jacket temperature frequencies, the tube is long enough to cause the outlet temperature to have the same amplitude as the jacket temperature.

7-6 Control of the Tubular Energy Exchanger

The open loop step-change response and the Bode plot suggest that there should not be too much trouble if feedback control is attempted. The large-diameter tubular energy exchanger has some idiosyncrasies but the total phase lag varies about 90° (why?—because, despite the distributed nature of the process, it is still basically first order).

This section starts by applying PI control to the large-diameter tubular exchanger where $L = 1$, $v = 1$, $\tau_T = 1$. Since the process gain is nominally unity, an initial proportional gain of unity was tried. This was increased to 3.0 by trial and error simulation using a Simulink model. Then the integral gain was increased slowly until a value of 2.0 was found to be satisfactory. Figure 7-11 shows the response to a unit step in the set point at time zero.

Implicit in this control scheme is the presence of a slave loop that will manipulate a valve so as to affect the steam jacket temperature set point which is the control output of the master loop that we are

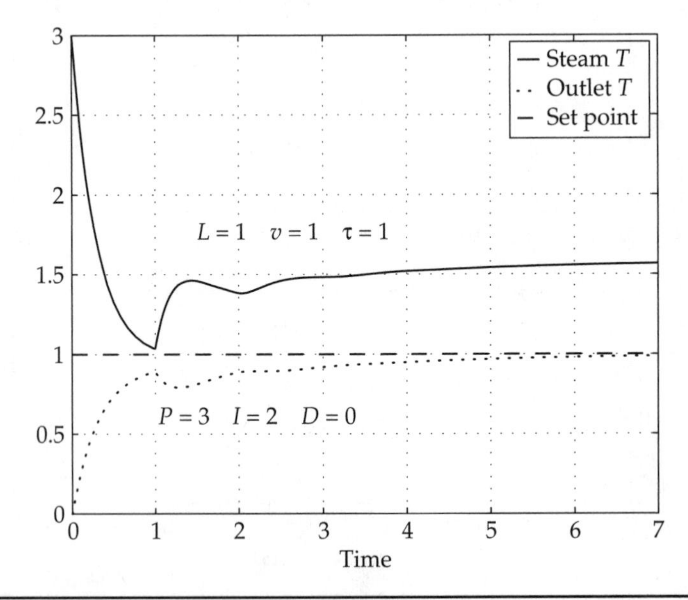

FIGURE 7-11 PI control of the large-diameter tubular heat exchanger.

going to key on. The dynamics of the slave loop are considered to be so much faster than those of the liquid temperature control loop that they can be neglected. This master/slave configuration is a cascade control structure and will be dealt with in Chap. 11.

At time zero, the step in the set point causes the proportional component of the control algorithm to jump to 3.0 (which makes sense because the proportional gain is 3.0). As the outlet temperature starts to respond, the proportional component backs off. At $t = 1.0$ the initial contents of the tube have passed through the exchanger. Although there is a delay for the liquid to pass through the tube, all of the liquid sees the step in the jacket temperature immediately so there is no dead-time effect. For $t > 1.0$ the tube will contain material that has entered the exchanger after the set point was stepped and the outlet temperature will back off slightly with an associated response of the proportional component. All this time, the integral component has been slowly working to bring the outlet temperature near the set point. The reader can try other control gain combinations by modifying the Matlab script that generated Fig. 7-11.

The same control gains are applied to the small-diameter tubular exchanger, where $L = 1$, $v = 1$, $\tau_T = 0.1$. The results are shown in Fig. 7-12.

Unlike the large-diameter tubular exchanger, there is no residence time effect here—rather, the process behaves similarly to a simple first-order process. Figure 7-12 suggests that the proportional gain is

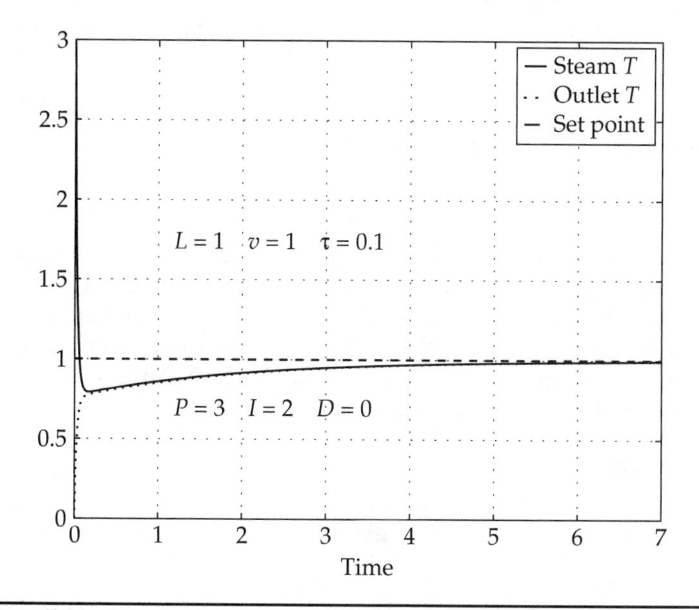

Figure 7-12 PI control of the small-diameter tubular heat exchanger.

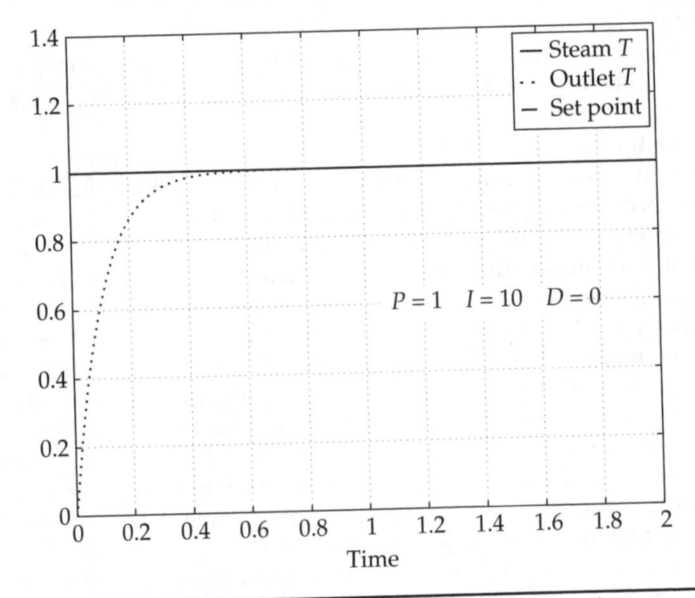

FIGURE 7-13 PI control of the small-diameter tubular heat exchanger (lower proportional gain).

unnecessarily high and the integral gain is too low. Trying $P = 1, I = 10$ produces the behavior shown in Fig. 7-13.

Curiously, these control gains cause the control output to be a straight line such that the steam jacket temperature takes on the value of 1.0 immediately upon the step in the set point and stays at the value indefinitely.

7-7 Lumping the Tubular Energy Exchanger

7-7-1 Modeling an Individual Lump

Often, process analysts like to approximate distributed models, described by partial differential equations, with lumped models, described by ordinary differential equations. The tubular exchanger could be approximated in this way. For example, consider Fig. 7-14 where the tubular exchanger is to be modeled by N tanks. The N-tank model has the following characteristics:

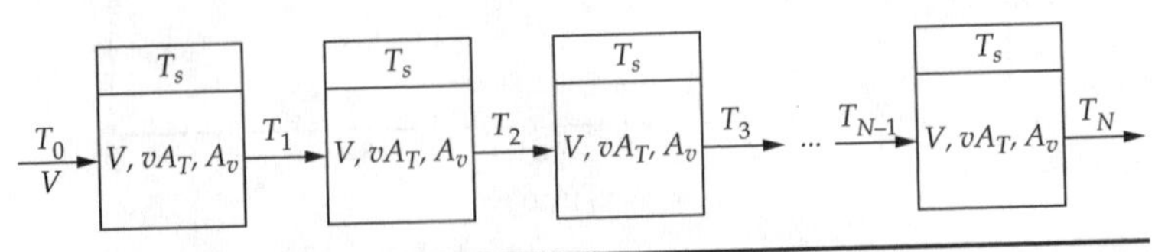

FIGURE 7-14 N-lump approximation to tubular exchanger.

1. Each tank is completely mixed in the sense that the exit temperature is the same as the temperature throughout the lump or tank. These lumps are often called *continuous stirred tanks* (CSTs).
2. Each tank is jacketed and is exposed to the jacket temperature T_s. Although this need not be case in general, each tank sees the same jacket temperature. That is, the jacket temperature does not vary from tank to tank.
3. The temperature leaving the kth tank is the inlet temperature for the k+1th tank.
4. Parameterwise, each tank is identical. This is not necessary but it does make the mathematics more manageable.

The mixing in each lump, mentioned in condition 1, will change the propagation of step changes significantly. The relationship between the number of lumps and the degree of axial mixing (or axial diffusion) will be examined later.

A dynamic energy balance over the kth tank yields the following:

In from $(k-1)$th tank: $vA_vC_p\rho T_{k-1}$
Out from kth tank: $vA_vC_p\rho T_k$
Into the kth tank from steam jacket: $UA_T(T_s - T_k)$
Accumulation in kth tank: $VC_p\rho\dfrac{dT_k}{dt}$

Where A_v is the cross-sectional area for flow into and out of the tank, A_T is the transfer area between the jacket and the tank, and V is the volume of the tank.

The balance becomes

$$VC_p\rho\frac{dT_k}{dt} = vA_vC_p\rho T_{k-1} - vA_vC_p\rho T_k + UA_T(T_s - T_k) \tag{7-16}$$

Equation (7-16) can be simplified slightly with the introduction of a time constant and two gains, as in

$$\tau\frac{dT_k}{dt} + T_k = g_iT_{k-1} + g_ST_s \tag{7-17}$$

$$\tau = \frac{VC_p\rho}{UA_T + vA_vC_p\rho} \qquad g_S = \frac{UA_T}{UA_T + vA_vC_p\rho} \qquad g_i = \frac{vA_vC_p\rho}{UA_T + vA_vC_p\rho}$$

Note that both gains, independent of the parameterization, have to be less than unity and, what's more, they add up to unity, as in

$$g_i + g_s = 1 \tag{7-18}$$

Although there appear to be eight parameters in Eq. (7-17), only three, τ, g_s (or g_i) and N are independent. Note that τ is not the same as τ_T. There are a variety of ways to choose the parameters. For example, one could specify that the area for energy transfer of each tank A_T could be the total transfer area of the tubular exchanger divided by N. The same approach could be taken for choosing the volume of the tank or the length. However, before diving into a simulation let's play some simple mathematical games with the tank equations and try to gain some insight.

7-7-2 Steady-State Solution

First, in steady state we would like the N-tank model to look something like the steady-state solution obtained in Sec. 7-1 which was

$$T(z) = T_0 e^{-\frac{z}{\psi}} + \left(1 - e^{-\frac{z}{\psi}}\right) T_s$$

The N-tank steady-state solution is quite simple since the derivative in Eq. (7-17) is zero, so

$$T_k = g_s T_s + g_i T_{k-1}$$

$$T_1 = g_s T_s + g_i T_0$$

$$T_2 = g_s T_s + g_i T_1 = g_s T_s + g_i (g_s T_s + g_i T_0) = g_s T_s (1 + g_i) + g_i^2 T_0$$

$$T_3 = g_s T_s + g_i T_2 = g_s T_s + g_i (g_s T_s + g_i T_1) = g_s T_s (1 + g_i + g_i^2) + g_i^3 T_0$$

$$\ldots$$

$$T_N = g_s T_s (1 + g_i + g_i^2 + \cdots + g_i^{N-1}) + g_i^N T_0 \tag{7-19}$$

When a quantity like g_i is less than unity, the sum of the geometric series, contained in the parentheses in Eq. (7-19), can be written compactly as

$$T_N = g_s T_s \frac{1 - g_i^N}{1 - g_i} + g_i^N T_0 \tag{7-20}$$

Remember that

1. g_s and g_i depend on the area for energy transfer A_T of each tank.
2. $g_s + g_i = 1$.

3. N increases A_T, the energy transfer area for the kth tank decreases so as to maintain the total energy transfer area constant.

Therefore, Eq. (7-20) becomes

$$T_N = T_0 g_i^N + \left(1 - g_i^N\right) T_s \tag{7-21}$$

This is to be compared with Eq. (7-3)

$$T(L) = T_0 e^{-\frac{L}{\psi}} + \left(1 - e^{-\frac{L}{\psi}}\right) T_s$$

This suggests that if the tube length L is divided up so that each tank has length $L/N = \Delta z$, then $g_i^N = e^{-L/\psi}$ or $g_i = e^{-L/N\psi} = e^{-\Delta z/\psi}$. Thus, $N \to \infty$, $g_s \to 0$, and $g_i \to 1$. This exercise suggests that the lumping approach is approximately similar to the continuous approach, at least in steady state.

7-7-3 Discretizing the Partial Differential Equation

An alternative approach to lumping returns to the partial differential equation in Eq. (7-6) and replaces the partial derivative with respect to axial distance z with a finite difference, as in

$$\begin{aligned}
&\frac{\partial T}{\partial t} + v\frac{\partial T}{\partial z} = \frac{1}{\tau_T}(T_s - T) \\
&\tau_T = \frac{D\rho C_p}{4U} = \frac{\psi}{v} \\
&T \Rightarrow T_k \qquad \frac{\partial T}{\partial z} \Rightarrow \frac{T_k - T_{k-1}}{\Delta z} \\
&\frac{dT_k}{dt} + v\frac{T_k - T_{k-1}}{\Delta z} = \frac{1}{\tau_T}(T_s - T_k) \\
&\frac{dT_k}{dt} = -\left(\frac{v}{\Delta z} + \frac{1}{\tau_T}\right) T_k + \frac{v}{\Delta z} T_{k-1} + \frac{1}{\tau_T} T_s
\end{aligned} \tag{7-22}$$

If the reader makes the following substitutions

$$A_v = \frac{\pi D^2}{4} \qquad A_T = \pi D \Delta z \qquad V = \frac{\pi D^2}{4}\Delta z \qquad \tau_T = \frac{D\rho C_p}{4U}$$

in Eq. (7-17), she will arrive at Eq. (7-22). As before, the small-diameter tube has $L = 1$, $v = 1$, and $\tau_T = 0.1$ and the large diameter tube has $L = 1$, $v = 1$, and $\tau_T = 1$.

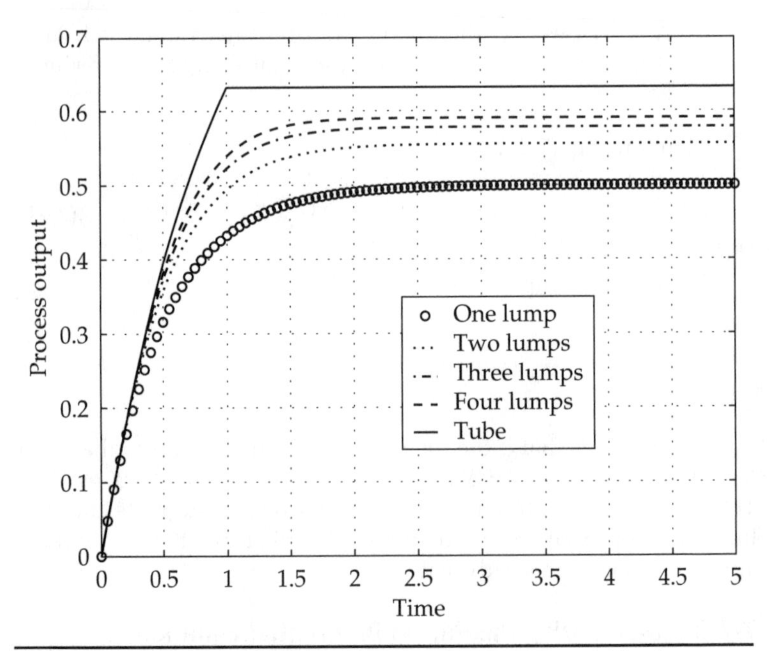

FIGURE 7-15 Response to steam temperature jacket T_s—matching large tubular reactor with lumped models, $\tau_t = 1.0$.

Figures 7-15 and 7-16 show one, two, three, and four tank approximations to the large-diameter and small-diameter tube energy exchangers, respectively. The approximation is poorer for the large-diameter tube exchanger probably due to the extensive mixing in the tank approximations as compared to none in the plug-flow model.

Equation (7-22) can also be solved for a step in the inlet temperature to the first tank. Figures 7-17 and 7-18 show the response for the two cases.

In Fig. 7-17, for the large-diameter tube, one can see a slow improvement in the approximation as the number of lumps increases. Note the unrealistically sharp response for the tube. In Fig. 7-18, for the small-diameter tube, there is the same progression. The 10-lump model is visually indistinguishable from the tube.

Therefore, one needs to be aware of the physical consequences of solving a partial differential equation by replacing one (or more) of the partial derivatives with a finite difference. In effect, you might be replacing a model that has no axial mixing or axial diffusion with a model that has extensive axial mixing. Section 7-8 takes a closer look at the relationship between lumping and axial transport.

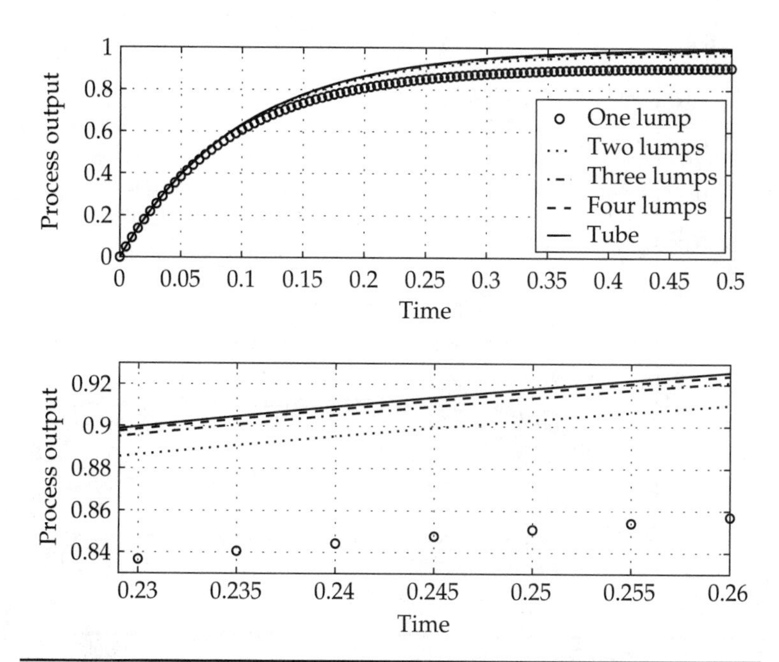

FIGURE 7-16 Response to steam jacket temperature T_s—matching small tubular reactor with lumped models, $\tau_t = 0.1$.

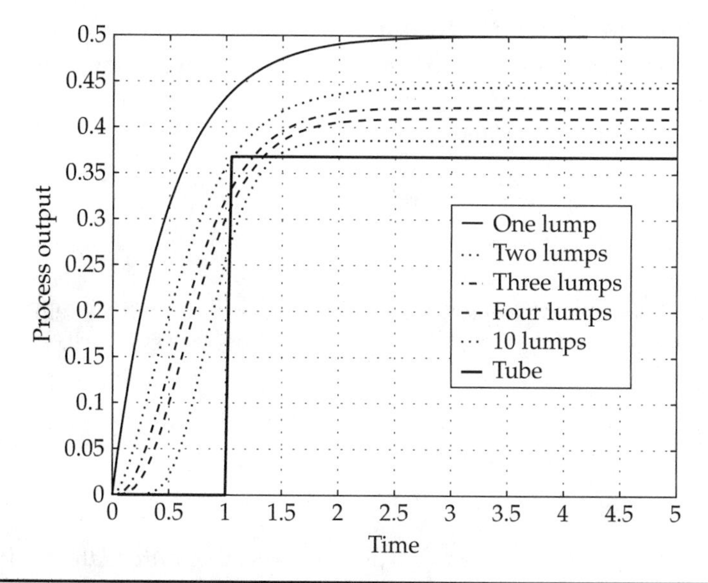

FIGURE 7-17 Response to inlet temperature jacket T_0—matching tubular reactor with lumped models, $\tau_t = 1.0$.

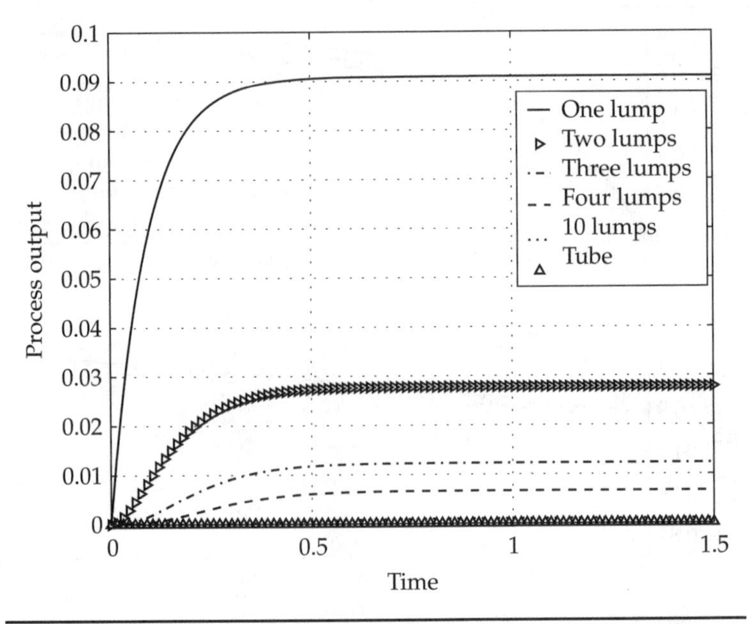

FIGURE 7-18 Response to inlet temperature jacket T_0—matching tubular reactor with lumped models, $\tau_t = 0.1$.

7-8 Lumping and Axial Transport

For reference, we repeat the partial differential equation that has energy transport by convection *and* axial diffusion:

$$\frac{\partial T}{\partial t} + v\frac{\partial T}{\partial z} = \frac{4U}{D\rho C_p}[T_s - T(z,t)] + \frac{k}{\rho C_p}\frac{\partial^2 T}{\partial z^2} \tag{7-23}$$

If the tubular exchanger model *without* axial diffusion is to be approximated by N lumps of length Δz, then each lump is described by

$$\frac{dT_k}{dt} + v\frac{T_k - T_{k-1}}{\Delta z} = \frac{1}{\tau_T}(T_s - T_k) \tag{7-24}$$

Using Taylor's series (see App. D), the temperature of the $k-1$th lump can be related to that of the kth lump by

$$T_{k-1} = T_k - \frac{\partial T}{\partial z}\Delta z + \frac{\partial^2 T}{\partial z^2}\frac{\Delta z}{2} + \text{h.o.t.}$$

or (7-25)

$$T_k - T_{k-1} = \frac{\partial T}{\partial z}\Delta z - \frac{\partial^2 T}{\partial z^2}\frac{(\Delta z)^2}{2} + \text{h.o.t.}$$

Replacing the $T_k - T_{k-1}$ term in Eq. (7-24) with the expression in Eq. (7-25) while ignoring the higher-order terms, one obtains

$$\frac{dT_k}{dt} + v\frac{\partial T}{\partial z} = \frac{1}{\tau_T}(T_s - T_k) + v\frac{\Delta z}{2}\frac{\partial^2 T}{\partial z^2}$$

or (7-26)

$$\frac{\partial T}{\partial t} + v\frac{\partial T}{\partial z} = \frac{4U}{D\rho C_p}[T_s - T(z,t)] + \frac{k}{\rho C_p}\frac{\partial^2 T}{\partial z^2}$$

$$\frac{k}{\rho C_p} = v\frac{\Delta z}{2} = v\frac{L}{N}$$

Since Eq. (7-26) follows from Eq. (7-24), one can conclude that lumping introduces an effective axial transport mechanism which is inversely proportional to the number of lumps in the approximation. This suggests that as N increases the sharpness in the propagation of steps through the tubular exchanger will increase because there will be less axial mixing or diffusion.

This analysis is consistent with the discussion we had in Sec. 5-5 about multitank processes and with the results obtained in Sec. 7.7.

Question 7-2 Can you make sense out of Fig. 7-18?

Answer As was shown in Section 7-4-2, Fig. 7-6, for the small diameter case the time constant $\tau_T = \dfrac{D\rho C_p}{4U}$ is 0.1 seconds, the dead time is 1.0 second, the attenuation factor $\exp(-t_D / \tau_T)$ equals 4.5e-5 and there is no axial transport other than convection. Note that the energy transfer coefficient U occurs in the factor and the relative smallness of τ_T suggests a large amount of energy transfer (and attendant loss of temperature). On the other hand, the one-tank approximation has an effective axial diffusion of (from Eq. 7-26) equal to vL/N. The one-tank approximation exhibits perfect mixing such that the inlet temperature step appears at the tube outlet in first order fashion. There is no axial mixing in the tubular exchanger so, although there is a small step at time $t = t_D$ because of the attenuation factor it is virtually undetectable. As the number of tanks increases there is less axial mixing. The ten-tank approximation is relatively close to tubular exchanger.

7-9 State-Space Version of the Lumped Tubular Exchanger

The equations describing the finite difference approximation in the Sec. 7-8 can be written as

$$
\frac{d}{dt}\begin{pmatrix} T_1 \\ T_2 \\ \cdots \\ T_{N-1} \\ T_N \end{pmatrix} = \begin{pmatrix} -\left(\frac{v}{\Delta z}+\frac{1}{\tau_T}\right) & 0 & \cdots & 0 & 0 \\ \frac{v}{\Delta z} & -\left(\frac{v}{\Delta z}+\frac{1}{\tau_T}\right) & \cdots & 0 & 0 \\ \cdots & \cdots & \cdots & \cdots & \cdots \\ 0 & 0 & \cdots & -\left(\frac{v}{\Delta z}+\frac{1}{\tau_T}\right) & 0 \\ 0 & 0 & \cdots & \frac{v}{\Delta z} & -\left(\frac{v}{\Delta z}+\frac{1}{\tau_T}\right) \end{pmatrix}\begin{pmatrix} T_1 \\ T_2 \\ \cdots \\ T_{N-1} \\ T_N \end{pmatrix}
$$

$$
+\begin{pmatrix} \frac{1}{\tau_T} & 0 & \cdots & 0 & 0 \\ 0 & \frac{1}{\tau_T} & \cdots & 0 & 0 \\ \cdots & \cdots & \cdots & \cdots & \cdots \\ 0 & 0 & \cdots & \frac{1}{\tau_T} & 0 \\ 0 & 0 & \cdots & 0 & \frac{1}{\tau_T} \end{pmatrix}\begin{pmatrix} T_{s_1} \\ T_{s_2} \\ \cdots \\ T_{s_{N-1}} \\ T_{s_N} \end{pmatrix} \tag{7-27}
$$

$$
\frac{d}{dt}X = AX + BU
$$

$$
Z = CX
$$

This form is different (and more general) in that the steam jacket temperatures for each lump have been specified. This is equivalent to a different physical situation where the tube is sectioned into N zones and where each zone's steam jacket temperature is adjustable.

Figures 7-19 and 7-20 show the step-change response of a 20-lump approximation of the tubular energy exchanger for the two cases of small- and large-diameter. All of the lump's steam temperatures were stepped in unison from 0.0 to 100.0. This is the first time we have tried

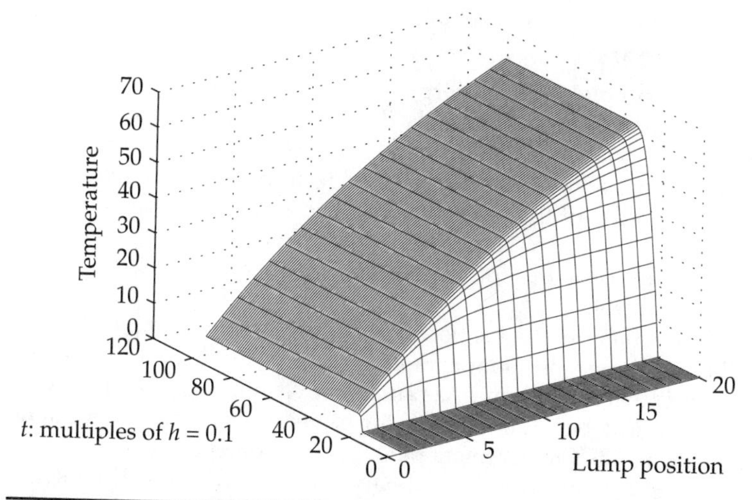

FIGURE 7-19 Steam step-change response for large-diameter case, $\tau_t = 1.0$, $N = 20$.

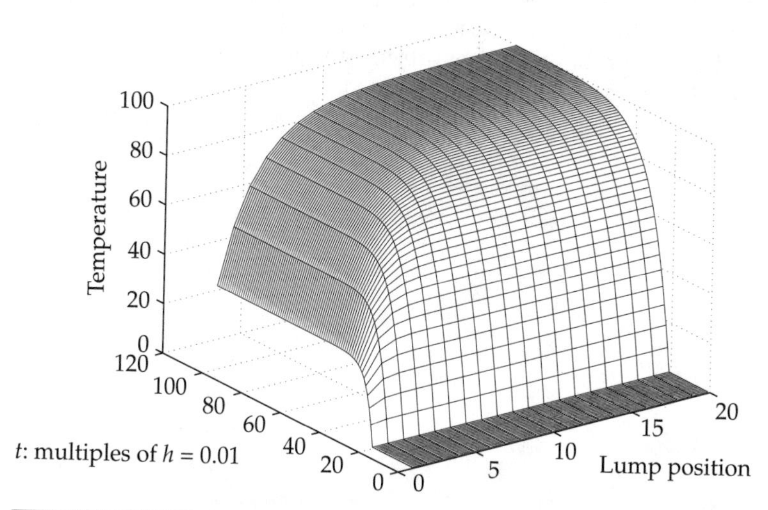

FIGURE 7-20 Steam step-change response for small-diameter case, $\tau_t = 0,1$ $N = 20$.

graphically presenting the spatial *and* temporal response of a process variable and these two figures may require some study. Think about why the responses are different.

In Chap. 10 this kind of multizoned tubular energy exchanger will be controlled via a state-space control algorithm.

7-10 Summary

With the simple tubular energy exchanger we have completed the introduction of process models. This model is described by a partial differential equation instead of an ordinary differential equation and has some idiosyncrasies, especially if the residence time is on the order of the time constant. The analysis in the frequency domain showed how some interesting resonances can appear. Still, the new process did not present any insuperable obstacles in the face of control.

The tubular exchanger is a particularly simple example of a distributed process. In more complicated cases the partial differential equation that might appear as an attractive model is often relatively difficult to solve. Lumping, using several sequential continuous stirred tanks as a replacement model, is sometimes attractive because the tanks are described by ordinary rather than partial differential equations. Continuous stirred tanks can be used to approximate the tubular exchanger but the effective axial mixing introduced by the lumping can lower the quality of approximation. On the other hand, the partial differential equation model does have the physically unrealistic characteristic of sharp step propagation so the axial mixing introduced by the lumping may be more realistic.

In the next chapter, a new subject, stochastic disturbances and the discrete time domain will be presented.

CHAPTER 8

Stochastic Process Disturbances and the Discrete Time Domain

Developing a successful control algorithm often requires proper identification of the disturbances. In Chap. 1, unautocorrelated process disturbances (white noise) and autocorrelated process disturbances were presented using the "large hotel water tank" example. In this chapter these terms and concepts will be revisited with a little more rigor using the autocorrelation, the line spectrum, the cumulative line spectrum, and the expectation operator. The ability of a PI controller to deal with different kinds of disturbances will be discussed.

Chapter 9 will revisit the discrete time domain and introduce the Z-transform.

8-1 The Discrete Time Domain

In the previous chapters, for the most part, the time domain was considered as continuous. Differential equations were derived based on this concept. Laplace transforms were used to solve these differential equations and also to provide a path to the frequency domain which was also considered continuous. In this chapter the time domain will be discrete in the sense that a data stream will now consist of a sequence of numbers usually sampled at a constant interval of time. For example, a data stream might consist of samples of a temperature $T(t)$, as is

$$T(t_1), T(t_2), \ldots, T(t_N)$$

or

$$T_1, T_2, \ldots, T_N$$

with the sample-instants in time being equally spaced, in the sense that

$$t_i = t_{i-1} + h \qquad i = 1, 2, \ldots$$

where h is the sampling interval, which will be assumed to be constant unless otherwise stated. The sampling frequency is $1/h$.

Instead of differential equations where the independent variable is continuous time, there will be algebraic equations with the independent variable being an index, such as i, to an instant of time. A simple example of an indexed equation would be a running sum of a data stream consisting of sampled values of the variable x, as in

$$S_i = S_{i-1} + x_i \qquad i = 2, 3, \ldots, N$$

The average of the $x_1, x_2, \ldots$ data after N samples would be

$$\bar{x}_N = \frac{1}{N} S_N$$

The sample average $\bar{x}_N$ is an estimator of the population mean μ, which we will discuss in more detail later in this chapter.

8-2 White Noise and Sample Estimates of Population Measures

Consider a data stream of infinite extent

$$w_1, w_2, \ldots$$

from which N contiguous samples have been taken. Figure 8-1 shows two views of the data stream. The infinite data stream represents a population having certain *population characteristics* and the subset of size N mentioned above is a sample of that population. The subset has certain *sample characteristics*, which can be used as estimates of the population characteristics. The data shown in Fig. 8-1 will soon be shown to be samples of "white noise." For now, we simply refer to it as a stochastic sequence. The word "stochastic," means "nondeterministic" in that the value at time t_i does not completely determine the value at time t_{i+1}. In the white noise stochastic sequence shown in Fig. 8-1, the value at t_i has no influence whatsoever on the value at t_{i+1}. In other non–white stochastic sequences to be covered later in the chapter, the value at t_{i+1} still is nondeterministic but the value at t_i does have an influence. Note that the two streams shown in Fig. 8-1 have different sample standard deviations.

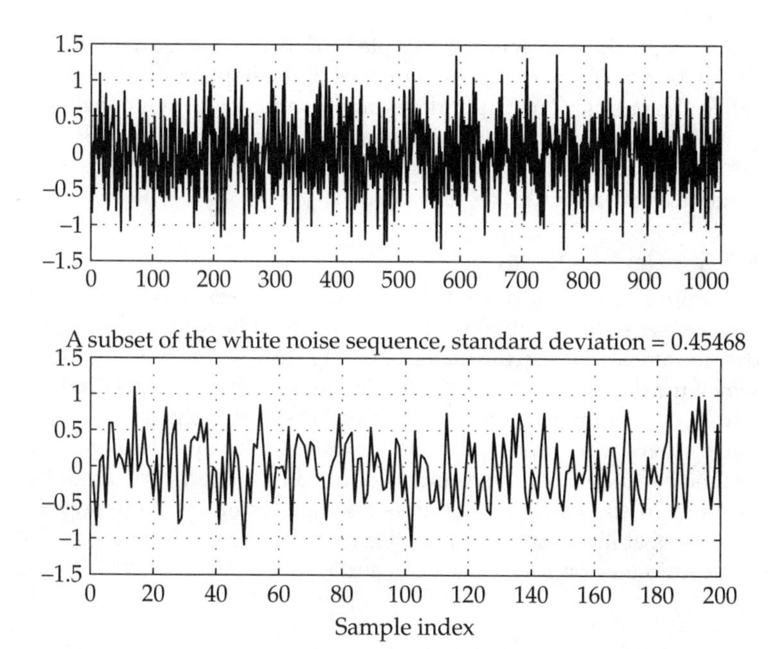

Figure 8-1 Two views of a white noise sequence with $\sigma = 0.515$.

8-2-1 The Sample Average

The sample average of the finite subset of the infinite stream of data is

$$\overline{w} = \frac{\sum_{k=1}^{N} w_k}{N} = \frac{w_1 + w_2 + \cdots + w_N}{N} \tag{8-1}$$

The average of many stochastic sequences used in this book will be removed from the data stream. If that were the case here, the subset would have a zero average. Equation (8-1) applies to any data stream, white noise, or otherwise. As mentioned in Sec. 8-1, the sample average is an estimate of the population mean μ which we will discuss later on in this chapter.

8-2-2 The Sample Variance

A measure of the strength of the variation of $w_1, w_2, \ldots$ about its average (which may be zero if the average has been removed) is the sample variance V_w, defined as

$$V_w = \frac{1}{N}\sum_{k=1}^{N}(w_k - \overline{w})^2 \tag{8-2}$$

As with the average, the sample variance is an estimate of the population variance σ^2. The square root of the population variance is called the population standard deviation and is often symbolized by σ. Note in Fig. 8-1 that "almost" all of the values lie between -3σ and 3σ. The sample variance and the sample standard deviation are often symbolized by s^2 and s, respectively.

8-2-3 The Histogram

The values in the data stream shown in Fig. 8-1 seem to cluster about the average of approximately zero. A picture of how these values are distributed is given by the histogram in Fig. 8-2 where the range over which the data stream varies is divided into 10 "bins" or cells and the number occurring in each bin is plotted versus the center of each bin. The histogram augments and extends the sample variance to give the analyst a feel for how the elements of the data stream vary about the average. In effect, the histogram is a sample estimate of the population's probability distribution. In this case, the population probability distribution is normal or Gaussian and is given by

$$p(x) = \frac{1}{\sigma\sqrt{2\pi}} e^{-\frac{(x-\mu)^2}{2\sigma^2}}$$

We will return to the probability distribution later on in this chapter. Figure 8-3 shows the shape of two normal probability distributions. Each has zero mean but one has a standard deviation of 1.0 while that for other is 1.5. Note how these curves qualitatively match that of the

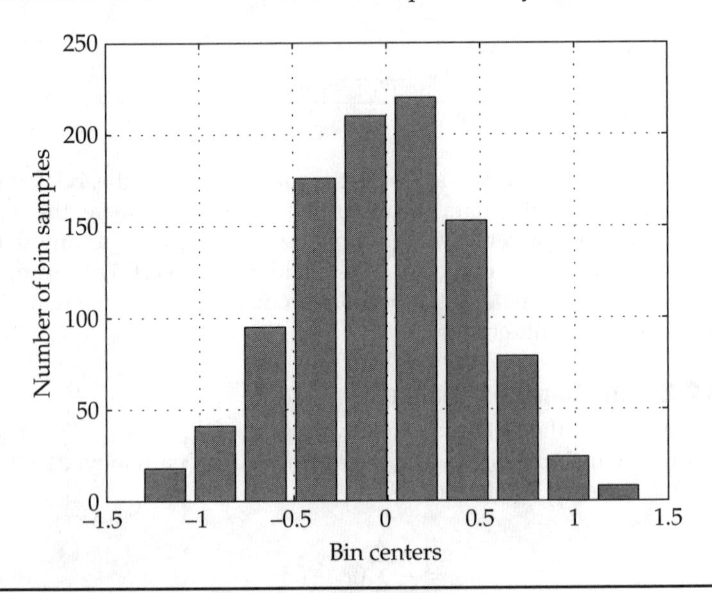

FIGURE 8-2 Histogram of a white noise sequence.

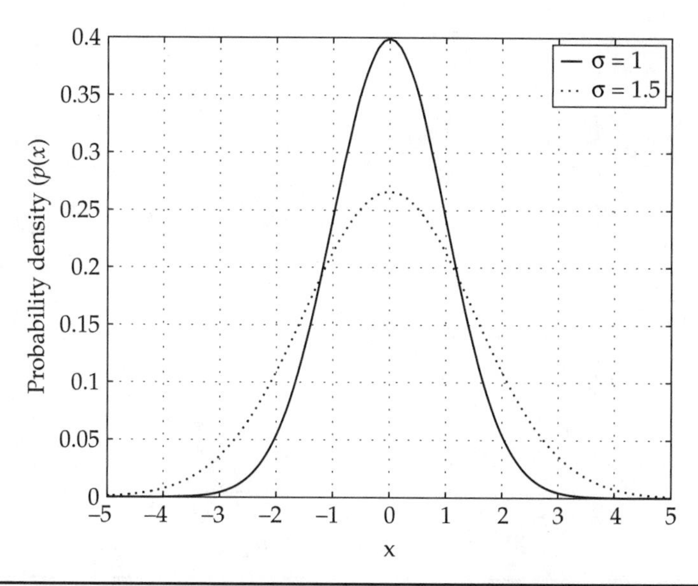

Figure 8-3 Two normal or Gaussian probability distributions.

histogram in Fig. 8-2. The histogram gives no insight into how the elements are interrelated in time. To gain some insight in that area we need other tools such as the autocorrelation and the line spectrum.

8-2-4 The Sample Autocorrelation

The white noise data stream in Fig. 8-1 is *unautocorrelated* because each sample, w_i, is independent of each and every other sample. That is, w_i is independent of w_k for every $k \neq i$. This condition could also be described using a lag index n, where samples w_i and w_{i+n} are considered uncorrelated in the sense that an average of the products of w_i and w_{i+n} taken over a set of N samples would be so close to zero as to be insignificant.

The sample estimate of the autocorrelation, $r_w(n)$, which uses the lag index n as a parameter, is one way to characterize this condition and is defined as follows:

$$r_w(n) = \frac{1}{(N-n)\,V_w} \sum_{k=1}^{N-n} (w_k - \overline{w})(w_{k+n} - \overline{w}) \tag{8-3}$$

The sample estimate of the autocorrelation is basically an average of the product of the lagged products over the available data set for all possible lags. As the lag size increases, the size of the data set available for the calculation in Eq. (8-3) becomes smaller and the estimate becomes less reliable.

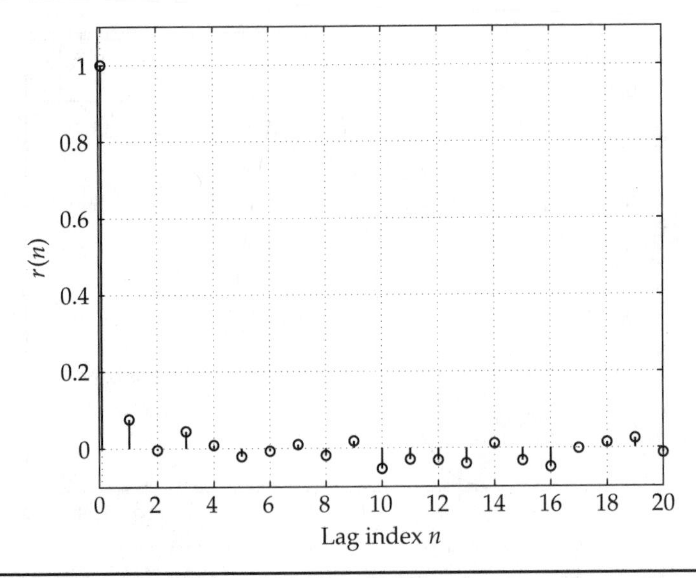

FIGURE 8-4 Autocorrelation of a white noise sequence.

If the data stream symbolized by w is unautocorrelated, $r_w(n)$ will be small for all n. On the other hand, if there is a periodic component in data stream then $r_w(n)$ will have a significant value for the value of n (and multiples of it) corresponding to the period of the oscillating component in the data stream.

The $r_w(n)$ of the white noise sequence plotted in Fig. 8-1 is shown in Fig. 8-4. Notice that the autocorrelation for a lag index of zero is unity because the ith sample is completely autocorrelated with itself. For the other lag indices the $r_w(n)$ bounces insignificantly around zero.

After adding a sine wave to the noisy data in Fig. 8-1, a new signal is created that also looks like white noise. This new signal is shown in Fig. 8-5. The histogram of this sequence is shown in Fig. 8-6. The autocorrelation of this second data sequence is shown in Fig. 8-7. The peaks show that there is a periodic component that appears to have a period of approximately six or seven samples. That is, samples spaced apart by 6 or 7 samples are autocorrelated. In fact, the sine wave buried in the white noise has a period of 6.5 sample intervals. The time domain plot of the data in Fig. 8-5 gives no hint as to the presence of a periodic component because of the background noise. However, the autocorrelation plot shows peaks because the averages of the lagged products tend to allow the noise to cancel out.

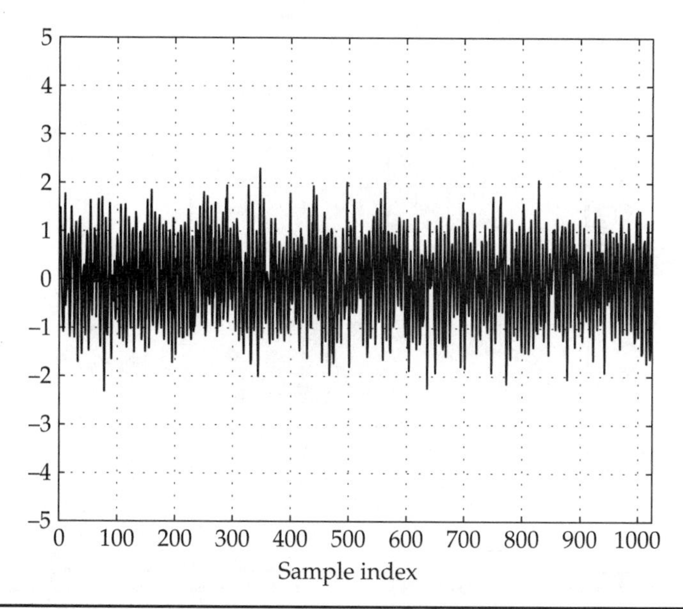

FIGURE 8-5 A white noise sequence containing a sine wave.

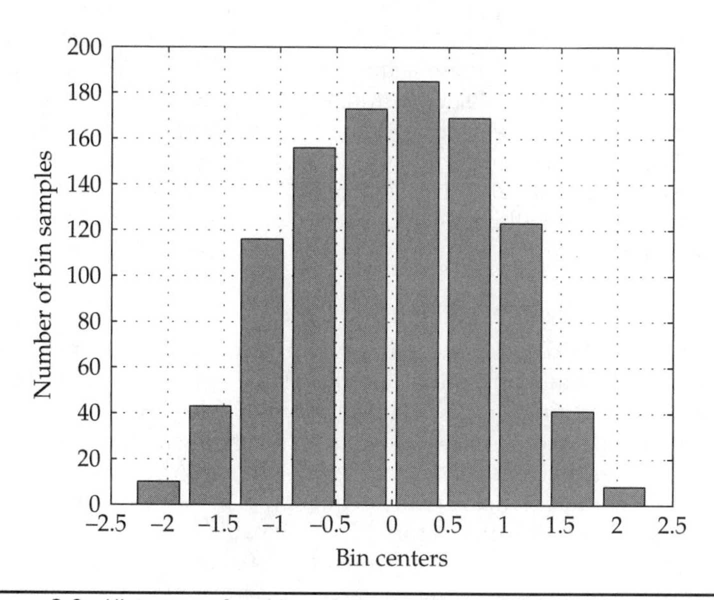

FIGURE 8-6 Histogram of a white noise sequence containing a sine wave.

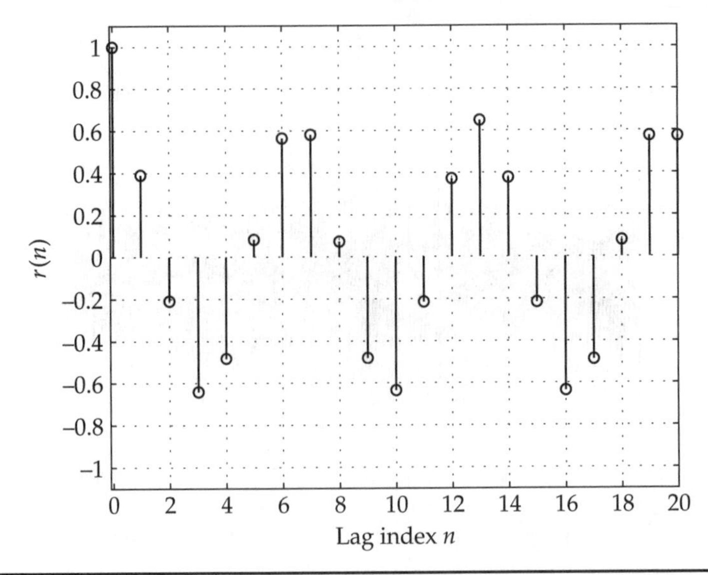

FIGURE 8-7 Sample autocorrelation of a white noise sequence containing a sine wave.

8-2-5 The Line Spectrum

In Chap. 2, the line spectrum was shown to be a handy tool for analyzing noisy processes. In App. C it is discussed in more detail. When applied to the white noise sequence in Fig. 8-1 the line spectrum shown in Fig. 8-8 results. Unfortunately, this spectrum does not give us much insight. Like the time domain sequence, it contains so many localized peaks that one could draw incorrect conclusions about hidden periodic signals. For the signal to be "white," its spectrum should contain power at all frequencies, that is, the spectrum should be flat. Figure 8-8 suggests otherwise, so is the signal really white noise?

8-2-6 The Cumulative Line Spectrum

To address this question, one uses the cumulative line spectrum, which is basically a running sum of the line spectrum. As we have suggested in App. A and will show later in this chapter, the operation of summing a sequence is analogous to integrating a function. As we will also see, both operations are low-pass filters. When the line spectrum is summed, many high-frequency stochastic variations are attenuated and the true nature of the signal is revealed. If the line spectrum of white noise is supposed to be flat (as in being constant) then the running sum (or the integral) of a straight flat line would be a ramp.

The cumulative line spectrum of the data in Fig. 8-1 is shown in Fig. 8-9. Note that the cumulative line spectrum behaves as a ramp and is well within the upper and lower K-S test limits shown in

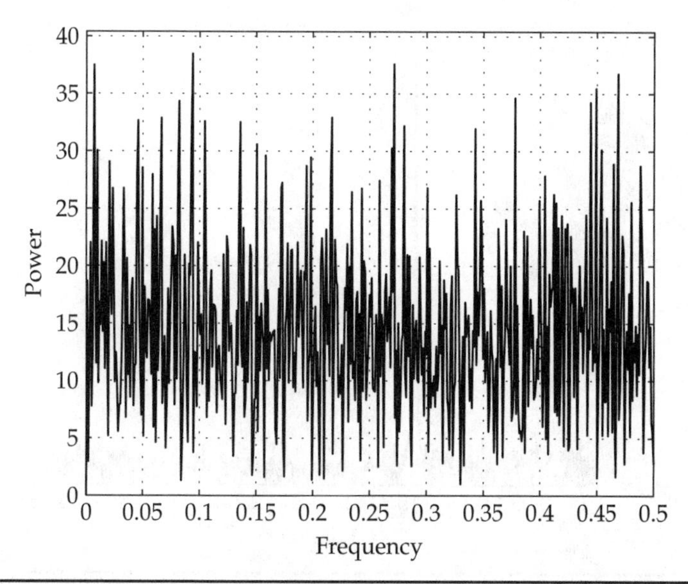

FIGURE 8-8 Line spectrum of a white noise sequence sampled at 1Hz.

Fig. 8-9. (If the cumulative line spectrum lies within the K-S limits there is a 99% probability that the associated stochastic sequence is white.) No more mention of the K-S, as in Kolmogorov-Smirnov, test limits will be made here. The interested reader can search the web and perhaps check out Kolmogorov-Smirnov in the Wikipedia.

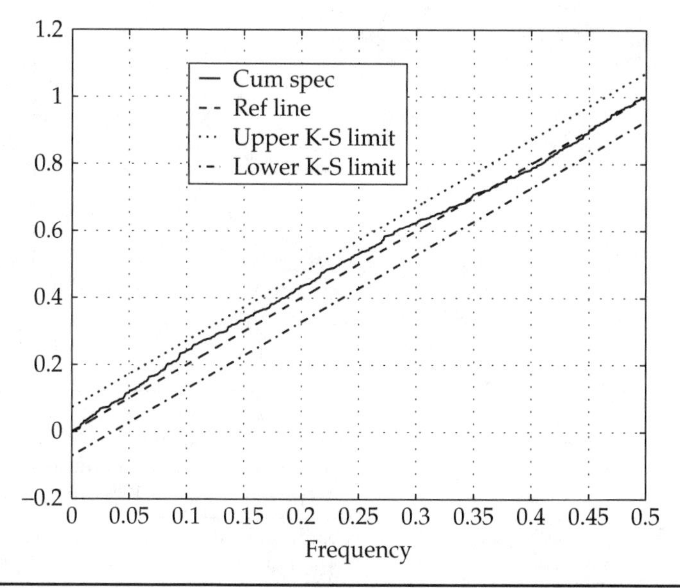

FIGURE 8-9 Cumulative line spectrum of a white noise sequence sampled at 1Hz.

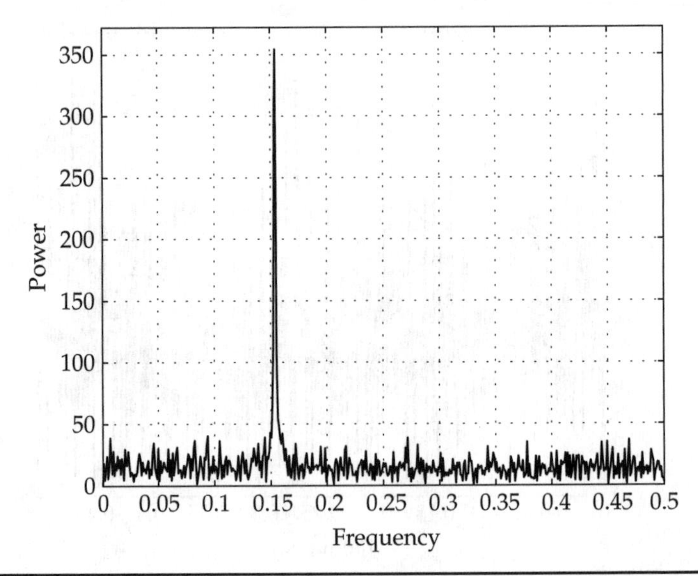

FIGURE 8-10 Line spectrum of signal with periodic component sampled at 1Hz.

For the sake of completeness, Figs. 8-10 and 8-11 show the line spectrum and cumulative line spectrum for the signal containing the periodic component presented in Fig. 8-5. Here, there is no need to resort to the cumulative line spectrum to convince the reader that

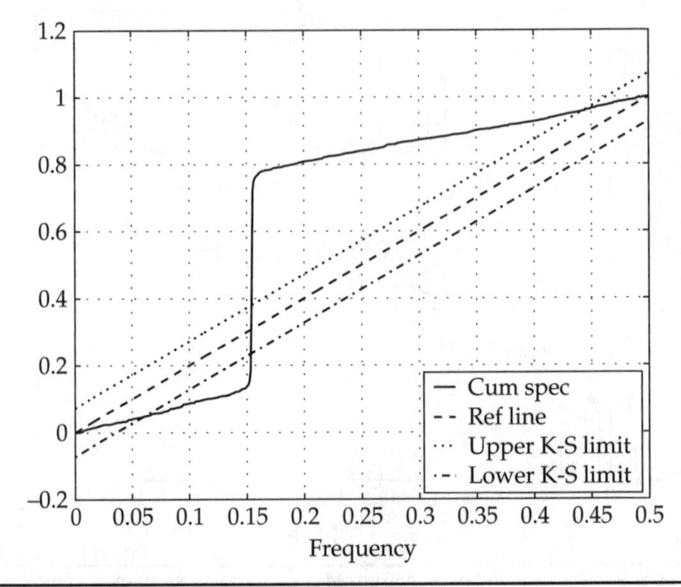

FIGURE 8-11 Cumulative line spectrum of signal with periodic component.

there is a periodic signal lurking in the sequence that looks like white noise. Note that the peak in Fig. 8-10 appears at a frequency of approximately 0.153 Hz which is consistent with the sine wave having a period of 6.5.

In summary, this section introduced the concept of white noise via the autocorrelation and the line spectrum. We used the sample estimate, the sample variance, and the histogram to help characterize white noise data streams.

8-3 Non-White Stochastic Sequences

8-3-1 Positively Autoregressive Sequences

Stochastic sequences that are autocorrelated can be generated by feeding white noise into various equations that will be shown in Chap. 9 to be discrete time filters. A simple example is the autoregressive filter, as in

$$y_k = a y_{k-1} + w_k \quad k = 1,2,3,\ldots \tag{8-4}$$

The input to the filter is the white noise sequence, w_k, $k = 1,2,\ldots$ and the output is y_k, $k = 1,2,\ldots$ with y_0 as an initial value. This sequence is termed autoregressive because it depends on its own previous values. The parameter a is the autoregressive parameter. Although the value of y_k is nondeterministic because of the impact of w_i, it is influenced by y_{k-1} because of the presence of a. In general, an Mth-order stochastic autoregressive sequence can be defined as

$$y_k = \sum_{m=1}^{M} a_m y_{k-m} + w_k \tag{8-5}$$

An example of Eq. (8-4) when $a = 0.9$ is shown in Fig. 8-12. Unlike the white noise sequence, this data stream tends to wander about with a low-frequency variation. In fact, it might even look as though it is periodic but that is not the case.

The histogram of this autoregressive sequence is shown in Fig. 8-13. Note that the histograms of white noise and the autoregressive sequence have the same overall shape, namely a "normal" or "Gaussian" or "bell" shape indicating that the values are distributed around the average with the most frequently occurring values being near the average.

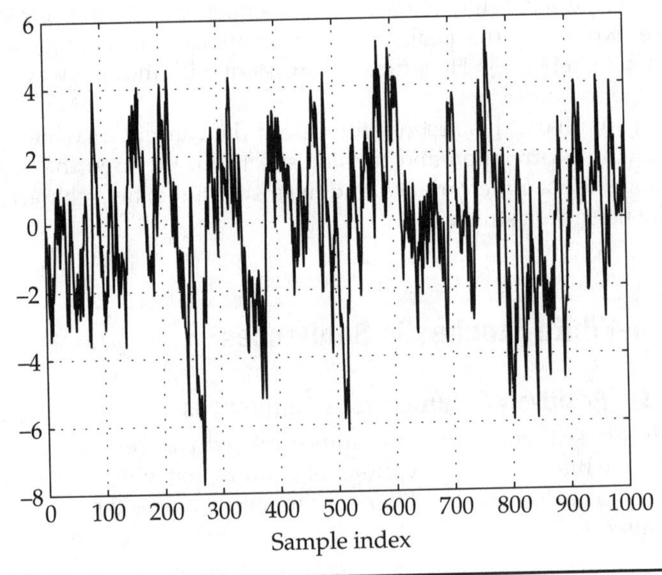

FIGURE 8-12 An autoregressive sequence, $a = 0.9$.

The autocorrelation, line spectrum, and cumulative line spectrum are shown in Figs. 8-14, 8-15, and 8-16. The shape of the autocorrelation curve makes sense because the height of each stem is approximately 90% of the height of its neighbor on the left. The line spectrum shows signal power in the lower frequencies consistent with apparent low-frequency variation shown in Fig. 8-12.

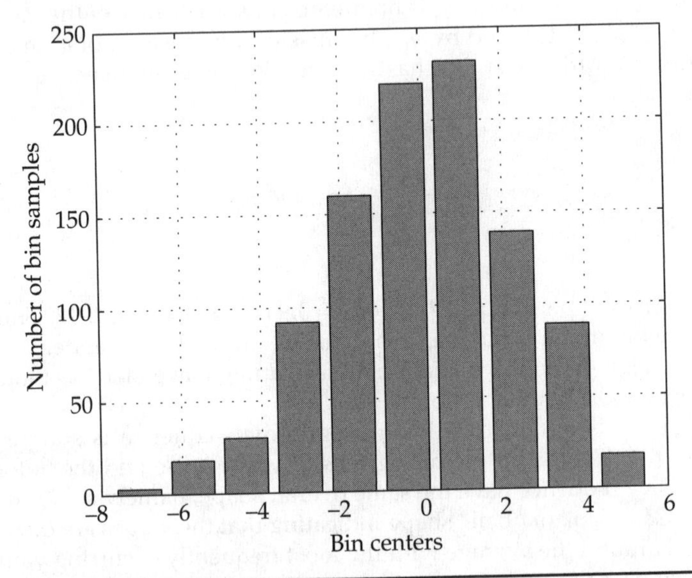

FIGURE 8-13 Histogram of an autoregressive sequence, $a = 0.9$.

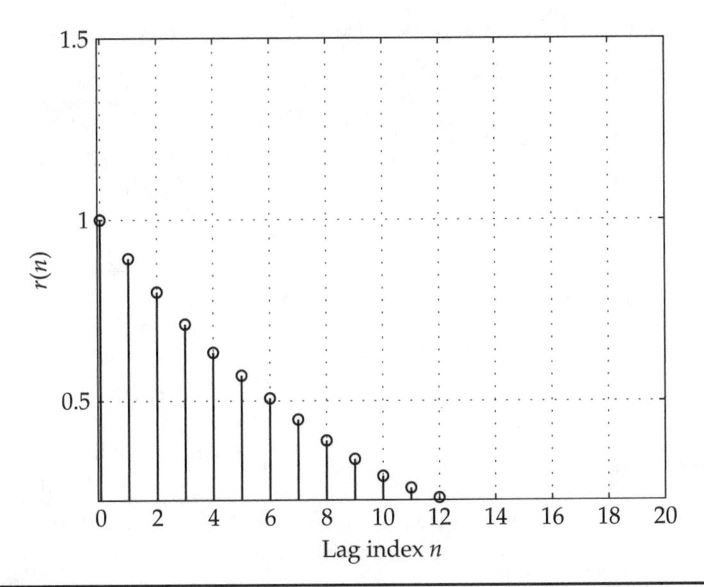

Figure 8-14 Autocorrelation of an autoregressive sequence, $a = 0.9$.

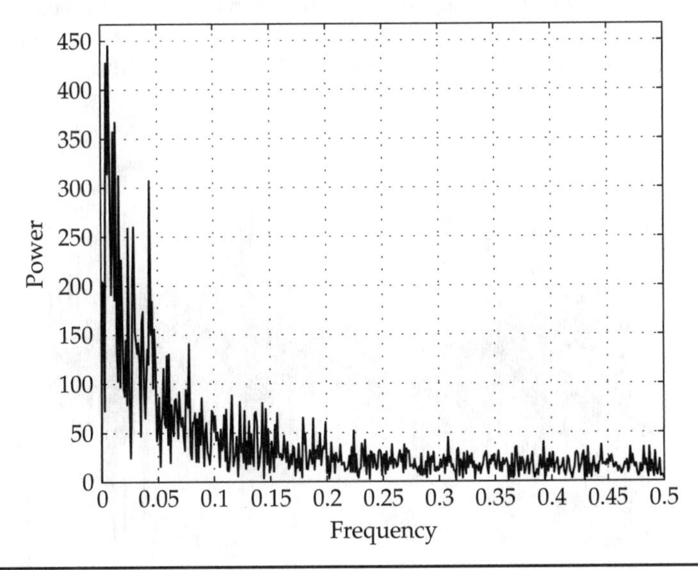

Figure 8-15 Line spectrum of an autoregressive sequence, $a = 0.9$.

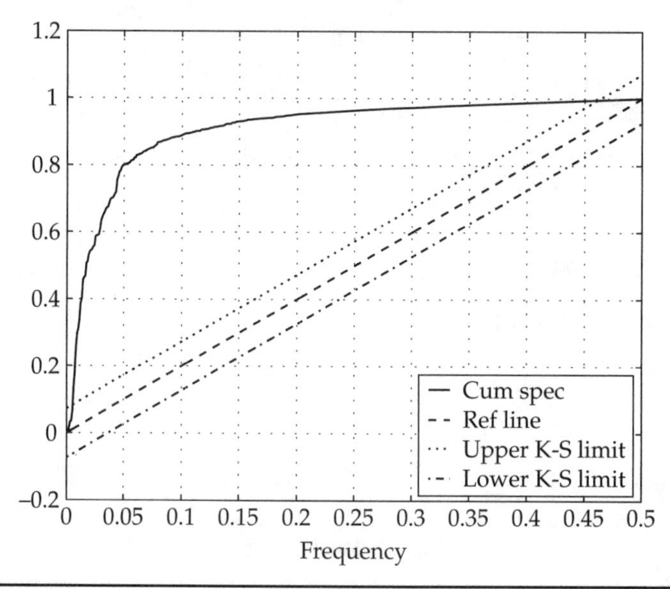

FIGURE 8-16 Cumulative line spectrum of an autoregressive sequence, $a = 0.9$.

8-3-2 Negatively Autoregressive Sequences

If the autoregressive parameter is changed to $a = -0.9$ an entirely different looking sequence results as is shown in Fig. 8-17. This sequence exhibits localized bursts of high-frequency variation rather than the low-frequency wandering shown in Fig. 8-12 for the positively

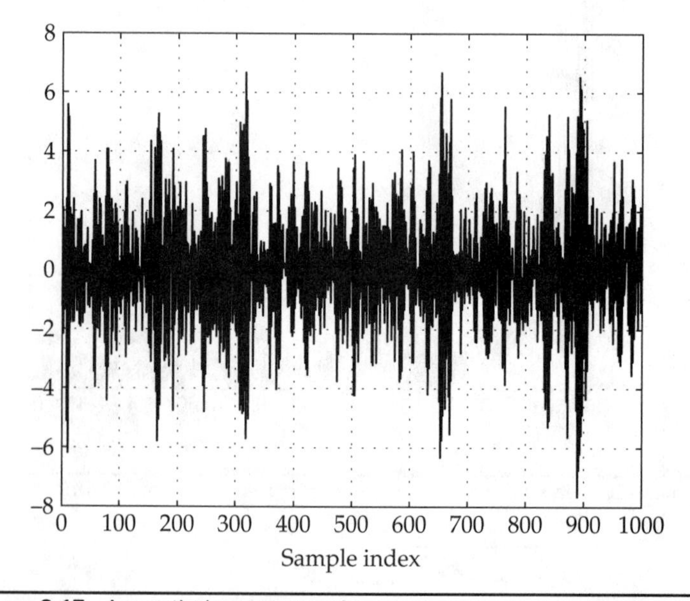

FIGURE 8-17 A negatively autoregressive sequence.

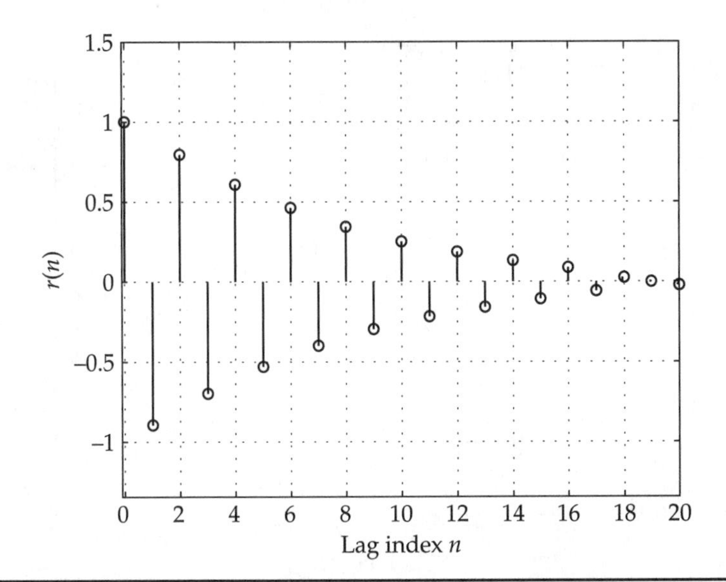

FIGURE 8-18 Autocorrelation of a negatively autoregressive sequence, $a = -0.9$.

autoregressive stochastic sequence. At first glance, it might not look much different than white noise, but the autocorrelation, line spectrum, and cumulative line spectrum, shown in Figs. 8-18, 8-19, and 8-20, point out the difference. Note the spectral power at the high frequencies in Fig. 8-19.

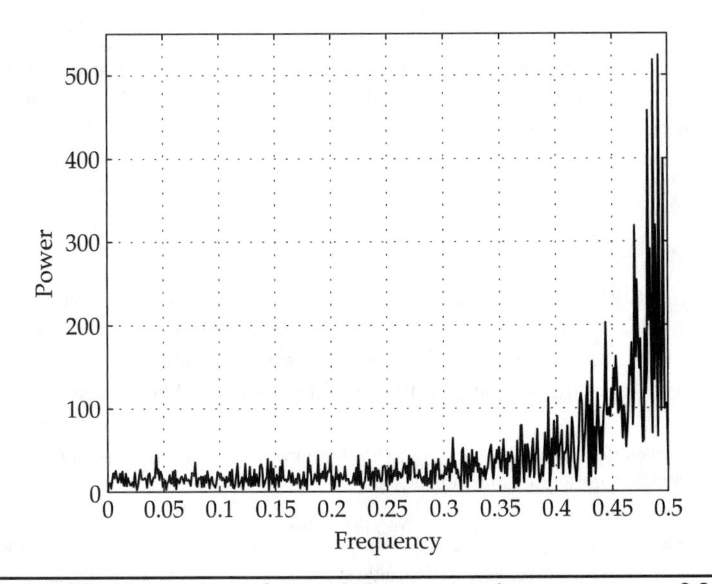

FIGURE 8-19 Line spectrum of a negatively autoregressive sequence, $a = -0.9$.

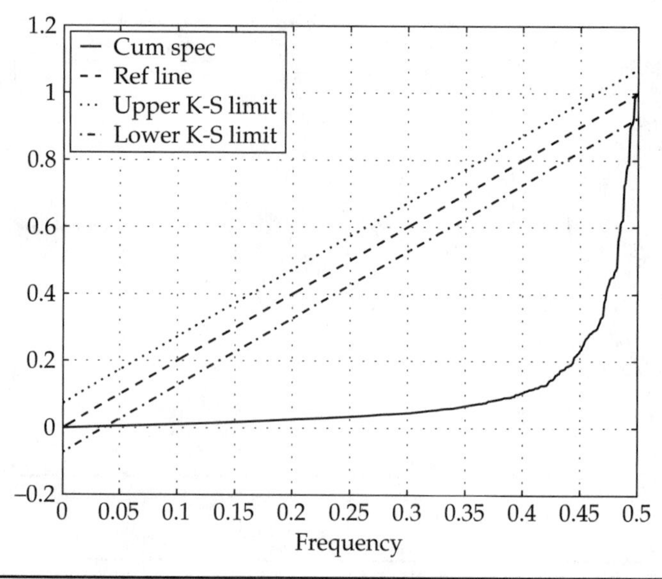

FIGURE 8-20 Cumulative line spectrum of a negatively autoregressive sequence, $a = -0.9$.

8-3-3 Moving Average Stochastic Sequences

The simplest moving average sequence is defined as

$$y_k = b_1 w_k + b_2 w_{k-1} \tag{8-6}$$

If $b_1 = b_2 = 0.5$ then Eq. (8-6) would be a common two-point moving average of a white noise stochastic sequence. In general, there can be M terms on the right-hand side as in

$$y_k = \sum_{m=1}^{M} b_m w_{k-m-1} \tag{8-7}$$

Figure 8-21 shows a two-point moving average stochastic sequence with $b_1 = b_2 = 0.5$. The autocorrelation of this sequence is shown in Fig. 8-22. Note that there is but one significant stem at lag one. In general, moving averages will have $M - 1$ stems of significance.

Question 8-1 Why does the autocorrelation of the two-point moving average have but one significant stem?

Answer The value of y_i, by definition, depends strongly on y_{i-1} but not on y_{i-2} or any other sample, hence the single stem.

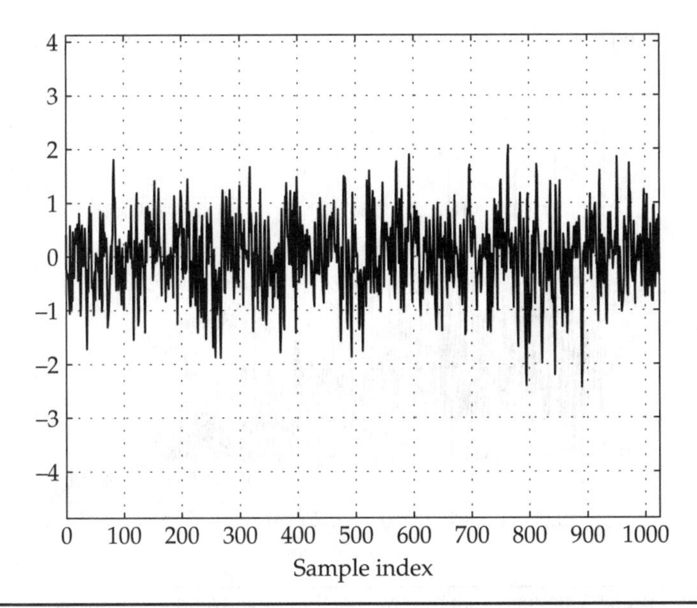

FIGURE 8-21 A two-point moving average stochastic sequence, $m = 2$.

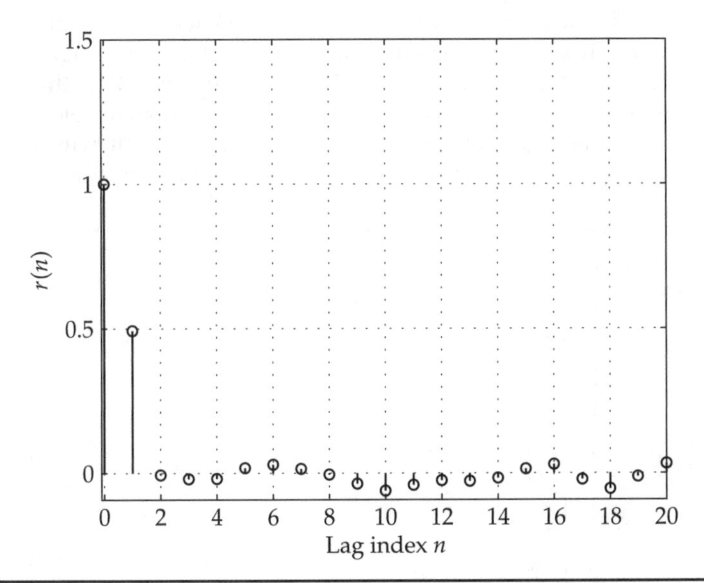

FIGURE 8-22 Autocorrelation of a two-point moving average stochastic sequence, $m = 2$.

The line spectrum of the moving average stochastic sequence is shown in Fig. 8-23. Here there is nothing of great distinction in the line spectrum so it appears that the autocorrelation is useful in revealing the presence of a moving average. However, note that when the input

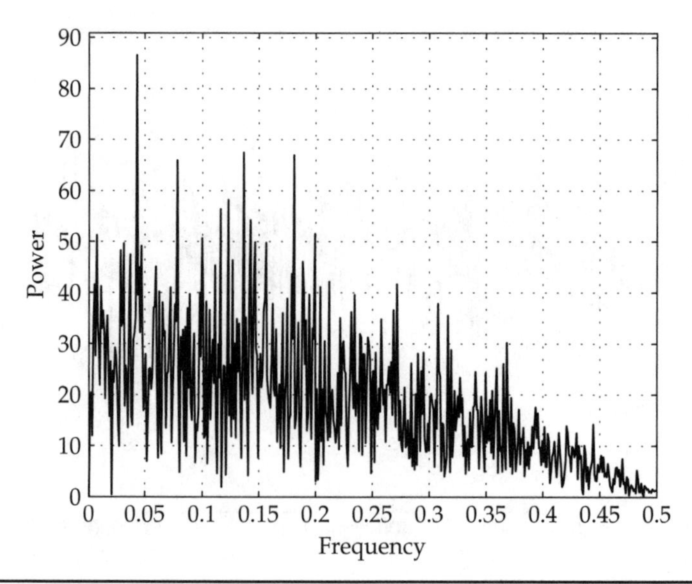

FIGURE 8-23 Line spectrum of a two-point moving average stochastic sequence, $m = 2$.

signal has a frequency of 0.5 (the folding frequency) the moving average output has zero power—more about this in the next paragraph.

Consider the case where $M = 3$ and $b_1 = b_2 = b_3 = 1/3$, that is, a common three-point moving average. The time domain plot looks about the same as previous stochastic sequences so it will not be shown. The autocorrelation in Fig. 8-24 has two significant stems

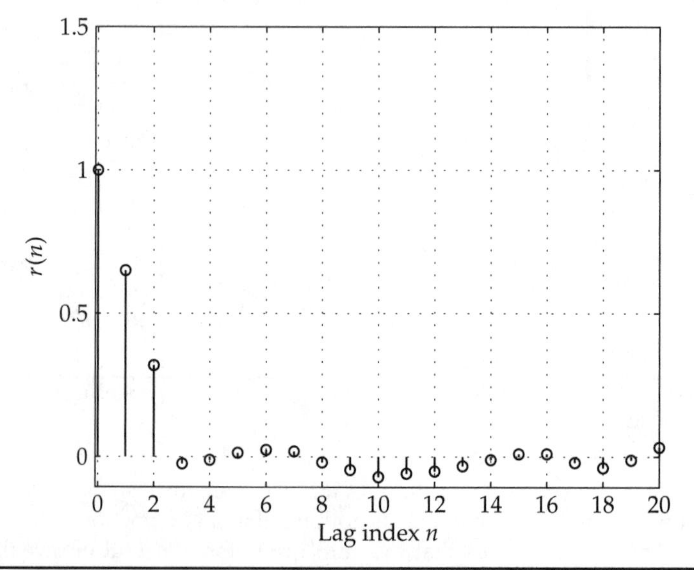

FIGURE 8-24 Autocorrelation of a three-point moving average stochastic sequence, $m = 3$.

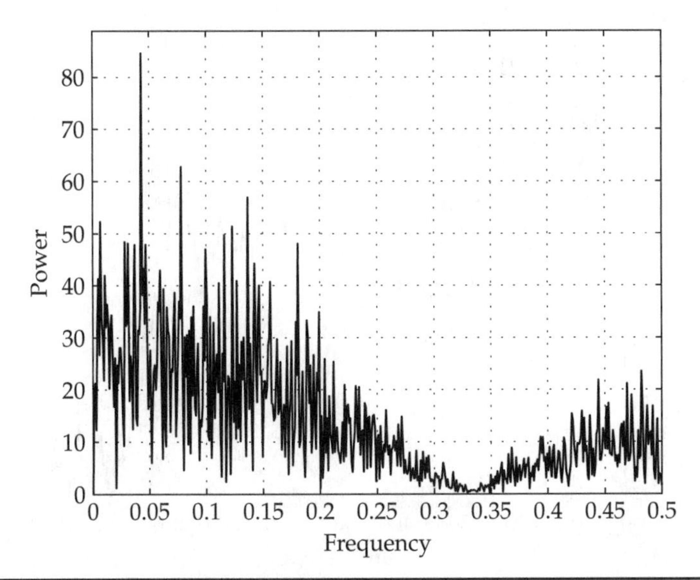

FIGURE 8-25 Line spectrum of a three-point moving average stochastic sequence, $m = 3$.

other than the one at a lag of zero. The line spectrum in Fig. 8-25 shows a valley centered at a frequency of 0.333 corresponding to a period of three samples. It makes sense that a sinusoid having a period of three samples would be completely obliterated by a moving average of three terms. For the first moving average sequence with $M = 2$, signals having a period of two samples would be obliterated and those signals would have a frequency of 0.5, which coincides with the folding frequency. Think about why this is the case.

This characteristic of obliterating a particular frequency tends to raise its ugly head in unexpected situations. Occasionally, I have come across line spectra of noisy processes sampled by a microprocessor that had strange steep valleys at certain frequencies. Subsequent detective work revealed that somewhere in the bowels of the microprocessor program that carried out the sampling there was some averaging. Therefore, one should be careful of doing anything to the raw data before carrying out a spectral analysis.

8-3-4 Unstable Nonstationary Stochastic Sequences

The variances of the stochastic sequences presented herein have always been finite. However, it is possible to have stochastic sequences that are not bounded. For example, consider a simple autoregressive stochastic sequence that has an autoregressive coefficient a where $|a| \geq 1$. Figure 8-26 shows a sequence that has $a = 1.002$. This sequence behaves as though its population mean were shifting in time. The behavior of the sequence suggests that it will continue to wander off with an unbounded variance. The autocorrelation becomes numerically

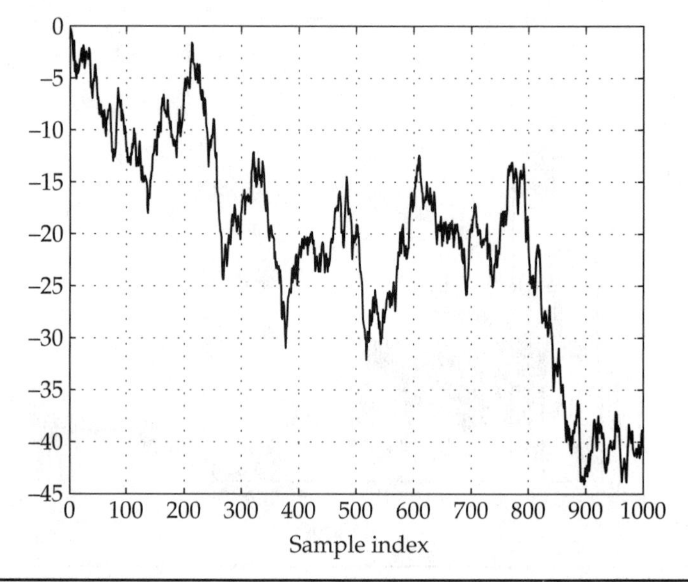

FIGURE 8-26 An unstable autoregressive sequence, $a = 1.002$.

unwieldy and the line spectrum in Fig. 8-27 shows excessive low-frequency power. In cases like these where the excessive low frequency may be washing out something in the data stream it behooves the analyst to *difference* the data and generate a new data stream from

$$y'_k = y_k - y_{k-1} \qquad k = 1, 2, \ldots, N$$

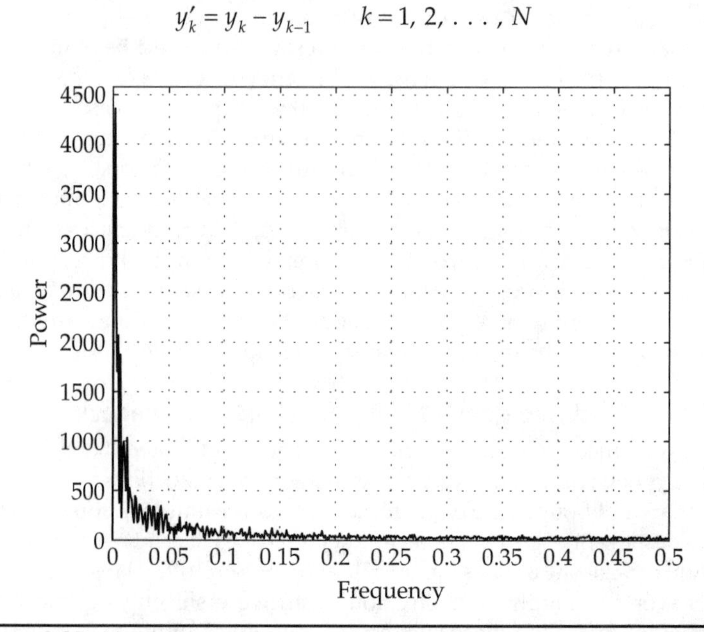

FIGURE 8-27 Line spectrum of an unstable autoregressive sequence, $a = 1.002$.

Differencing, which is the discrete time analog of differentiating, removes low-frequency power in the signal, making it a bounded signal that might yield to further analysis.

8-3-5 Multidimensional Stochastic Processes and the Covariance

Consider the two stochastic sequences shown in Fig. 8-28. One way of describing these two sequences is the covariance matrix, which contains the auto- and crosscorrelations of the sequences, as in

$$R_{xy}(i) = \begin{pmatrix} r_{xx}(i) & r_{xy}(i) \\ r_{yx}(i) & r_{yy}(i) \end{pmatrix} = \begin{pmatrix} \frac{1}{N}\sum_{k=1}^{N-i}(x_k - \bar{x})(x_{k+i} - \bar{x}) & \frac{1}{N}\sum_{k=1}^{N-i}(x_k - \bar{x})(y_{k+i} - \bar{y}) \\ \frac{1}{N}\sum_{k=1}^{N-i}(x_{k+i} - \bar{x})(y_k - \bar{y}) & \frac{1}{N}\sum_{k=1}^{N-i}(y_k - \bar{y})(y_{k+i} - \bar{y}) \end{pmatrix}$$

Later on, when dealing with the estimation of the state in Chap. 10, we will use $R_{xy}(0)$. For the time being, note that if the two sequences were interdependent then cross terms in the covariance matrix $r_{yx}(i)$ would be significant.

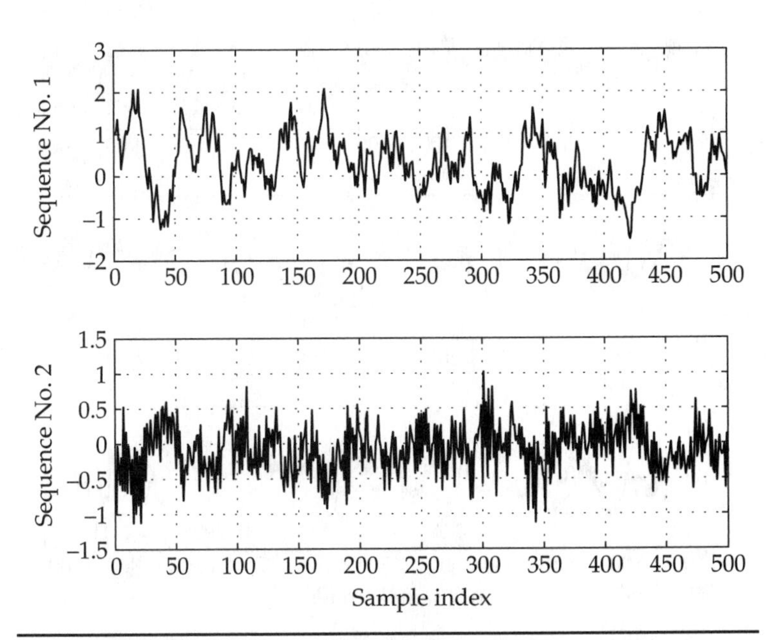

Figure 8-28 Two crosscorrelated stochastic sequences.

8-4 Populations, Realizations, Samples, Estimates, and Expected Values

8-4-1 Realizations

Each of the preceding examples is a finite sequence that is a *realization* of a population. For example, Fig. 8-29 shows three realizations of an autoregressive stochastic sequence that has an autoregressive parameter of 0.95 and a 0.0 mean. They all are described by

$$y_i = 0.95 y_{i-1} + w_i \qquad i = 1, 2, \ldots, N \tag{8-8}$$

Each is driven by a white noise sequence, $w_i, i = 1, \ldots, N$, whose average is approximately zero and whose variance is approximately unity. However, the three white noise sequences used to generate the sequences in Fig. 8-29 are three different realizations drawn from a population of white noise having zero mean and unity variance. After N samples have been realized, estimates of the population mean, called the average, and the population variance, called the sample variance, can be made. The averages of the three autoregressive sequences shown in Fig. 8-29 are –0.841, +0.957, and –0.548,

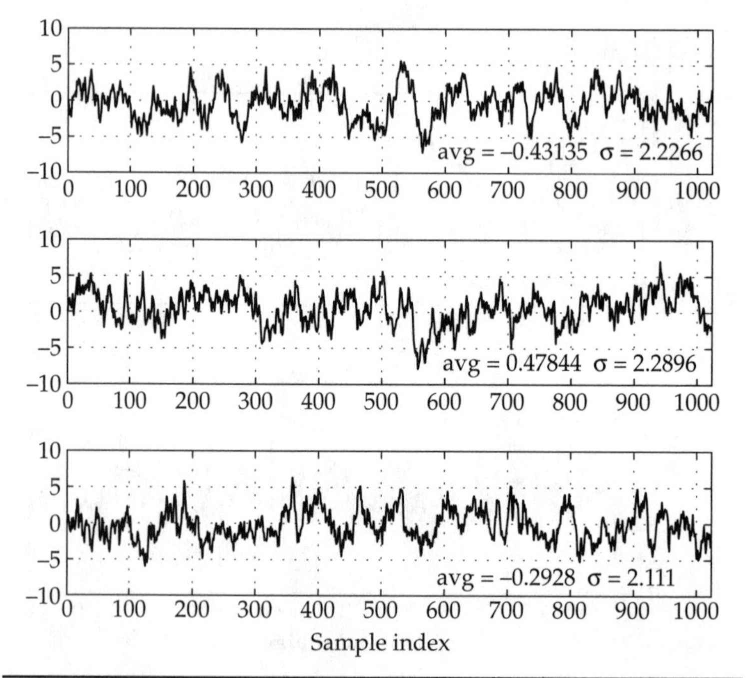

FIGURE 8-29 Three realizations of an autoregressive stochastic sequence.

while the population mean is zero. On the other hand, the three sample estimates of the variance are 8.96, 11.02, and 8.07, while the population variance is 10.26 (we will show how to arrive at this population parameter in the next section).

Implicitly, we have introduced the concept of a population as an infinite data stream from which finite length samples can be taken. These sample data streams of length N can be used to calculate estimates of the population parameters. This approach suggests that conceptually one could obtain, say, the population mean by averaging the infinite population.

At the other extreme, let's say you started at sample point number 500 in the first realization shown in Fig. 8-29 and averaged over samples 500, 501, and 502. Alternatively, one could have started at sample point number 500 in *each* of the three realizations in Fig. 8-29 and computed an average from those three values. That is, one could have averaged *over* the three realizations at one point in time. Hopefully, the reader will see the difference between these two approaches. Extending this concept, one could say that the population mean at sample point 500 is the average over *all* the realizations of the stochastic sequence sampled at point 500.

8-4-2 Expected Value

Following up on the latter approach, consider an infinite sequence of values, $T(t_1), T(t_2), \ldots, T(t_i), \ldots$ defining a stochastic sequence. At any sample point in time t_i there is a probability *density* $p(T_i, t_i)dT$ that gives the probability that T will have a value between $T(t_i)$ and $T(t_i) + dT$. This probability density is the same quantity that was introduced in Sec. 8-2-3. By definition, the integral of the probability density over all of the possible values of T, at t_i, would equal unity. That is,

$$\int_{-\infty}^{\infty} p(T, t_i) dT = 1 \tag{8-9}$$

This is just saying that the probabilities of each of the possible values of T at sample point t_i have to add up to unity. The integration is done over all realizations at time t_i, *not* over all time. Since $T(t_i)$ varies continuously, we can use an integral rather than a sum.

The *expected value* of T at sample point t_i would be the weighted integral of all the possible values of $T(t_i)$ at the sample point t_i, as in

$$\int_{-\infty}^{\infty} p(T, t_i) T(t_i) dT = E\{T, i\} = \mu_T(i) \tag{8-10}$$

In a crude sense, Eq. (8-10) is a weighted sum of all the possible temperatures at time t_i with the weight being the probability density.

The expected value is also the population mean, sometimes symbolized by μ.

Compare this with the average of T over one realization of length N, as in

$$\bar{T} = \frac{1}{N}\sum_{k=1}^{N} T_k \tag{8-11}$$

The average is taken over N discrete samples of one realization while the expected value is taken over *all* of the realizations at one particular point in time.

8-4-3 Ergodicity and Stationarity

If the population parameters such as the mean can be determined by taking the average of a data stream of infinite extent then the population is said to be *ergodic*. This assumes that, as one passes through all of the infinite number of values in the data stream, one will also pass over all the possible values in all of the realizations. This word *ergodic* will be of absolutely no use to you or your control engineer associates but it does have name-dropping value.

Another more useful term is *stationarity*. If the population parameters do not change with time and the population is bounded in value, the population is considered *stationary*. In this case we could suppress the appearance of t_i in the probability density $p(T(t_i),t_i)$.

8-4-4 Applying the Expectation Operator

Extending the idea of the expected value, one can define the population variance at the instant t_i as the expected value of deviations of T from its mean, as in

$$\sigma_T^2(i) = E\{[T_i - \mu_T(i)]^2\} = \int_{-\infty}^{\infty} p(T,t_i)[T_i - \mu_T(i)]^2 dT \tag{8-12}$$

Compare this with the sample estimate of the variance

$$V_T = \frac{1}{N}\sum_{k=1}^{N}(T_i - \bar{T})^2 \tag{8-13}$$

The autocorrelation function for a population is defined as

$$R_T(n) = E\{(T_i - \mu_T)(T_{i+n} - \mu_T)\} = \int_{-\infty}^{\infty} p(T,t_i)(T_i - \mu_T)(T_{i+n} - \mu_T)dT \tag{8-14}$$

which can be compare to the sample estimate of

$$r_T(n) = \frac{1}{(N-n)\,V_w} \sum_{k=1}^{N-n} (T_k - \bar{T})(T_{k+n} - \bar{T})$$

We can use the expectation operator $E\{\ldots\}$ to calculate the means, population variances, and other values. For example, the autoregressive sequence mentioned in Eq. (8-8) has a population mean determined by applying the expected value operator as follows:

$$\begin{aligned} y_k &= a y_{k-1} + w_i \qquad i = 1, 2, \ldots \\ \mu_Y &= E\{y_k\} = aE\{y_{k-1}\} + E\{w_i\} \end{aligned} \tag{8-15}$$

The expected value of y_k should be the same as that of y_{k-1} if the population is stationary (which we assume it to be) and the expected value of the white noise sequence w_i is zero. Therefore, Eq. (8-15) can be continued as follows

$$\begin{aligned} \mu_Y &= a\mu_Y + 0 \\ \mu_Y &= 0 \end{aligned} \tag{8-16}$$

The variance of that stochastic sequence can be calculated by applying the expectation operator to the square of the sequence as follows.

$$\begin{aligned} V_y &= E\{y_k^2\} = E\{a^2 y_{k-1}^2 + 2ay_{k-1}w_k + w_k^2\} \\ &= E\{a^2 y_{k-1}^2\} + E\{2ay_{k-1}w_k\} + E\{w_k^2\} \end{aligned}$$

The white noise sequence w_k, by definition, is not correlated with y_{k-1}, so $E\{y_{k-1}w_k\} = 0$. Furthermore, $E\{y_k^2\} = E\{y_{k-1}^2\} = \sigma_y^2$ because of stationarity and $E\{w_k^2\} = \sigma_w^2$ by definition, so

$$\begin{aligned} \sigma_y^2 &= a^2\sigma_y^2 + \sigma_w^2 \\ &= \frac{\sigma_w^2}{1-a^2} \end{aligned} \tag{8-17}$$

Question 8-2 Why is w_k not correlated with y_{k-1}?

Answer y_{k-1} occurs at sample time t_{k-1} and w_k, which occurs at time t_k, is independent of, or not correlated with, anything that took place earlier. Therefore, $E\{y_{k-1}w_k\} = 0$.

Question 8-3 What is the variance of the moving average stochastic sequence?

Answer Apply the expectation operator to the square of Eq. (8-7):

$$E\{y_k^2\} = E\left\{\left(\sum_{m=1}^{M} b_m w_{k-m-1}\right)^2\right\}$$

Since white noise is not autocorrelated all of the cross terms, as in $E\{w_k w_j\}$ will be zero, that leaves

$$\sigma_y^2 = E\{y_k^2\} = \sum_{m=1}^{M} (b_m)^2 \sigma_w^2 = \sigma_w^2 \sum_{m=1}^{M} (b_m)^2$$

For the case where it is just a conventional moving average, $b_m = 1/M$ and

$$\sigma_y^2 = \frac{\sigma_w^2}{M} \quad \text{or} \quad \sigma_y = \frac{\sigma_w}{\sqrt{M}} \tag{8-18}$$

which is a formula dear to the hearts of many statisticians.

8-5 Comments on Stochastic Disturbances and Difficulty of Control

8-5-1 White Noise

How well can a process be controlled when it is subject to white noise only? This is an interesting question because many statisticians will immediately throw up their hands and make some condescending comment suggesting that control engineers should keep their hands off processes subject to white noise because any attempts to control such a process only causes troubles.

Part of that answer, minus the condescension, is correct, at least in my opinion. Consider the case where you are the controller and you observe samples of the process output whose average has been satisfactorily close to set point and that suffers only from white noise disturbances. Should you make an adjustment to the control output upon observing a sample of the process output that is not on set point? If the *average* of the process output is indeed nearly at the set point then any deviation, if it is really white or unautocorrelated, will be completely independent of the previous value of the control output and it will have no impact on subsequent disturbances. Therefore, if you should react to such a deviation, you would be wasting your time because the next observation will contain another deviation that has nothing to do with the previous deviation on which you acted. You, in fact, may make things worse.

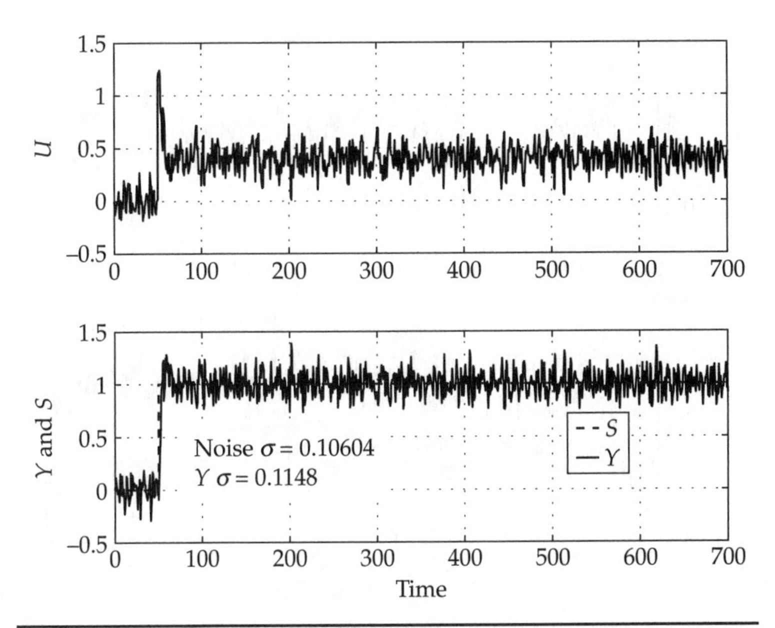

FIGURE 8-30 Controlling in the face of white noise.

But what happens when you want to change the set point? Consider Fig. 8-30 where the process output is subject to white noise (whose standard deviation is 0.103), the process is first-order with a time constant of 10.0 time units and a PI controller, conservatively tuned, is active. Note the activity of the control output as the noise on the process output feeds through the controller. At 50 time units, the set point is stepped and the controller satisfactorily drives the average value of the process output to the new set point. However, the standard deviation of the process output about the set point is 0.115. So, the controller has amplified the noise. Ha! The statistician is smirking. A feedback controller cannot decrease the standard deviation of the white noise riding on the process output. At best it can keep the average on set point. The catch is that in most industrial situations one needs the controller actively watching and controlling the process in case there are set-point changes and in case some non–white noise disturbance appears. To quote a famous control engineer, "life is not white noise." This subject will be revisited in Chap. 11.

8-5-2 Colored Noise

Now, let's make the noise riding on the process output autocorrelated with an autoregressive coefficient of $a = 0.95$. This noise wanders around its nominal average value of zero. Because of the autoregressive nature, it occasionally spends some time near its last deviation. That is, each sample does indeed depend somewhat on the

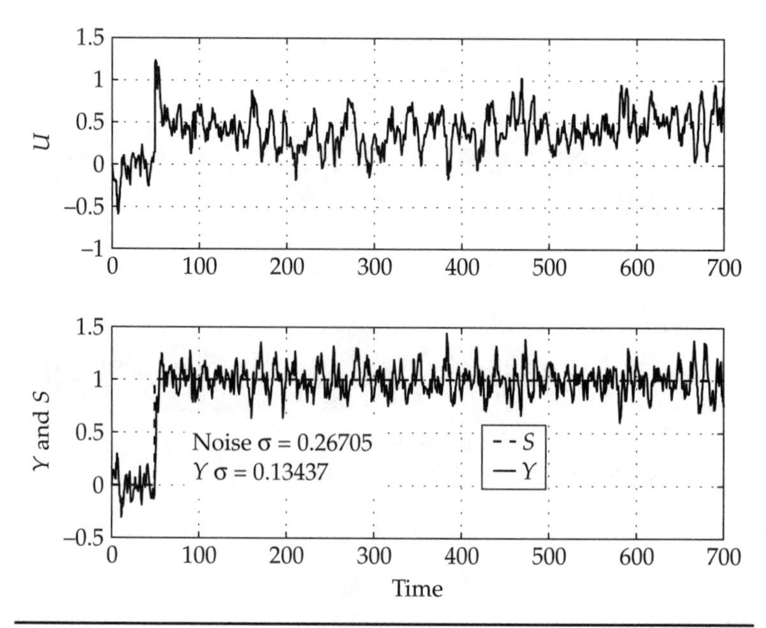

FIGURE 8-31 Controlling in the face of autoregressive noise ($a = 0.95$).

previous sample. This suggests that it might be beneficial to go after the deviation because it has some persistence. Figure 8-31 shows what happens in this case. Here the autocorrelated noise riding on the process output has a standard deviation of 0.267 but the standard deviation of the controlled process output is 0.134. The controller has actually lowered the noise level. Note the different nature of the controller output as it moves around to deal with noise that is not unautocorrelated (double negatives, sorry). The reader may see a trend here.

Consider the next situation where the autoregressive parameter is $a = 1.001$ in Fig. 8-32. Now we have a nonstationary noise sequence riding on the process output. If left uncontrolled, the process output would wander off into the boonies. The same controller, however, turns the noise standard deviation of 0.68 into a noise standard deviation of 0.13. Note how much movement the control output has to make to keep the process output on set point. By the way, the value of 0.68 comes from the autocorrelated noise added to the process output before it is fed to the controller during the period of simulation. In fact, the population standard deviation of the added noise is unbounded.

Figure 8-33 shows the error transmission curve for the first-order process with PI. Note that only power in the frequency range from 0.0 to 0.05 Hz is attenuated. Fortunately, that is where the nonstationary power of the noise resides, as was shown in Fig. 8-27.

As mentioned previously, we will touch on this subject again in Chap. 11 when we do the summing up.

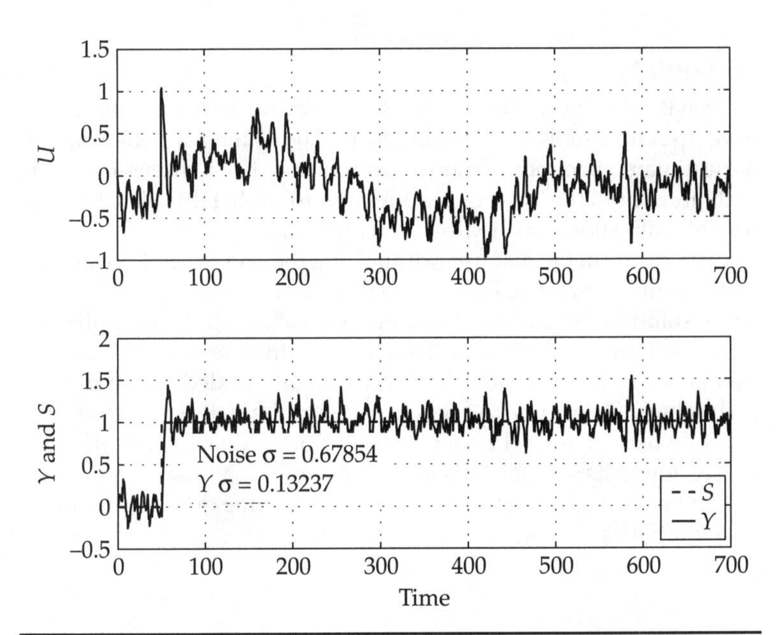

FIGURE 8-32 Controlling in the face of nonstationary noise ($a = 1.001$).

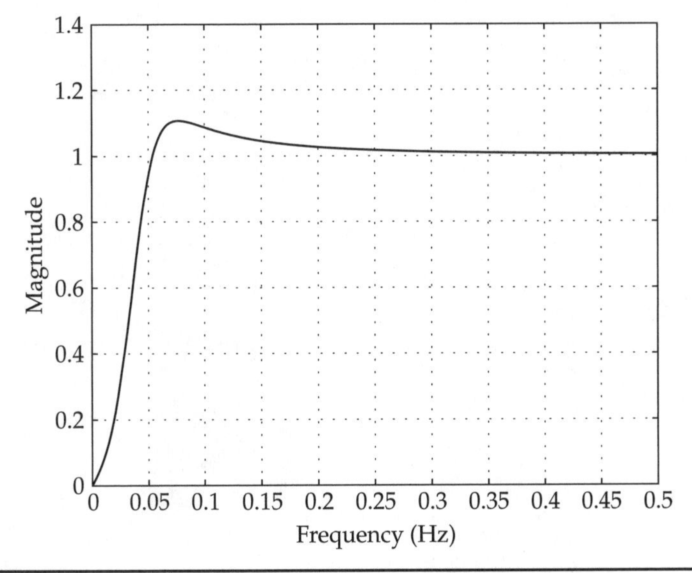

FIGURE 8-33 Error transmission curve for first-order process with PI.

8-6 Summary

We have introduced several kinds of stochastic disturbances and have presented tools for detecting them. The difference between sample estimates and population estimates has been discussed along with an admittedly painful introduction to probability density, concept of realizations, and expectation operator.

The nature of the disturbance affecting the process is often ignored by the control engineer. Depending on its nature there may be a variety of solutions, some not including feedback control. When the disturbance is primarily white, feedback controllers cannot lower its standard deviation. Controllers still may be needed to handle setpoint changes and unexpected changes in the nature of the disturbance. When the disturbance takes on color, feedback controllers can indeed lower its standard deviation.

The next chapter will stay in the discrete time domain and introduce the Z-transform.

CHAPTER 9

The Discrete Time Domain and the Z-Transform

The discrete time domain is important because (1) most data collected during a process analysis consists of samples at points in time separated by a constant interval, (2) most custom control algorithms are implemented digitally, and (3) concepts like white noise and the delta function for pulses are physically realizable in this domain. (Remember how the Dirac delta function in the continuous time domain had no specific shape and had to be defined in terms of an integral.) In the previous chapter stochastic processes defined in the discrete time domain were introduced. Here, several familiar continuous time model equations will be discretized and the Z-transform will be introduced. Just as the Laplace transform aided and abetted our attempts to solve problems and gain insight in the continuous time domain, the Z-transform will be used in the discrete time domain.

There are a couple of ways to introduce the Z-transform: (1) using the backshift operator in a manner similar to using the Laplace operator s to replace derivatives, or (2) deriving the Z-transform from the Laplace transform of a sampled time function. The latter approach is quite elegant and more general but I think it is best placed in App. I. Therefore, in Secs. 9.1 and 9.2 the backshift operator approach will naturally fall out of the discretization of the first-order model. With the new tool in hand, several other models, algorithms, and filters will be recast and studied in the Z-transform domain. As with the Laplace transform, there will be a transition to the frequency domain where more insight will be gained. The chapter closes with a discussion of fitting discrete time domain data to models.

This will perhaps be the longest chapter in the book, so you might want to break your reading plan into four parts. In the first part, you will learn about the Z-transform. In the second part, you will see how several unconventional control algorithms can be designed using both the Z-transform and the Laplace transform. In the third part, we will move from the Z-transform domain to the frequency domain. The fourth part deals with the discrete time domain data-fitting problem. Take your time.

9-1 Discretizing the First-Order Model

This and the following three sections are busy. The first-order process model is studied for a special case where the process input is a series of steps. The describing equation will be modified slightly when time is discretized. The result will be rewritten using the backshift operator which will lead to the Z-transform. This necessitates a discussion of sampling and holding. The discretized unity-gain first-order model is then reinterpreted as a discrete time filter.

Back in Chap. 3 we presented the first-order model in the continuous time domain via the differential equation

$$\tau\frac{dy}{dt}+y=gU(t) \tag{9-1}$$

In the Laplace domain we wrote the transfer function between process input and output as

$$\frac{\tilde{Y}(s)}{\tilde{U}(s)}=G_p(s)=\frac{g}{\tau s+1} \tag{9-2}$$

When Eq. (9-1) or (9-2) was solved for the case where the process input U is a step change we obtained

$$y(t)=y_0e^{-\frac{t}{\tau}}+gU_c\left(1-e^{-\frac{t}{\tau}}\right) \tag{9-3}$$

Before moving to the discrete time domain, let's apply Eq. (9-3) for one time increment of size h over which $U(t)$ is held constant at U_0.

$$y(h)=y_0e^{-\frac{h}{\tau}}+gU_0\left(1-e^{-\frac{h}{\tau}}\right) \tag{9-4}$$

Equation (9-4) moves information at $t=0$, namely y_0 and U_0, to $t=h$ to produce $y(h)$. The information at $t=h$ can be moved to $t=2h$ by reapplying Eq. (9-4) suitably modified, as in

$$y(2h)=y(h)e^{-\frac{h}{\tau}}+gU_1\left(1-e^{-\frac{h}{\tau}}\right) \tag{9-5}$$

Note that the value of $U(t)$ over the interval $h \le t < 2h$ is held constant at U_1 which may be different from U_0. We have effectively moved to the discrete time domain by breaking the time variable up into samples spaced apart by h sec. This means that time is described by

$$t \Rightarrow t_i = hi \qquad i = 0, 1, 2, \ldots$$

The process output becomes

$$y(t) \Rightarrow y_i \qquad i = 0, 1, 2, \ldots$$

The process input becomes

$$U(t) \Rightarrow U_i \qquad i = 0, 1, 2, \ldots$$

With this in mind, Eqs. (9-4) or (9-5) can be written as

$$y_i = y_{i-1} e^{-\frac{h}{\tau}} + gU_{i-1}\left(1 - e^{-\frac{h}{\tau}}\right) \qquad i = 0, 1, 2, \ldots \tag{9-6}$$

Spend a few moments thinking about Eq. (9-6). It is the same as Eq. (9-4) except that it is applied over the time interval from t_{i-1} to t_i during which U_{i-1} is held constant. Figure 9-1 shows how a first-order process with a time constant of 5.0 sec and a gain of 1.1 responds to a

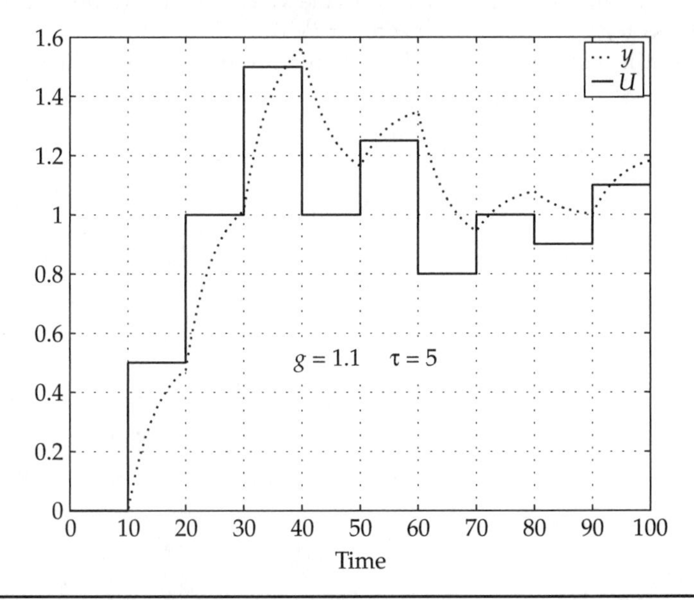

Figure 9-1 Response of first-order model to a series of steps in the process input.

series of steps in U lasting 10 sec ($h = 10$) during which U is held constant. Eq. (9-6) gives the values of the process output at the end of each time period over which U was held constant. It provides no information for $y(t)$ during $t_i < t < t_{i+1}$.

The stepwise behavior of $U(t)$ could be the result of a digital-to-analog converter that is activated every 10 sec.

9-2 Moving to the Z-Domain via the Backshift Operator

The quantity y in Eq. (9-6) can be converted to its Z-transform counterpart $\hat{y}(z)$ by introducing the backshift operator z^{-1} defined as follows

$$y(t-h) = z^{-1}\, y(t) \tag{9-7}$$

where h is the sampling interval. I have not added the "hat" to the variable $y(t)$ because Eq. (9-7) is in a kind of limbo between the time and the z domains. It is perhaps better to proceed with the actual definition of the Z-transform which is

$$\hat{y}(z) = Z\{y(t)\} = \sum_{k=0}^{\infty} z^{-k} y(k) \tag{9-8}$$

Note that this is a weighted sum over all of the sampled values of $y(t)$ as compared to the Laplace transform which is a weighted integral over all the continuously variable values of $y(t)$. In both cases, values of $y(t)$ are considered zero for $t < 0$.

If the variable $y(t)$ is shifted in time one sample, Eq. (9-8) would be

$$\begin{aligned} Z\{y(t-1)\} &= \sum_{k=0}^{\infty} z^{-k} y(k-1) \\ &= \sum_{p=-1}^{\infty} z^{-(p+1)} y(p) \\ &= z^{-1} \sum_{p=-1}^{\infty} z^{-p} y(p) \\ &= z^{-1} \sum_{p=0}^{\infty} z^{-p} y(p) \\ &= z^{-1} Z\{y(t)\} \\ &= z^{-1} \hat{y}(z) \end{aligned} \tag{9-9}$$

The summation index in the third line of Eq. (9-9) is changed to zero because, like the Laplace transform, $y(t)$ is assumed to be zero for $t < 0$. The manipulations in Eq. (9-9) should convince the reader that z^{-1} is a backshift operator.
With this in mind, Eq. (9-6) becomes

$$
\begin{aligned}
y_i &= y_{i-1}e^{-\frac{h}{\tau}} + gU_{i-1}\left(1-e^{-\frac{h}{\tau}}\right) \qquad i = 0, 1, 2, \ldots \\
\hat{y}(z) &= e^{-\frac{h}{\tau}}z^{-1}\hat{y}(z) + g\left(1-e^{-\frac{h}{\tau}}\right)z^{-1}\hat{U}(z)
\end{aligned}
\tag{9-10}
$$

As with the Laplace transform, Eq. (9-10) is algebraic and can be solved for $\hat{y}$ as in

$$
\hat{y}(z) = \frac{g\left(1-e^{-\frac{h}{\tau}}\right)z^{-1}}{1-e^{-\frac{h}{\tau}}z^{-1}}\hat{U}(z) \tag{9-11}
$$

or

$$
\hat{y}(z) = G(z)\hat{U}(z) \tag{9-12}
$$

where $G(z)$ is the transfer function in the Z-domain for the first-order process model. For this multiplication in Eq. (9-12) to be valid, the time domain variable $U(t)$ has to behave as in Fig. 9-1.

9-3 Sampling and Zero-Holding

In Eq. (9-10), $U(t)$ is a series of steps, as if it were the output of a digital/analog (D/A) on a microprocessor. Alternatively, and more elegantly, one can say that $U(t)$ has been put through a sampler and a *zero-order hold device* which samples the value of $U(t)$ at time t_i and holds it for a period of h sec. The device releases it at time t_{i+1} at which time the device samples the new value of $U(t)$ and holds it. There is no zero-order hold device associated with $y(t)$ so it is considered as a sequence of isolated sampled values that exist only at time t_i, $i = 0, 1, 2, \ldots$. The sampled variable $y(t)$ could also be considered as a train of spikes with the height of each spike equal to the value of $y(t_i)$ as depicted in Fig. 9-2. Figure 9-3 shows how the zero-order hold device is introduced. Note the samplers that act on $U(t)$ and on $Y(t)$.

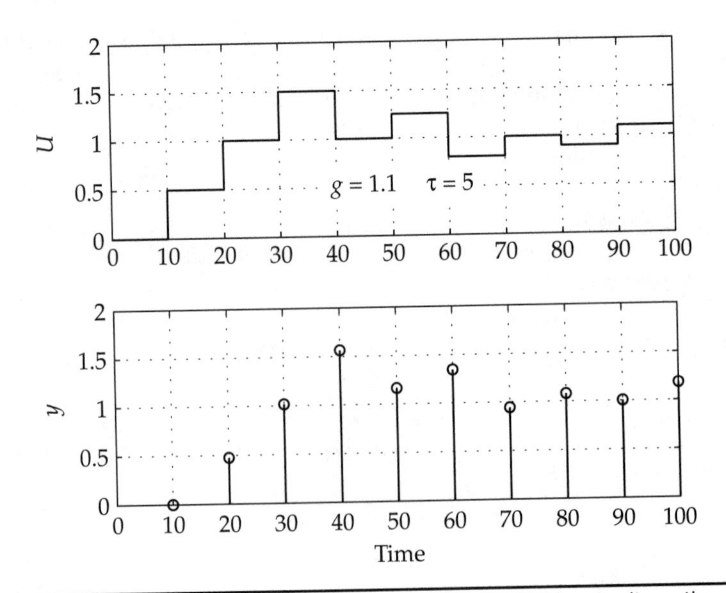

FIGURE 9-2 Response to a series of steps in the process input, alternative view.

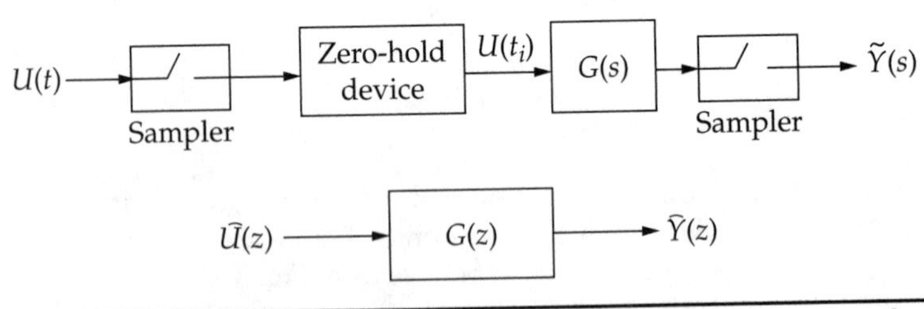

FIGURE 9-3 The sampler zero-hold device as part of the Z-transform transfer function.

In App. I, the zero-order hold is studied in detail and it is shown that the transfer function in Eq. (9-11) can be written as

$$\hat{Y}(z) = \frac{g\left(1-e^{-\frac{h}{\tau}}\right)z^{-1}}{1-e^{-\frac{h}{\tau}}z^{-1}}\hat{U}(z) = Z\{\Pi_h(s)G(s)\}\hat{U}(z) \tag{9-13}$$

where

$$\Pi_h(s) = \frac{1-e^{-sh}}{s} \qquad G(s) = \frac{g}{\tau s+1}$$

That is, the Z-transform transfer function can be obtained from the Laplace transform of the zero-order hold applied to the Laplace transform transfer function for the first-order process. Equation (9-13) represents a great leap of faith and unless the reader is familiar with the Z-transform she should spend some time reading App. I where the origin of $\Pi_h(s)$ is explained. In this chapter, we have taken the path that moves directly from the discrete time domain to the Z-transform domain via the backshift operator. Appendix I goes through the analysis of the sampling and holding functions via the Laplace transform. For our needs, either approach is sufficient but it will greatly help the reader to study both. Therefore, the Z-transform transfer function relating the sampled and zero-held process input to the sampled process output is

$$G(z) = \frac{g\left(1 - e^{-\frac{h}{\tau}}\right)z^{-1}}{1 - e^{-\frac{h}{\tau}}z^{-1}}$$

This transfer function can come from the discrete time domain with a stepped process input augmented by the backshift operator or from the Laplace transform domain where transfer functions in s are converted to Z-transforms in z.

If $y(t)$ turns out to be the input to another stage as in Fig. 9-4, then the Z-transform cannot be used without some consideration. Note that no sampler is applied to $y(t)$ before it becomes an input to the box represented by $H(s)$.

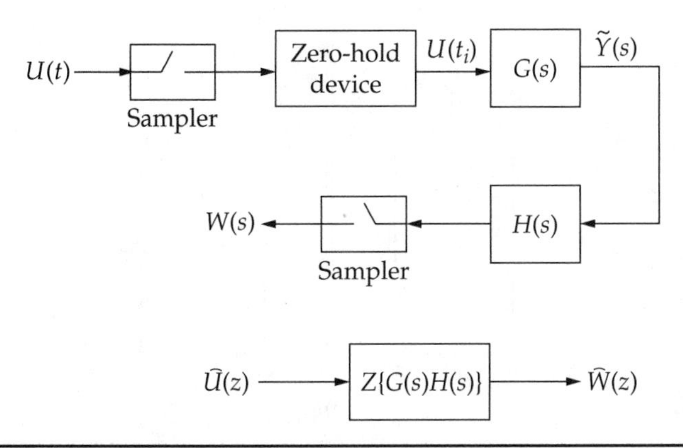

FIGURE 9-4 The zero-hold device as part of the Z-transform.

The correct expression to yield $\widehat{W}(z)$ would be

$$\widehat{W}(z) = Z\{\Pi_h(s)G(s)H(s)\}\widehat{U}(z) \tag{9-14}$$

However, if a sampling device were to be placed between stages $G(s)$ and $H(s)$ then the expression for $\widehat{W}(z)$ would be

$$\begin{aligned}\widehat{W}(z) &= Z\{\Pi_h(s)G(s)\}\,Z\{H(s)\}Z\{\tilde{U}(s)\} \\ &= G(z)H(z)\widehat{U}(z)\end{aligned} \tag{9-15}$$

To complete the concept of sampling and holding, consider the case where there is a sampler applied to the input but there is no hold as is shown in Fig. 9-5. Now the process is responding to a train of spikes and the response in Fig. 9-6 is quite different from that in Fig. 9-1. Note the first-order response to each of the U spikes (or un-zero-held samples).

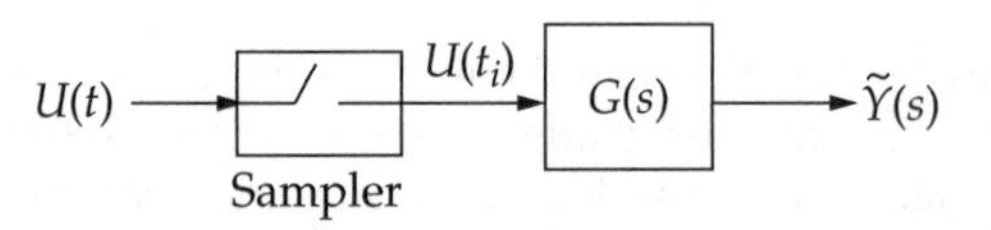

Figure 9-5 Removing the zero-hold device.

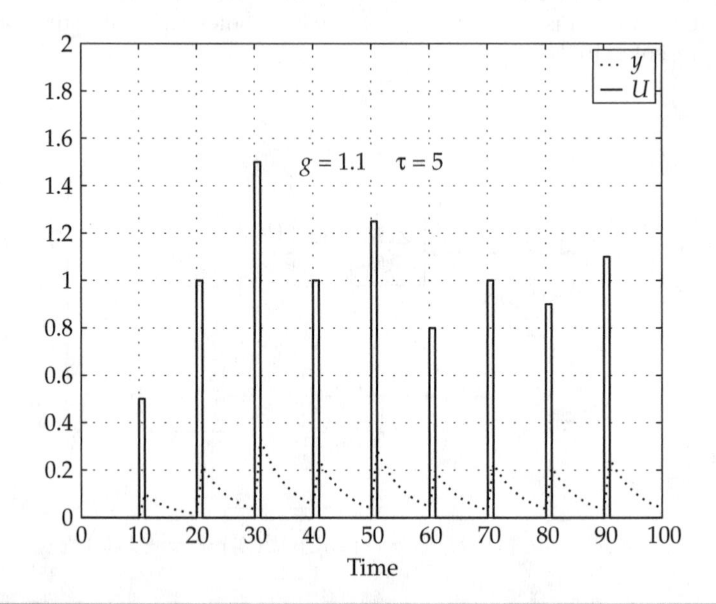

Figure 9-6 Response to a series of spikes in the process input.

9-4 Recognizing the First-Order Model as a Discrete Time Filter

If in Eq. (9-6) we set the gain to unity, then

$$y_i = y_{i-1}e^{-\frac{h}{\tau}} + U_{i-1}\left(1 - e^{-\frac{h}{\tau}}\right) \qquad i = 0, 1, 2, \ldots$$

can be written as

$$y_i = ay_{i-1} + bU_{i-1} \qquad i = 0, 1, 2, \ldots$$
$$a = e^{-\frac{h}{\tau}} \qquad \tau = -\frac{h}{\ln a} \qquad b = 1 - a \tag{9-16}$$

Equation (9-16) is the widely used unity-gain discrete first-order filter with a time constant τ (also sometimes called the "exponential filter").

For the special case where the input U is white noise, one can apply the expected value operator to the square of Eq. (9-16) and obtain the variance reducing property of the discrete first-order filter, as in

$$E\{y^2\} = V_y = E\{(ay_i + bU_{i-1})^2\}$$
$$= E\{a^2y_i^2 + 2aby_{i-1}U_{i-1} + b^2U_{i-1}^2\}$$
$$= a^2V_y + b^2V_U$$

$$V_y = \frac{b^2}{1-a^2}V_U = \frac{(1-a)^2}{(1-a)(1+a)}V_U = \frac{1-a}{1+a}V_U \tag{9-17}$$

Note that the cross-term in Eq. (9-17), $2abE\{y_{i-1}U_{i-1}\}$, disappears because y_{i-1} and U_{i-1} are not cross-correlated. From the definition of the parameter a in Eq. (9-16), it is clear that as the time constant is increased, the variance gain, $b^2/1-a^2$, is decreased. In Sec. 9-12, after we have moved from the discrete time domain to the frequency domain, the increased lag of the filter output will become apparent. Alternatively, one could look as the first-order process model as a filter acting on the process input U. The engineer can use Eq. (9-16) as a filter to reduce noise. The amount of noise reduction, if the noise is white, is given by Eq. (9-17). As the parameter a increases (remember it has to be < 1), the filter time constant τ increases, as does the noise reduction. The increase in the filter time constant may have adverse effects if the filter resides in a feedback loop. Therefore, there is a balance between the benefits of noise reduction and the disadvantage of increased lag.

9-5 Descretizing the FOWDT Model

In Chap. 4, the FOWDT model was presented as

$$\tau\frac{dy}{dt}+y=gU(t-D) \tag{9-18}$$

In the Laplace domain, the FOWDT model transfer function was written as

$$\frac{\tilde{y}(s)}{\tilde{U}(s)}=G_p(s)=e^{-sD}\frac{g}{\tau s+1} \tag{9-19}$$

In Chap. 4, we rushed to the frequency domain without solving Eq. (9-18) for a constant process input. To obtain the solution in the discrete time domain, adjust the indices of U, as in

$$\begin{aligned} y_i &= y_{i-1}e^{-\frac{h}{\tau}}+g\left(1-e^{-\frac{h}{\tau}}\right)U_{i-1-n} \qquad i=0,1,2,\ldots \\ \hat{Y}(z) &= e^{-\frac{h}{\tau}}z^{-1}\hat{Y}(z)+g\left(1-e^{-\frac{h}{\tau}}\right)z^{-1-n}\hat{U}(z) \end{aligned} \tag{9-20}$$

where dead time is $D=nh$. For the time being, assume that the dead time D can be exactly divided up into n increments, each of size h.

The transfer function between y and U follows from Eq. (9-20):

$$\frac{\hat{Y}(z)}{\hat{U}(z)}=G(z)=\frac{g\left(1-e^{-\frac{h}{\tau}}\right)z^{-1-n}}{1-e^{-\frac{h}{\tau}}z^{-1}} \tag{9-21}$$

where, again, the process input is passed through a sampler and a zero-order hold.

9-6 The Proportional-Integral Control Equation in the Discrete Time Domain

The proportional-integral (PI) control equation, introduced in Chap. 3, can be written as

$$U(t)=U_0+ke(t)+I\int_0^t du\,e(u) \tag{9-22}$$

If Eq. (9-22) is evaluated at discrete points in time, the result is

$$
\begin{aligned}
e(t_i) &= S_i - y_i && \text{(error)} \\
Q_i &= Q_{i-1} + he_i && \text{(integral component)} \\
U(t_i) &= U_0 + ke(t_i) + IQ_i && \text{(PI algorithm)}
\end{aligned}
\tag{9-23}
$$

The transition from Eq. (9-22) to (9-23) bears some discussion. The integral of the error is approximated simply (and crudely) by

$$\int_0^t du\, e(u) \cong \sum_{k=0}^{N} he_k \qquad t = Nh$$

But, to put this approximation directly into Eq. (9-22) would pose implementation problems when it comes to evaluating the sum (try it with some practice code in any language such as BASIC). It is simpler to use three equations, as in Eq. (9-23), and update the integral of the error separately. Failure to break the PI control equation into these parts has been the bane of many a microprocessor programmer. Several times in my career, by separating the equation into three parts, I was able to help out a junior engineer, who just couldn't get his code to work.

Equation (9-23) presents the *positional* form of the PI control algorithm. Alternatively, the PI control algorithm can be implemented as

$$\Delta U_{\text{tent}} = k\Delta e(t_i) + I h e_i$$

$$
\begin{aligned}
&\text{If } \Delta U_{\text{tent}} > \Delta U_{\max} && \text{then} && \Delta U = \Delta U_{\max} \\
&\text{If } \Delta U_{\text{tent}} < -\Delta U_{\max} && \text{then} && \Delta U = -\Delta U_{\max} \\
&\text{else } \Delta U = \Delta U_{\text{tent}}
\end{aligned}
\tag{9-24}
$$

$$
\begin{aligned}
&U_{\text{tent}} = U_{i-1} + \Delta U \\
&\text{if } U_{\text{tent}} > U_{\max} && \text{then} && U_i = U_{\max} \\
&\text{if } U_{\text{tent}} < U_{\min} && \text{then} && U_i = U_{\min} \\
&\text{else } U_i = U_{\text{tent}}
\end{aligned}
$$

where the first line of Eq. (9-24) is the *incremental* form of the PI control algorithm. A new symbol Δ is the differencing operator and it operates on a variable to generate a difference, as in

$$\Delta E = E_i - E_{i-1}$$

The differencing operator can be related to the backshift operator as

$$\Delta E = E_i - E_{i-1}$$
$$= (1 - z^{-1})E_i$$

Note that the minimum and maximum allowed values of U and ΔU are included in Eq. (9-24). This is preferable to Eq. (9-23) because the clipping of U and ΔU is easier in the sense that one does not have to figure out how to clip the integral component Q in Eq. (9-23).

A common problem occurs when a programmer "talks the algorithm out" before coding it. He might say, "I will compute the error between the set point and the process output, multiply it by a proportional control gain, and add it to the old value of the control output and send it on its way via a D/A converter." In doing so, the following algorithm is being implemented:

$$U_i = U_{i-1} + K(S - Y_i)_i \tag{9-25}$$

Equation (9-25) is an integral-only control algorithm. As you might already suspect, and as we will show later in this chapter, this algorithm adds an immediate phase lag of 90° and can be deadly if the process under control already contributes 90° or more of phase lag. Many junior engineers consider Eq. (9-25) to be a proportional control algorithm and are mystified when the controlled system oscillates.

Finally, there is often a temptation, when programming the PI control algorithm in a microprocessor, to take a short cut as in

$$\begin{aligned} \Delta U(t_i) &= Ae(t_i) + Be_{i-1} \\ U_i &= U_{i-1} + \Delta U(t_i) \end{aligned} \tag{9-26}$$

where

$$A = k + Ih \quad \text{and}$$
$$B = -k$$

Equation (9-26) appears to be equivalent to Eq. (9-24). It also appears to save one computational step. But, if the control interval h and the integral gain I are small (and they often are) then A and $-B$ might be numerically quite close, especially if the microprocessor has a relatively short word length. This can cause serious round-off errors. I always recommend avoiding Eq. (9-26).

9-7 Converting the Proportional-Integral Control Algorithm to Z-Transforms

To move to the Z-transform domain, one merely applies the backshift operator to Eq. (9-24), as in

$$(1-z^{-1})\hat{U} = k(1-z^{-1})\hat{E} + Ih\hat{E}$$

$$\hat{U} = \frac{k(1-z^{-1})\hat{E} + Ih\hat{E}}{(1-z^{-1})} \tag{9-27}$$

$$= \left(k + \frac{Ih}{1-z^{-1}}\right)\hat{E}$$

In the development of Eq. (9-27), the reader should have perceived that there are some matches between discrete and continuous time operators as in

Derivative : $\quad s \quad \Leftrightarrow \quad$ Difference: $\quad 1-z^{-1}$

Integral: $\quad \frac{1}{s} \quad \Leftrightarrow \quad$ Sum: $\quad \frac{1}{1-z^{-1}}$

9-8 The PIfD Control Equation in the Discrete Time Domain

The PIfD algorithm was introduced as a tool for controlling the underdamped process in Chap. 6. In the Laplace transform domain the PIfD algorithm can be written as

$$\frac{\tilde{u}(s)}{\tilde{e}(s)} = G_c = P + \frac{I}{s} + D\frac{s}{\tau_f s + 1} \tag{9-28}$$

The discrete time realization of this algorithm can be constructed as follows. First, construct the proportional-integral-derivative (PID) without the filter, as in

$$E_i = S_i - y_i$$

$$\Delta U_i = P\Delta E_i + IhE_i + D\frac{\Delta^2 E_i}{h}$$

$$U_i = U_{i-1} + \Delta U_i \tag{9-29}$$

where

$$\Delta E_i = E_i - E_{i-1}$$

$$\Delta^2 E_i = E_i - 2E_{i-1} + E_{i-2}$$

Note that $\Delta^2 E_i$ can be obtained by a repeated application of the differencing operator Δ, as in

$$\begin{aligned}\Delta^2 E_i &= \Delta(\Delta E_i) = \Delta(E_i - E_{i-1})\\ &= \Delta E_i - \Delta E_{i-1}\\ &= (E_i - E_{i-1}) - (E_{i-1} - E_{i-2})\\ &= E_i - 2E_{i-1} + E_{i-2}\end{aligned}$$

Also, note that the h appears in Eq. (9-29) because the derivative is being approximated by a difference, as in

$$\frac{dE_i}{dt} \cong \frac{\Delta E_i}{h}$$

Two modifications should be made immediately. First, the set point should be removed from the derivative term because, if there be a step in the set point, the derivative of a step change is a spike which would disrupt the controller behavior.

$$\begin{aligned}E_i &= S_i - y_i\\ \Delta U_i &= P\Delta E_i + IhE_i - D\frac{\Delta^2 y_i}{h}\\ U_i &= U_{i-1} + \Delta U_i\end{aligned}$$

where

$$\begin{aligned}\Delta E_i &= E_i - E_{i-1}\\ \Delta^2 y_i &= y_i - 2y_{i-1} + y_{i-2}\end{aligned}$$

Second, the derivative term should be separated from the main equation and put through our newly introduced first-order filter, as in

$$\begin{aligned}q_i &= \frac{y_i - y_{i-1}}{h}\\ q_i^f &= (1-a)q_i + a q_{i-1}^f\\ a &= e^{-\frac{h}{\tau_f}}\end{aligned}$$

where q_i^f is the filtered difference. The PIfD algorithm becomes

$$
\begin{aligned}
E_i &= S_i - y_i \\
q_i &= \frac{y_i - y_{i-1}}{h} \\
q_i^f &= (1-a)q_i + a q_{i-1}^f \\
\Delta U_i &= P\Delta E_i + IhE_i - D\Delta q_i^f \\
U_i &= U_{i-1} + \Delta U_i
\end{aligned}
\tag{9-30}
$$

The Z-transform version of Eq. (9-30) can be written by inspection as

$$(1-z^{-1})\hat{U} = P(1-z^{-1})\hat{E} + Ih\hat{E} - D(1-z^{-1})\hat{q}^f$$

$$[1-(1-a)z^{-1}]\hat{q}^f = a\frac{(1-z^{-1})}{h}\hat{y}$$

or (9-31)

$$\hat{U} = P\hat{E} + \frac{Ih}{(1-z^{-1})}\hat{E} - \frac{D}{h}\frac{a(1-z^{-1})}{(1-(1-a)z^{-1})}\hat{y}$$

9-9 Using the Laplace Transform to Design Control Algorithms—the Q Method

We have put this subject off until we could carry out the design in both the continuous and discrete time domains. The former is more straightforward because there are no samplers or zero-order holds.

9-9-1 Developing the Proportional-Integral Control Algorithm

We will start with the first-order process model and derive a control algorithm that will accomplish a desired response of the process output to a step in the set point. This means that the transfer function relating the process output to the set point will be the same as that given in Eq. (3-42) in Chap. 3, namely,

$$\frac{\tilde{Y}}{\tilde{S}} = \frac{G_p G_c}{1 + G_p G_c} \tag{9-32}$$

where, in this case, the process transfer function $G_p(s)$ is known to be

$$G_p = \frac{g}{\tau_p s+1} \tag{9-33}$$

The transfer function for the controller $G_c(s)$ is unknown and is to be determined.

We specify that the response of the process output to a step in the set point should be first order with a specified time constant τ_d as in

$$\frac{\tilde{Y}}{\tilde{S}} = Q = \frac{1}{\tau_d s+1} \tag{9-34}$$

Equation (9-34) means that the controller will drive the process output to the set point in a first-order fashion described by a time constant τ_d. You might expect that we would want $\tau_d < \tau_p$. The gain is unity in Eq. (9-34) because we want the process output to line out at the set point.

Eliminating $\tilde{Y}/\tilde{S}$ between Eqs. (9-32) and (9-34) gives

$$\frac{G_p G_c}{1+G_p G_c} = Q$$

or, after solving for G_c,

$$\boxed{\frac{\tilde{U}}{\tilde{E}} = G_c = \frac{Q}{1-Q}\frac{1}{G_p}} \tag{9-35}$$

which is the desired transfer function for the controller. All that remains is to replace Q and G_p, which we will do in the following:

$$\begin{aligned}\frac{\tilde{U}}{\tilde{E}} = G_c &= \frac{Q}{1-Q}\frac{1}{G_p} \\ &= \frac{\dfrac{1}{\tau_d s+1}}{1-\dfrac{1}{\tau_d s+1}}\frac{1}{\dfrac{g}{\tau_p s+1}} \\ &= \frac{\tau_p}{g\tau_d} + \frac{1}{g\tau_d s}\end{aligned}$$

This is nothing more than the PI control algorithm! This can be seen by continuing the algebra

$$\tilde{U} = \frac{\tau_p}{g\tau_d}\tilde{E} + \frac{1}{g\tau_d}\frac{\tilde{E}}{s}$$

$$= P\tilde{E} + I\frac{\tilde{E}}{s}$$

$$\boxed{P = \frac{\tau_p}{g\tau_d} \qquad I = \frac{1}{g\tau_d}} \tag{9-36}$$

Therefore, given the process time constant τ_p and the process gain g, the proportional control gain and the integral control gain can be adjusted by choosing the desired time constant τ_d. Equation (9-36) shows that as τ_d decreases the tuning gets more aggressive. It is comforting that the PI control algorithm is the natural outcome of asking the response of the first-order process model output to behave in a first-order manner to a step in the set point.

The practical use of Eq. (9-36) is important. Many processes behave *approximately* as first-order processes. Therefore, one would apply a step change to the process input while the controller is inactive and collect the response. Visually or otherwise, one would then estimate the process gain and the process time constant. Then the controller gains could be obtained from Eq. (9-36) after the desired time constant is specified. My experience suggests that $\tau_p / \tau_d = 2$ is a good start.

Many PI controllers are designed to fit the following format:

$$\tilde{U} = K_c\left(1 + \frac{1}{\tau_I}\right)\tilde{E} \tag{9-37}$$

where τ_I is often called the "reset time" and K_c is an overall control gain. In this case the tuning rules are slightly different.

$$\tilde{U} = \frac{\tau_p}{g\tau_d}\tilde{E} + \frac{1}{g\tau_d}\frac{\tilde{E}}{s}$$

$$= K_c\left(1 + \frac{1}{\tau_I}\right)\tilde{E}$$

$$\boxed{K_c = \frac{\tau_p}{g\tau_d} \qquad \tau_I = \tau_d} \tag{9-38}$$

Now the reset time is directly equal to the desired time constant and one would only adjust K_c to tune the algorithm. In this case, one adjustable parameter τ_d, easily visualized, can be used to determine values for two controller parameters.

By the way, note that the controller gain is inversely proportional to the process gain. This makes sense because when the process gain is high the controller gain should be low and vice versa.

In general, when the process is not first order, the controller designed by this method will not be PI and in fact may be difficult to implement in the continuous time domain.

9-9-2 Developing a PID-Like Control Algorithm

Having gone through the painful algebra for the simple first-order case, we will quickly develop a control algorithm for the underdamped process discussed in Chap. 6. First, the process transfer function is given as

$$G_p(s) = \frac{g\,\omega_n^2}{s^2 + 2\zeta\omega_n s + \omega_n^2}$$

Next, the transfer function for the desired second-order response of the process output to a step in the set point is specified as

$$Q = \frac{\omega_d^2}{s^2 + 2\zeta_d\omega_d s + \omega_d^2}$$

Once again, the gain is unity because we want the process output to line out at the set point. Note that ω_d and ζ_d are specified by the control designer.

These two transfer functions are inserted into Eq. (9-35), the algebraic crank is turned and out pops the following algorithm:

$$G_c = \frac{\left(s^2 + 2\zeta\omega_n s + \omega_n^2\right)}{\omega_n^2 g s}\frac{\omega_{nd}^2}{s + 2\zeta_d\omega_d}$$

which, after some further manipulation, becomes

$$G_c = \frac{1}{g}\frac{\zeta}{\zeta_d}\frac{\omega_d}{\omega_n}\left(1 + \frac{\omega_n}{2\zeta s} + \frac{s}{2\zeta\omega_n}\right)\frac{1}{\dfrac{s}{2\zeta_d\omega_d} + 1} \tag{9-39}$$

Equation (9-39) is seen to be a PID control algorithm, namely,

$$\frac{1}{g}\frac{\zeta}{\zeta_d}\frac{\omega_d}{\omega_n}\left(1 + \frac{\omega_n}{2\zeta s} + \frac{s}{2\zeta\omega_n}\right) \Leftrightarrow K_c\left(1 + \frac{1}{\tau_I s} + \tau_D s\right)$$

which is put through the following first-order filter:

$$\frac{1}{\dfrac{s}{2\zeta_d\omega_d}+1} \Leftrightarrow \frac{1}{\tau_f s+1}$$

In tuning this algorithm one would perhaps choose a desired damping factor such that ζ / ζ_d is less than unity (the desired response would not be as underdamped as the uncontrolled process) and a desired natural frequency such that ω_d / ω_n was greater unity (the desired natural frequency would be faster than the uncontrolled process). It is interesting that a four-parameter PID control algorithm (including the filter time constant) ends up requiring just two parameters, both of which are easily visualized:

$$\boxed{K_c = \frac{1}{g}\frac{\zeta}{\zeta_d}\frac{\omega_d}{\omega_n} \qquad \tau_I = \frac{2\zeta}{\omega_n} \qquad \tau_D = \frac{1}{2\zeta\omega_n} \qquad \tau_f = \frac{1}{2\zeta_d\omega_d}} \tag{9-40}$$

For both the example algorithms in this section, the reader should see that knowledge of the process parameters is critical in tuning the controllers.

9-10 Using the Z-Transform to Design Control Algorithms

To design a discrete time domain controller we start again with a feedback control loop but we insert a sampler and a zero-order hold as in Fig. 9-7. The block diagram algebra is similar to that for the Laplace transform except that one has to ensure that the location of the samplers and zero-order holds make sense. Here, the controller error is formed, sampled, and fed to the controller (which is probably implemented digitally) as a train of pulses. The

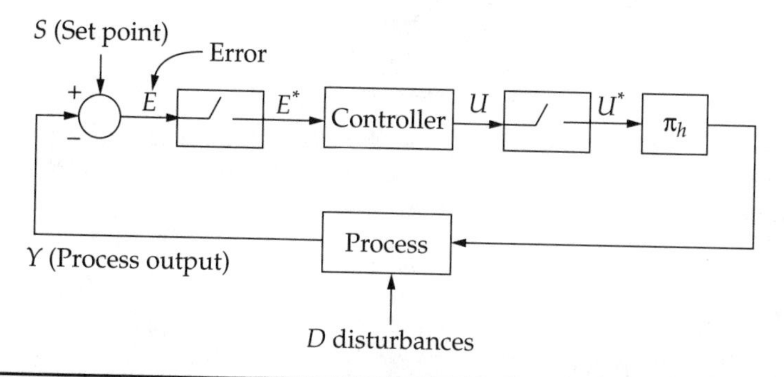

FIGURE 9-7 A feedback controller with a sampler and a zero-order hold.

output of the controller, U, is sampled, held and fed to the process as a sequence of steps.

$$\begin{aligned}
\hat{Y} &= Z\{\Pi_h(s)G_p(s)\}\hat{U}(z) \\
\hat{U} &= G_c\hat{E} \\
\hat{Y} &= Z\{\Pi_h(s)G_p(s)\}G_c(z)\hat{E} \\
\hat{E} &= \hat{S} - \hat{Y} \\
\hat{Y} &= Z\{\Pi_h(s)G_p(s)\}G_c(z)(\hat{S} - \hat{Y}) \\
\hat{Y} &= \frac{Z\{\Pi_h(s)G_p(s)\}G_c(z)}{1 + Z\{\Pi_h(s)G_p(s)\}G_c(z)}\hat{S} = H\hat{S}
\end{aligned} \tag{9-41}$$

This look messy, but the only difference between Eq. (9-41) and Eq. (3-42) is the additional factor of the zero-order hold $\Pi_h(z)$ which is dealt with in detail in App. I.

As in Sec. 9.9, the desired relationship between the process output and the set point is specified as

$$\frac{\hat{Y}}{\hat{S}} = Q(z) \tag{9-42}$$

Equations (9-41) and (9-42) are combined as before giving the expression for the controller

$$\frac{\Pi_h(z)G_p(z)G_c(z)}{1 + \Pi_h(z)G_p(z)G_c(z)} = Q(z)$$

or

$$\boxed{\frac{\hat{U}}{\hat{E}} = G_c(z) = \frac{Q(z)}{1 - Q(z)}\frac{1}{\Pi_h(z)G_p(z)}} \tag{9-43}$$

If we are dealing with first-order models, the desired response would be characterized by the discrete first-order unity gain transfer function

$$\hat{Y}(z) = \frac{\left(1 - e^{-\frac{h}{\tau_d}}\right)z^{-1}}{1 - e^{-\frac{h}{\tau_d}}z^{-1}}\hat{S}(z) = Q(z)\hat{S}(z) \tag{9-44}$$

Continuing the first-order approach, the expression for $\Pi_h(z)G_p(z)$ would be obtained from Eq. (9-13):

$$\widehat{Y}(z) = \frac{g\left(1-e^{-\frac{h}{\tau_p}}\right)z^{-1}}{1-e^{-\frac{h}{\tau_p}}z^{-1}}\widehat{U}(z) = Z\{\Pi_h(s)G(s)\}\widehat{U}(z) \tag{9-45}$$

The only difference between Eqs. (9-44) and (9-45) is the gain and the subscript on the time constant.

The algebraic crank-turning yields

$$G_c(z) = \frac{\left(1-e^{-\frac{h}{\tau_d}}\right)\left(1-z^{-1}e^{-\frac{h}{\tau_p}}\right)}{g\left(1-e^{-\frac{h}{\tau_p}}\right)(1-z^{-1})} \tag{9-46}$$

The control algorithm is therefore

$$\widehat{U}(z) = \frac{\left(1-e^{-\frac{h}{\tau_d}}\right)}{g\left(1-e^{-\frac{h}{\tau_p}}\right)}\frac{\left(1-z^{-1}e^{-\frac{h}{\tau_p}}\right)}{(1-z^{-1})}\widehat{E}(z) = \frac{B_d}{gB}\frac{(1-Az^{-1})}{(1-z^{-1})}$$

where (9-47)

$$A = e^{-\frac{h}{\tau_p}} \qquad A_d = e^{-\frac{h}{\tau_d}} \qquad B = 1-A \qquad B_d = 1-A_d$$

A little (actually, a lot of) algebraic manipulation of Eq. (9-47) gives

$$\widehat{U}(z) = \frac{B_d}{gB}\widehat{E}(z) + \frac{B_d}{g}\frac{z^{-1}}{1-z^{-1}}\widehat{E}(z) \tag{9-48}$$

The first term on the right-hand side of Eq. (9-48) is the proportional component and the second term is the integral component with an extra delay of one sample in the numerator. In practice, one would remove the extra delay because, from a common sense point of view, it adds nothing to the performance.

Therefore, converting to the discrete time domain we have a PI control algorithm:

$$\Delta U_i = \frac{B_d}{gB}\Delta E_i + \frac{B_d}{g}E_i$$
$$= P\Delta E_i + IhE_i \quad (9\text{-}49)$$
$$U_i = U_{i-1} + \Delta U_i$$

with the following tuning rules.

$$\boxed{P = \frac{B_d}{B} = \frac{1 - e^{-\frac{h}{\tau_d}}}{g\left(1 - e^{-\frac{h}{\tau}}\right)} \qquad I = \frac{1}{h}\frac{B_d}{g} = \frac{1 - e^{-\frac{h}{\tau_d}}}{gh}} \quad (9\text{-}50)$$

Question 9-1 If the control interval h is decreased to an infinitesimal value, will the tuning rules in Eq. (9-50) evolve into those of Eq. (9-36)?

Answer Yes, they would and I will leave it to the reader as an exercise. Sorry, unless you trust me you will have to work it out on your own.

As with the tuning rules given in Eq. (9-36), these in Eq. (9-50) are practical. When the digitally implemented PI controller has a control or sampling interval that is quite small relative to the dominant process time constant, these two sets of tuning rules are virtually identical and I would recommend using the former.

9-11 Designing a Control Algorithm for a Dead-Time Process

Before we get started, we have to understand that there really is no panacea for controlling processes with a dead time. The controllability of a process with a dead time can never be as good as that for a process without dead time, no matter how fancy the control algorithm is. Consider the following sequence of events. A disturbance causes the process output to deviate from the set point. You, acting as the controller, immediately initiate a control move to address the deviation. There will be no response to that control move until the dead time has elapsed. During that dead-time period, more nasty things can happen to the process output but you still haven't seen the effect of your initial action. This situation often leads to impatient and aggressive moves that cause more trouble than the original disturbance.

The discrete time domain is perhaps a better place to concoct a control algorithm for processes with dead times because to be somewhat effective the algorithm will have to store past control outputs or process outputs. A delay vector, digitally implemented, is probably the best way to deal with this problem.

The FOWDT process model from Eq. (9-21) will be used.

$$\frac{\hat{Y}(z)}{\hat{U}(z)} = G(z) = \frac{gBz^{-1-n}}{1-Az^{-1}}$$
$$A = e^{-\frac{h}{\tau}} \qquad B = 1-A \tag{9-51}$$

The desired response to a change in the set point will be

$$\frac{\hat{Y}(z)}{\hat{S}(z)} = Q = \frac{B_d z^{-1-n}}{1-A_d z^{-1}}$$
$$A_d = e^{-\frac{h}{\tau_d}} \qquad B_d = 1-A_d \tag{9-52}$$

This desired response contains the dead time. Therefore, we are conceding that our desired response cannot occur until the dead time has elapsed and only then can we specify a desired time constant τ_d. Combining Eqs. (9-51) and (9-52) results in

$$\frac{\hat{U}(z)}{\hat{E}(z)} = G_c = \frac{(1-Az^{-1})B_d}{Bg(1-A_d z^{-1} - z^{-1-n} + z^{-1-n}A_d)} \tag{9-53}$$

To get Eq. (9-53) into a better form we add z^{-1} to and subtract z^{-1} from both the numerator and the denominator (try it!) and get

$$(1-z^{-1})\hat{U}(z) = B_d(z^{-1-n} - z^{-1})\hat{U}(z) + \frac{B_d}{gB}(1-z^{-1})\hat{E}(z) + \frac{B_d}{g}z^{-1}\hat{E}(z) \tag{9-54}$$

This may be off-putting at first but after a closer look the reader will see the change in the controller output on the left-hand side of Eq. (9-54), namely, $(1-z^{-1})\hat{U}(z)$, equals a difference of delayed controller outputs, $(z^{-1-n} - z^{-1})\hat{U}(z)$. Next, there are two terms that look just like the PI algorithm obtained in Sec. 9.10 if one is willing to remove, once again, the backshift operator in the last term on the right-hand side of Eq. (9-54). In fact, the P and I control gains in these last two terms are the same as those arrived at in Sec. 9-11. The first term on the right-hand side of Eq. (9-54)

$$B_d(z^{-1-n} - z^{-1})\hat{U}(z)$$

can be realized by using a delay vector of the controller outputs.

In the discrete time domain the algorithm looks like

$$U_i = U_{i-1} + B_d(U_{i-1-n} - U_{i-1}) + \frac{B_d}{gB}(E_i - E_{i-1}) + \frac{B_d}{g}E_i \qquad (9\text{-}55)$$

Therefore, the control of a FOWDT process can be accomplished via an augmented PI control algorithm.

Figures 9-8 and 9-9 show the control of a dead-time process using the augmented algorithm and using regular PI. The process has a time constant of 10.0 time units and a gain of 2.0. The control interval is 1 time unit. The dead time is 5 time units. The desired time constant is 5 time units. The extra term in Eq. (9-55), containing the delayed control outputs, causes the PI controller to behave as though there were no dead time. Therefore, in Fig. 9-8 the control output responds to the step in the set point as though there were no dead time. Note how the control output immediately starts to back off after the first step associated with the set-point change. This backing off occurs before the process starts to respond.

Figure 9-9 shows the same process under regular PI control with the same P and I control gains (admittedly too high). Note how the integral component ramps up after the initial proportional jump at the time of the set-point change. This ramping occurs because the process has not yet responded and the error is constant.

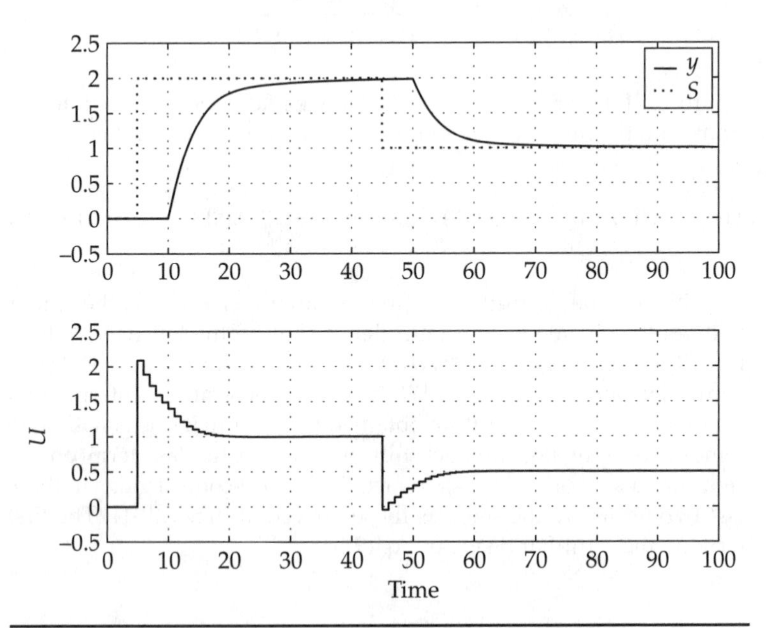

FIGURE 9-8 Control of a FOWDT process with compensated PI.

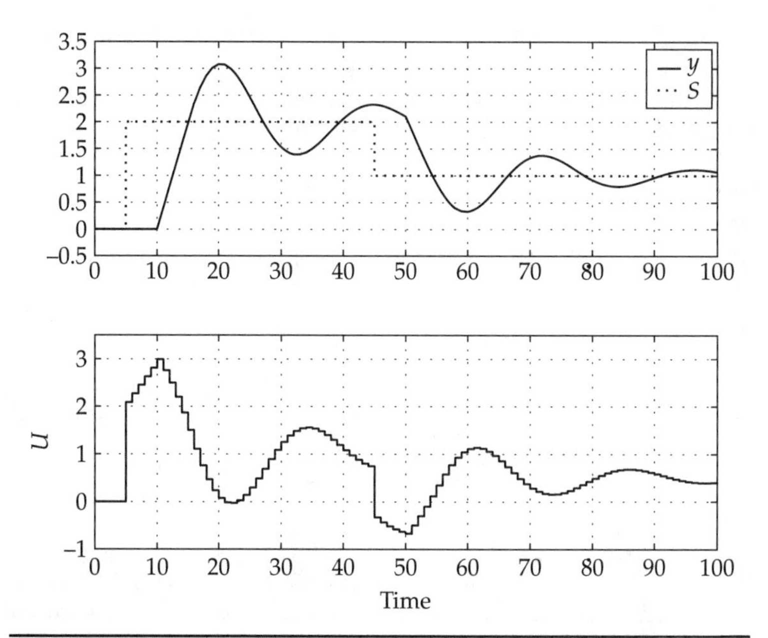

FIGURE 9-9 Control of a FOWDT process with PI.

In this example the process model parameters matched those of the actual process perfectly. In realistic cases there will be a mismatch and the performance will not be as good. However, I have used this type of control algorithm with reasonable success in a variety of industrial settings and I rarely chose the desired time constant to be less than half the process time constant. Anything smaller will likely be too aggressive and accentuate the differences between the model and the actual process. Just in case you come across the Smith Predictor, the algorithm in Eq. (9-55) is quite similar in concept and application.

Finally, note that this algorithm requires fetching values of U from a delay vector. If the dead time is variable, the algorithm discussed in Sec. 4-6 would be required.

9-12 Moving to the Frequency Domain

In the continuous time domain, the Laplace transform provided a path from differential equations to algebraic equations and then to the frequency domain via the substitution of $s \to j\omega$ or $s \to j2\pi f$. In the discrete time domain there is a similar path. The Z-transform converts discrete time index equations into algebraic equations. The transition to the frequency domain is provided by the substitution

$$z = e^{sh} \Rightarrow e^{j\omega h} = e^{j2\pi f h} = e^{j\Omega} = \cos\Omega + j\sin\Omega$$

where Ω varies from 0 to 2π, ω varies from 0 to $2\pi/h$, and f varies from 0 to $1/h$ (later on you will see that the range for f can be quite a bit wider). As can be seen from App. I, this substitution logically follows from the sampled Laplace transform.

9-12-1 The First-Order Process Model

Consider a first-order process with a time constant of 5.0 min and a gain of 2.0. Upon applying the above substitution to Eq. (9-11), one gets, after some serious algebra, which I don't think you are interested in,

$$G(e^{j\Omega}) = \frac{g(1-A)}{\sqrt{1-2A\cos(2\pi f h)+A^2}} e^{j\theta}$$
$$\theta = -\tan^{-1}\left(\frac{\sin(2\pi f h)}{\cos(2\pi f h)-A}\right) \tag{9-56}$$

The Bode plot for this transfer function is shown in Fig. 9-10 for three different values of the sampling interval h. The plot also gives the magnitude and phase for the continuous first-order model with the same time constant and gain.

This figure bears some study. First, note that as h decreases (or as the sampling frequency increases) the curves for the discrete models approach that of the continuous model. Second, note that for each h

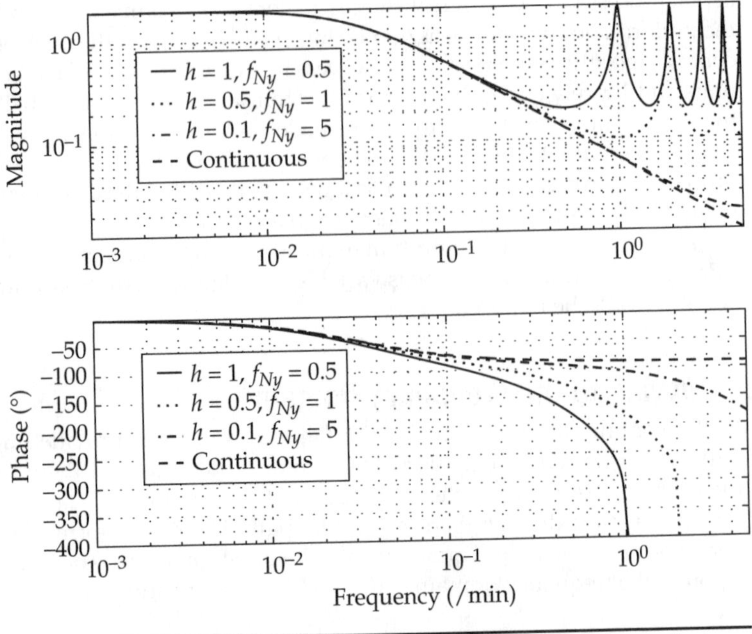

FIGURE 9-10 Bode plots for sampled first-order process models, $\tau = 5$.

there is a different folding frequency f_{Ny}. The frequency range in Fig. 9-10 is [0, 5] which includes the folding frequency for the highest sampling frequency of the three cases. For the two examples that have smaller folding frequencies, the magnitudes show a strange ripple that will be explained later.

For the continuous first-order model, the phase curves show that the maximum phase lag is 90°. The phases for the discrete model drop off precipitously once the folding frequency has been exceeded.

The sampling interval acts as a dead time with a value equal to one half the sampling interval. Consider the case when the process input is stepped. If the step happens to occur exactly at the sampling instant then it will appear as though there were no dead time. At the other extreme, if the step happens just after the sampling instant there will be an effective dead time of one sampling interval. On the average, the step will occur in the middle of the interval, hence the effective dead time of one-half sampling interval. This means that even the idealized dead-time-less first-order model becomes potentially unstable when discretized.

9-12-2 The Ripple

Now, what about those ripples in the magnitude curves? Before addressing this issue, we replot the magnitude curves with a linear frequency axis, in Fig. 9-11. In addition, we plot the continuous

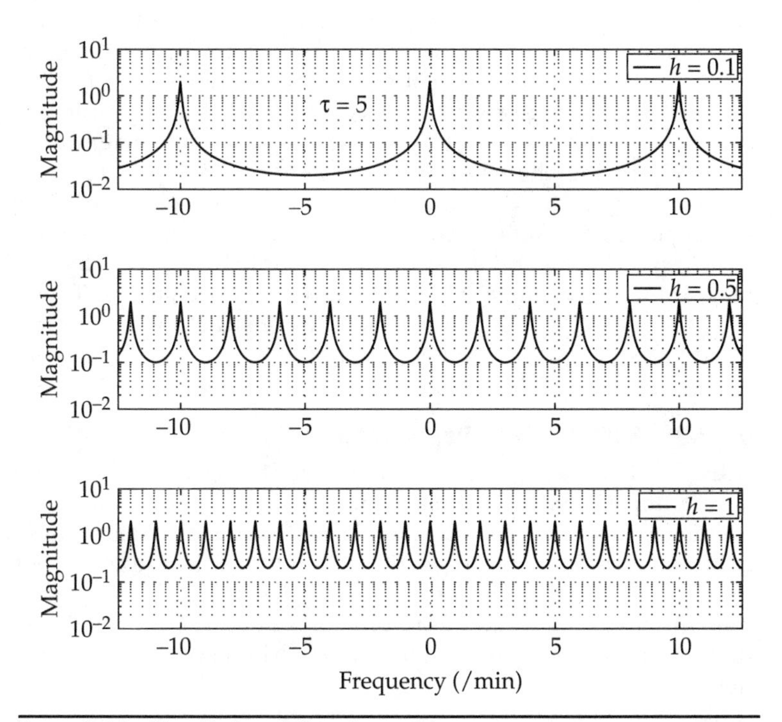

FIGURE 9-11 Magnitudes of discrete models.

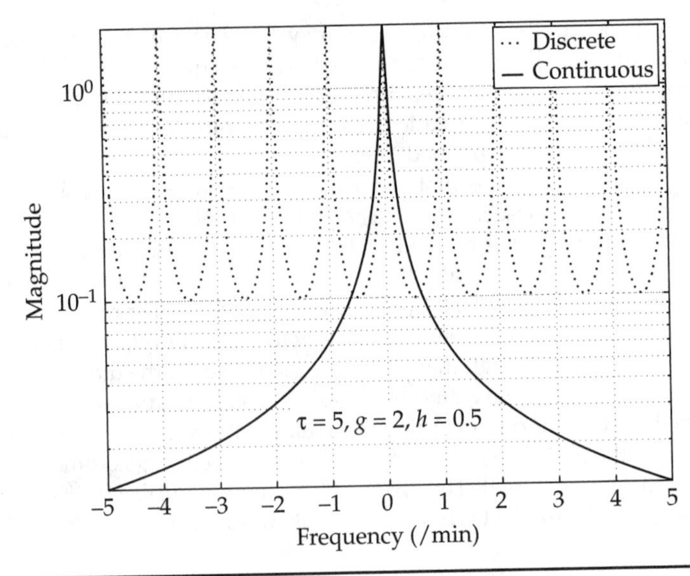

FIGURE 9-12 Discrete and continuous magnitude plot.

and discrete version of the magnitude in Fig. 9-12. Three things should be apparent. First, the act of sampling causes the continuous magnitude to be replicated at frequency intervals equal to the sampling frequency. For example, when $h = 0.5$ or $f_s = 2.0$, the continuous magnitude curve, centered at $f = 0.0$ is replicated at $f = -4, -2, 0, 2, 4, \ldots$. Second, the discrete time magnitudes overlap causing the discrete magnitude at, say, $f = 1.0$, to be greater than continuous magnitude. This is a consequence of the folding of higher frequencies into the Nyquist interval as aliases (see App. C for more discussion of aliasing). Third, note that the magnitude is symmetrical about the folding frequency. Therefore, there is really not much reason to construct the Bode plots for any frequency range other than $[0\ f_{Ny}]$.

9-12-3 Sampling and Replication

Although I am not going to demonstrate it, in general, sampling in one domain, say, the time domain, causes replication in the frequency domain, as in the case here. Conversely, sampling in the frequency domain causes replication in the time domain. For example, in App. C, it was suggested that the discrete Fourier transform was the result of sampling a continuous Fourier transform in the frequency domain and that periodicity resulted in the time domain data.

9-13 Filters

9-13-1 Autogressive Filters

We have mentioned filters in passing several times in earlier parts of the book. In the continuous time domain, the first-order low-pass filter

$$\frac{1}{\tau s+1} \tag{9-57}$$

was used to smooth out the derivative in the PIfD control algorithm. The Bode plot for first-order processes in Chap. 4 supports the moniker of "low pass."

Earlier in this chapter the first-order discrete model

$$\begin{aligned} &y_i = ay_{i-1} + bU_{i-1} \qquad i = 0, 1, 2, \ldots \\ &a = e^{-\frac{h}{\tau}} \qquad b = 1 - a \\ &\frac{\hat{y}}{\hat{U}} = \frac{bz^{-1}}{1 - az^{-1}} \end{aligned} \tag{9-58}$$

was suggested as a low-pass filter. In general, any operation on data that changes its spectral content can be considered a filter. Most industrial processes are low-pass filters.

The choice of the indices is somewhat arbitrary because Eq. (9-58) can also be written as

$$\begin{aligned} &y_i = ay_{i-1} + bU_i \qquad i = 0, 1, 2, \ldots \\ &a = e^{-\frac{h}{\tau}} \qquad b = 1 - a \\ &\frac{\hat{y}}{\hat{U}} = \frac{b}{1 - az^{-1}} \end{aligned} \tag{9-59}$$

In general, Eq. (9-59) can be written as

$$\begin{aligned} &y_i = a_1 y_{i-1} + a_2 y_{i-2} + \cdots + a_n y_{i-n} + b_0 U_i + b_1 U_{i-1} + \cdots + b_m U_{i-m} \\ &y_i = \sum_{k=1}^{n-1} a_k y_{i-k} + \sum_{k=1}^{m} b_k U_{i-k} \\ &\frac{\hat{y}}{\hat{U}} = \frac{\sum_{k=1}^{m} b_k z^{-k+1}}{\sum_{k=1}^{n-1} a_k z^{-k}} \end{aligned} \tag{9-60}$$

Note that Eq. (9-60) contains additional nonautoregressive terms in U_i that, in the next section will be considered as moving average components.

Question 9-2 If the first-order process model evolves into a discrete time first-order filter, what would the second-order model, say,

$$\frac{d^2y}{dt^2}+2\zeta\omega_n\frac{dy}{dt}+\omega_n^2 y = g\omega_n^2 U$$

$$\omega_n=\sqrt{\frac{k}{m}} \qquad \zeta=\frac{B}{2\sqrt{km}}$$

look like after discretization?

Answer The second-order process model would become a second-order discrete time model. There are several ways to approach this, none of which will be shown here, but using the following Matlab script is one of the simpler paths:

```
w=1; % natural frequency
z=.1; % damping coefficient
[n d]=ord2(w,z);
sys=tf(n,d);
disp('display Laplace transform')
sys
disp('convert to discrete')
dsys=c2d(sys,1);
disp('display Z-transform')
dsys
```

which generates the following screen output:

```
display Laplace transform

Transfer function:
1
---------------
s^2 + 0.2 s + 1

convert to discrete
display Z-transform

Transfer function:
0.431 z + 0.4023
-----------------------
z^2 - 0.9854 z + 0.8187

Sampling time: 1
```

This transfer function equals $\hat{Y}/\hat{U}$ or

$$(z^2-0.985z+0.819)\,\hat{Y}=(0.431z+0.4023)\,\hat{U}$$

Multiplying by z^{-2} gives

$$(1-0.985z^{-1}+0.819z^{-2})\,\hat{Y}=(0.431z^{-1}+0.4023z^{-2})\,\hat{U}$$

Knowing that z^{-1} is a backshift operator allows us to move to the time domain directly and get

$$y_k=0.985y_{k-1}-0.819y_{k-2}+0.431u_{k-1}+0.4023u_{k-2}$$

which is a second-order filter of the type shown in Eq. (9-60).

9-13-2 Moving Average Filters

An alternative to the first-order discrete filter is the moving average filter already alluded to in Chap. 8. For the case of two "taps," we have

$$\begin{aligned} y_k &= b_1U_k + b_2U_{k-1} \\ \hat{y} &= (b_1+b_2z^{-1})\hat{U} \end{aligned} \tag{9-61}$$

where often the coefficients would be chosen to effect a two-sample conventional average, as in

$$b_1=b_2=\frac{1}{2}$$

In general, the moving average filter is

$$y_k=\sum_{m=1}^{M} b_m U_{k-m-1} \tag{9-62}$$

Note that the so-called autoregressive filter in Eq. (9-60) contains a moving average component. Therefore, Eq. (9-60) really represents an autoregressive-moving average filter.

Consider the two simple cases, a first-order filter with a time constant of 10 time units and a moving average filter with 10 "taps."

$$\begin{aligned} & y_i = ay_{i-1}+bU_{i-1} \qquad i=0,1,2,\ldots \\ & a=e^{-\frac{1}{10}} \qquad b=1-a \\ & y_k=\frac{1}{10}\sum_{m=1}^{10} U_{k-m-1} \end{aligned} \tag{9-63}$$

Figure 9-13 shows the step-change responses of the two filters. Note how the moving average filter ramps up to the final value and flattens out at the tenth sample. On the other hand, the first-order filter reaches 63% of its final value at about the tenth sample. You should think about the response of the moving average filter and convince yourself that it makes sense.

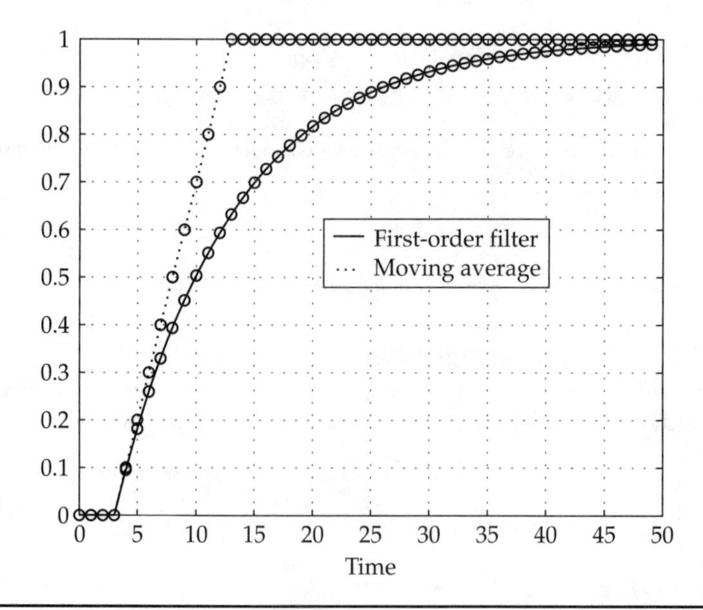

FIGURE 9-13 Moving average versus first-order filters, step-changer response.

Figure 9-14 shows the Bode plot magnitude of the two filters. Note how the moving average filter actually obliterates signals that have frequencies that are multiples of 1/10. Signals with these frequencies line up with the sampling such that positive values cancel

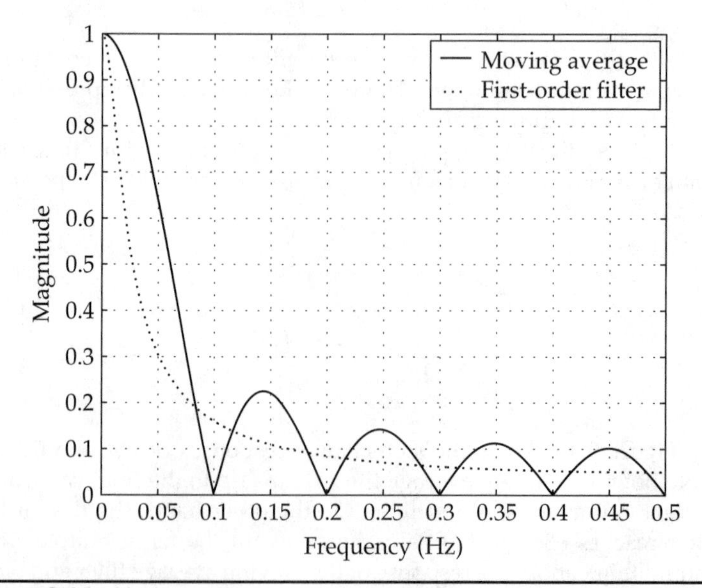

FIGURE 9-14 Moving average versus first-order filters, Bode plot.

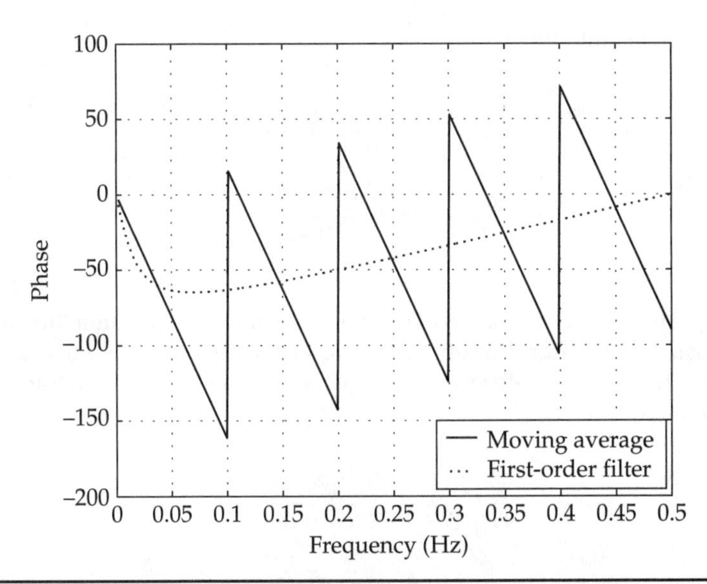

FIGURE 9-15 Ten point moving average versus 10 sec time constant first-order filters.

out negative values and the net output of the filter is zero. Figure 9-15 shows the phases of these two filters. Note the linearity of the phase for the moving average filter.

9-13-3 A Double-Pass Filter

As the reader can tell from the Bode plots, the first-order filter carries out attenuation, which is usually a desirable thing. However, it also adds phase lag to the output and this can sometimes be a problem when the data is analyzed graphically. For example, you might want to plot the raw data over the filtered data and you would probably want the filtered and unfiltered streams to be in phase. To address this problem, the filter is sometimes applied twice; once in the forward direction and once in the backward direction, as in

$$\begin{aligned} y_i &= ay_{i-1} + bU_i \qquad i = 1, 2, \ldots, N \qquad y_0 = U_0 \\ w_{i-1} &= aw_i + by_{i-1} \qquad i = N, N-1, \ldots, 1 \qquad w_N = y_N \end{aligned} \tag{9-64}$$

In the forward (left to right) direction there is a phase lag introduced but in the reverse direction (right to left) the additional phase cancels the phase lag of the first pass.

In the Z-transform domain, Eq. (9-64) becomes

$$\hat{y} = \frac{b}{1 - az^{-1}}\hat{U}$$

$$\hat{w} = \frac{z^{-1}b}{z^{-1} - a}\hat{y}$$

or, after combining

$$
\begin{aligned}
\widehat{w} &= \frac{bz^{-1}}{(z^{-1}-a)}\frac{b}{(1-az^{-1})}\widehat{U} \\
\frac{\widehat{w}}{\widehat{U}} &= \frac{b^2}{1-a(z^{-1}+z)+a^2}
\end{aligned}
\tag{9-65}
$$

This filter is now a second-order filter and in the denominator, the term $z^{-1}+z$, after the substitution $z = e^{j\Omega} = \cos(\Omega) + j\sin(\Omega)$, is simply $2\cos(\Omega)$. Therefore, in the frequency domain, the filter's transfer function is real, as in

$$
\frac{\widehat{w}(j\Omega)}{\widehat{U}(j\Omega)} = \frac{b^2}{1-2a\cos(\Omega)+a^2} \tag{9-66}
$$

and the phase is zero—meaning no lag. So, the filtered signal is completely in phase with the input signal.

Figures 9-16 and 9-17 compare the first-order filter (which would have phase lag) and the double-pass filter. The first plot compares the frequency domain magnitudes and the second compares how they filter noisy data.

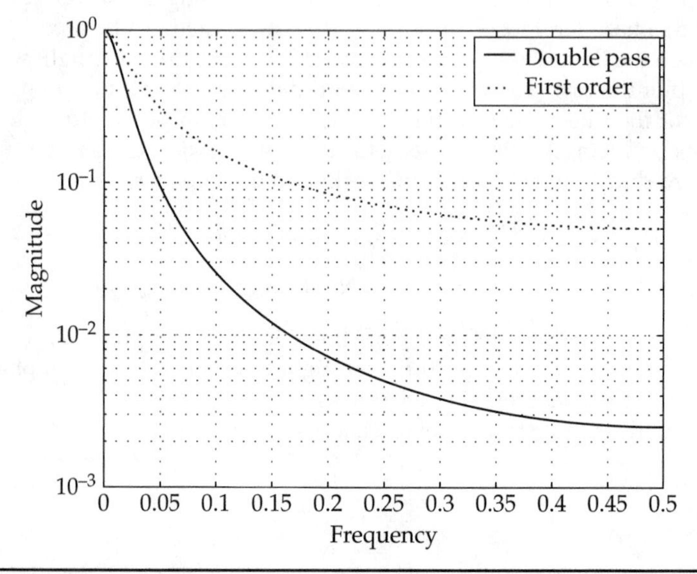

FIGURE 9-16 First-order versus double-pass filter, Bode plot.

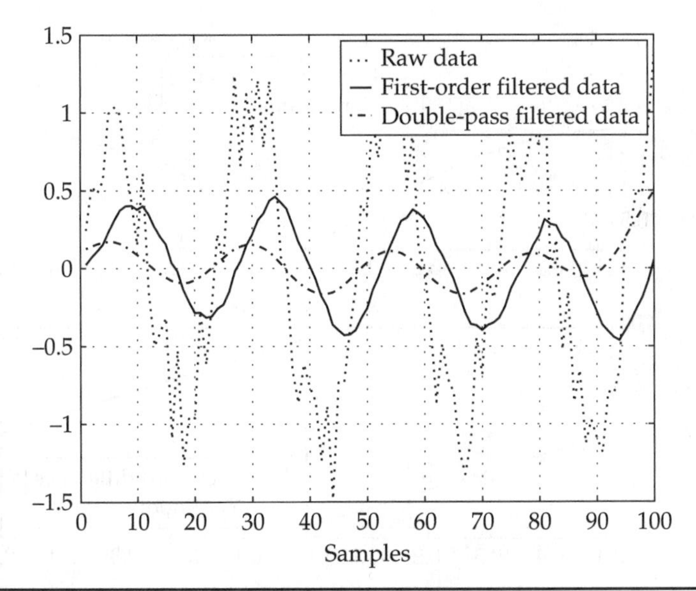

Figure 9-17 First-order versus double-pass filter, time domain.

There is more attenuation for the latter because it is a second-order filter. Also, the time domain plot in Fig. 9-17 shows that the double-pass filter keeps the filtered signal in phase with the raw data.

9-13-4 High-Pass Filters

The differentiator filter in the Laplace domain would be simply s. Its implementation is not obvious but you can probably conceive of a way to hook up a capacitor and a resistor to do it. If not, don't worry; the digital approach is probably simpler and more direct.

In the discrete time domain, differentiation is approximated by the difference filter, the simplest of which is the backward differencer:

$$
\begin{aligned}
y_i &= \frac{U_i - U_{i-1}}{h} \\
\frac{\hat{y}}{\hat{U}} &= \frac{1 - z^{-1}}{h} \\
\frac{\hat{y}(\Omega)}{\hat{U}(\Omega)} &= \frac{1}{h}\sqrt{2 - 2\cos(\Omega)}\, e^{j\theta} \\
\theta &= \tan^{-1}\left(\frac{\sin\Omega}{1 - \cos\Omega}\right)
\end{aligned}
\tag{9-67}
$$

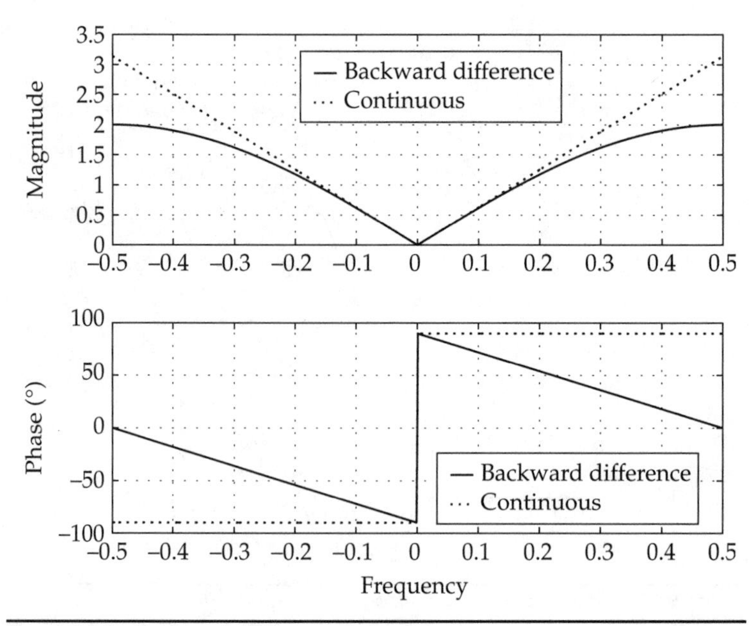

FIGURE 9-18 Backward difference Bode plot.

Figure 9-18 shows the magnitude and phase of the backward differencer. Note how the filter gain increases with frequency and how the phase starts at +90° and slowly decreases. The magnitude and phase for the theoretical continuous differentiator is also shown in Fig. 9-18. Since the gain at zero frequency (the frequency of a constant signal) is zero, the differencer is often used to remove offsets in data. Should analyst then do a spectral analysis on the data, say, looking for periodic components, they should be aware of the high frequency power that will appear in the spectrum.

Because the backward differencer amplifies noise, the more so the higher the frequency, a central difference is sometimes used.

$$y_i = \frac{U_i - U_{i-2}}{2h} \tag{9-68}$$

which can be considered as the average of two backward differences, as in,

$$\frac{1}{2}\left(\frac{1-z^{-1}}{h} + \frac{z^{-1}-z^{-2}}{h}\right) = \frac{1-z^{-2}}{2h}$$

The Bode plot for this difference operator is shown in Fig. 9-19. Note that both the backward and central differences have zero gain at zero frequency but the central difference has zero gain at the folding frequency, also.

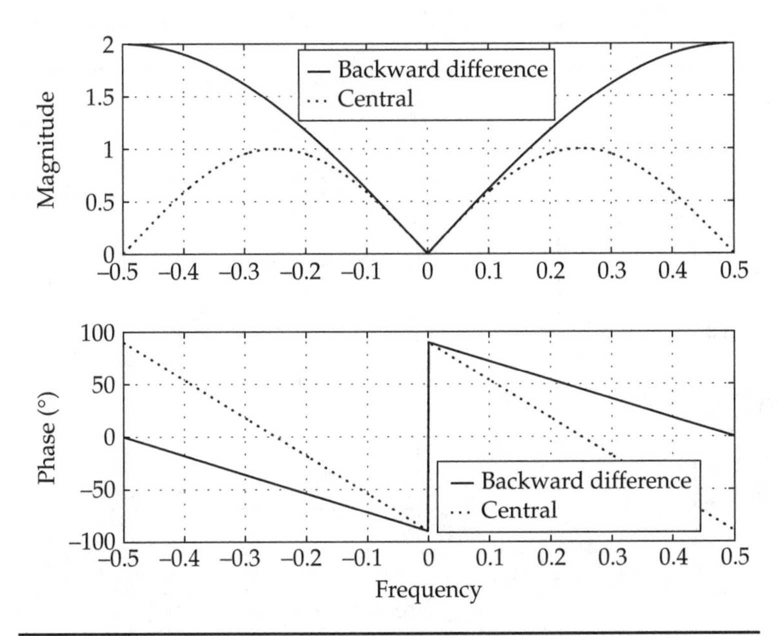

FIGURE 9-19 Backward and central differences, Bode plot.

9-14 Frequency Domain Filtering

There are a plethora of sophisticated computer-aided design techniques for moving average and autoregressive filters. A simple alternative uses frequency domain filtering where the Fourier transform of the data is multiplied by a magnitude factor which removes part of the spectrum. The modified transform is inverted back to the time domain yielding the filtered data. For example, should the analyst wish to remove a band of frequencies from the data, he might apply the factor shown in Fig. 9-20 to the transformed data in the frequency domain. This factor suggests that components in a signal having frequencies greater than about 0.21 Hz would be removed while those with frequencies less than 0.21 Hz would be passed unattenuated. For the readers who have read App. C, it might be worth noting that when multiplying in the frequency domain, one convolves in the time domain. However, that is a detail that is a little bit beyond the scope of this section.

Finally, note that the factor is symmetrical about the folding frequency of 0.5 Hz. Figure 9-21 shows the factor (after scaling to make it more presentable) and the spectrum of the signal to be filtered (two sinusoids, one in the pass band and one not). Figure 9-22 shows the result of applying the filter. One needs only the fast Fourier transform (and some code) to use this filtering method. The following is a crude Matlab script that carries out frequency-domain filtering.

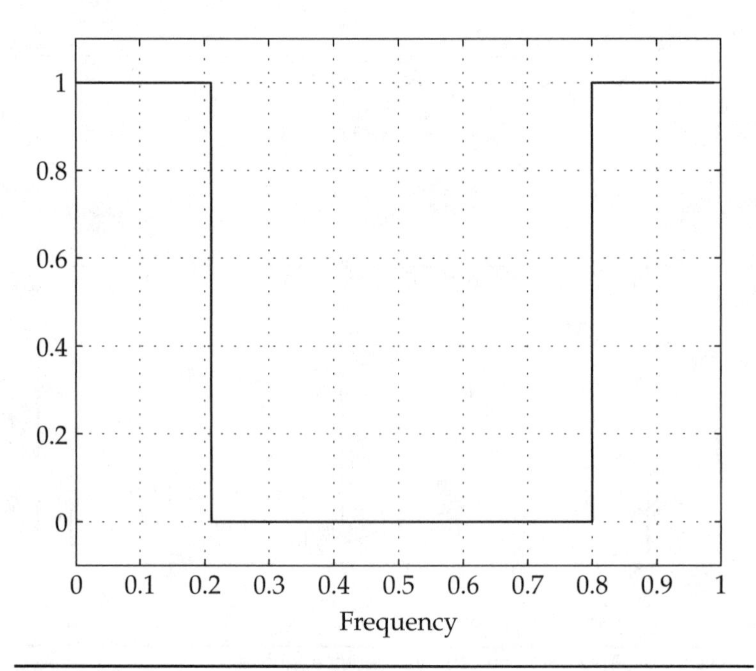

FIGURE 9-20 Factor to be applied to signal spectrum.

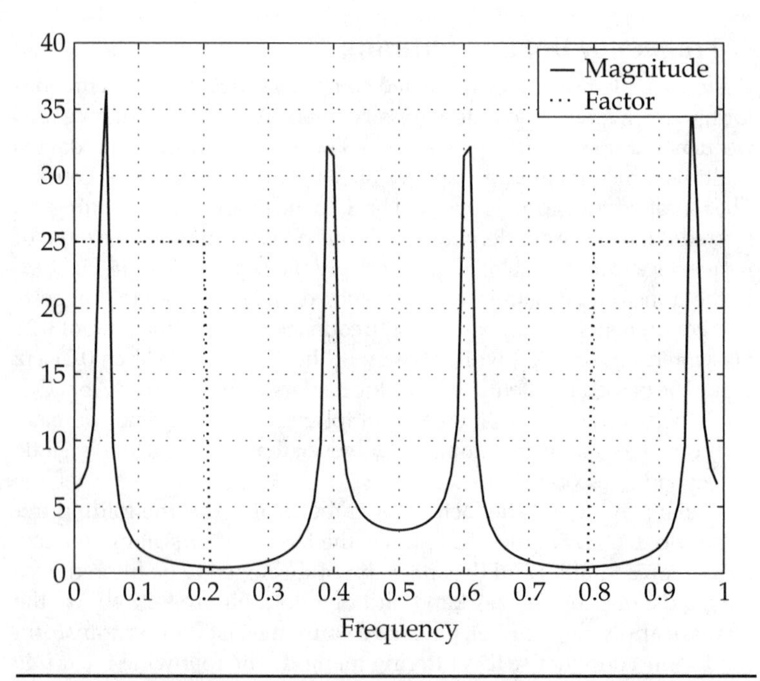

FIGURE 9-21 Signal spectrum (magnitude) and filtering factor (scaled by 25).

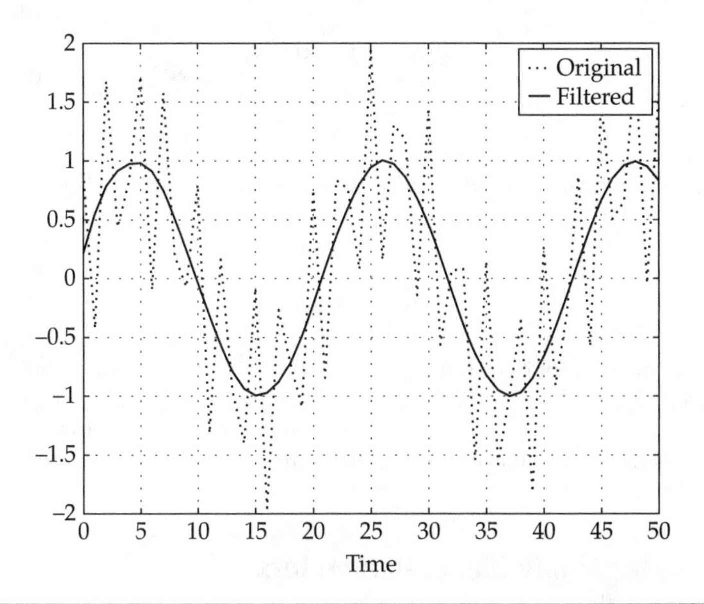

FIGURE 9-22 Original and filtered signal.

```
N=100; % generate test signal having two sinusoids
t=1:N;
y21=sin(2*pi*t/21.75);
y11=sin(2*pi*t/2.533);
ysig=y11+y21;
fNy=.5;  % set up frequencies
delf=1/N;
f=0:delf:1-delf;
Ysig=fft(ysig); % transform signal
i1=22*N/100;  % set up indices for factor
i2=51*N/100;
n=N;
nf=n/2+1;
fac(1:nf)=ones(1,nf);      % construct the factor: start
with all ones
fac(i1:i2)=zeros(1,i2-i1+1);  %zero out the removal region
fac(n:-1:n/2+2)=fac(2:n/2);
fac(n:-1:n/2+2)=fac(2:n/2); % duplicate the other half
of factor
Yfil=fac.*Ysig; % apply the 0-1 factor in frequency domain
yfil=ifft(Yfil); % transform back to time domain
```

9-15 The Discrete Time State-Space Equation

In Chap. 5 the continuous time state-space formulation was presented. It consists of two equations, as in

$$\frac{d}{dt}X = AX + BU \qquad (9\text{-}69)$$
$$y = Cx$$

The discrete time version has a similar form

$$x_i = \Phi x_{i-1} + \Lambda U_{i-1} \qquad (9\text{-}70)$$
$$y_i = Cx_i$$

A method for determining the matrices Φ and Λ from A and B is given in App. H. Note that Eq. (9-70) has the same form as Eq. (9-16) except that the quantities Φ and Λ, which are analogous to a and b, are matrices. Extensive use of this form of the state-space equation will be made in Chap. 10.

9-16 Determining Model Parameters from Experimental Data

One of the control algorithm tuning approaches presented in this chapter requires knowledge of the time constant and gain from a first-order model that best fits the actual process. Often, these parameters can be determined numerically from experimental data, say, resulting from applying step changes to the process.

9-16-1 First-Order Models

The discrete form of the first-order model is

$$y_i = ay_{i-1} + bU_{i-1} \qquad i = 0, 1, 2, \ldots$$
$$a = e^{-\frac{h}{\tau_p}} \qquad b = g(1-a) \qquad (9\text{-}71)$$
$$\tau_p = -\frac{h}{\ln a} \qquad g = \frac{b}{1-a}$$

Assuming that reader knows how to apply least squares regression to find a_1 and a_2 in $Y_i = a_1X_{1i} + a_2X_{2i}$, it should not be too much of a stretch for him to believe that least squares regression could be applied to $y_i = ay_{i-1} + bU_{i-1}$ to find the model parameters a and b by the following comparison:

$$Y_i = a_1X_{1i} + a_2X_{2i}$$
$$Y_i \Leftrightarrow y_i \qquad X_{1i} \Leftrightarrow y_{i-1} \qquad X_{2i} \Leftrightarrow U_{i-1}$$
$$y_i = ay_{i-1} + bU_{i-1}$$

where X_{1i} is equivalent to y_i shifted by one step. Using Eq. (9-71), the time constant and gain would follow from a and b and the PI control parameters would be

$$P = \frac{\tau_p}{g\tau_d} \qquad I = \frac{1}{g\tau_d}$$

Figure 9-23 shows how this would work for the case of slightly noisy data. Close inspection of Fig. 9-23 reveals that, due to the additive noise, the process output bounces slightly around the theoretical noiseless process output (to a degree almost undetectable with the naked eye). Furthermore, the plotted model determined by the least squares analysis is visually indecipherable from the theoretical model. The noiseless model used to generate the data (and to which the noise was added) was

$$y_i = 0.6065y_{i-1} + 0.4328U_{i-1} \qquad i = 0, 1, 2, \ldots$$

or

$$\frac{\hat{Y}(z)}{\hat{U}(z)} = \frac{0.4328z^{-1}}{1-0.6065z^{-1}}$$

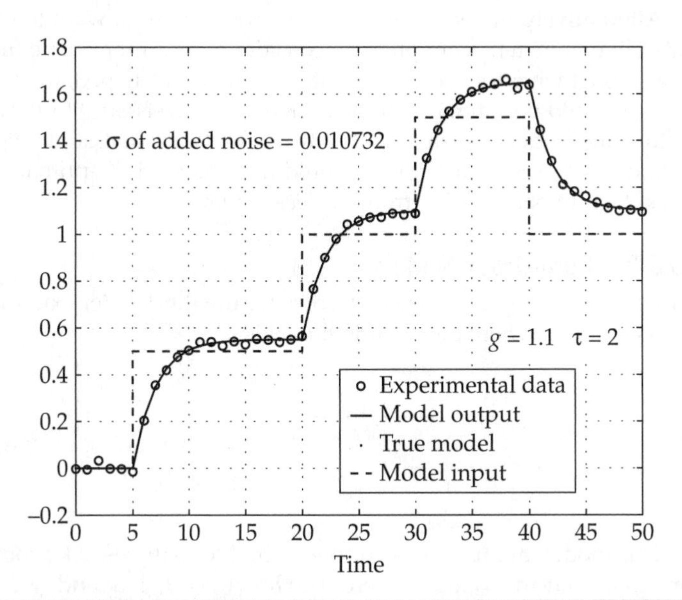

FIGURE 9-23 Fitting a slightly noisy first-order process with a first-order model.

This model has a z-plane pole at $z = 0.6065$ and the sampling interval h is 1.0. The model is a discrete time version of a first-order model

$$\frac{g}{\tau s+1} \qquad g = 1.1 \qquad \tau = 2$$

The pole of this model is at $s = -1/\tau = -.5$. As was shown in App. I, the s- and z-plane poles are related by

$$z = e^{sh} \qquad s = \frac{\ln z}{h} \qquad e^{-0.5} = 0.6065$$

The least squares model determined by using the Matlab `arx` function was

$$y_i = 0.6064 y_{i-1} + 0.4326 U_{i-1} \qquad i = 0, 1, 2, \ldots$$

(In the absence of the Matlab System Identification Toolbox, these parameters could be determined by any regression program—I used to write my own back in the QUICKBASIC era.) One could use the parameters from the least squares model, 0.6064 and 0.4326, to back-calculate the time constant and the gain and then the control parameters. Note that the pole of the numerically determined model is quite close to that of the theoretical noiseless model.

Alternatively, and sometimes preferably, you might want to estimate the parameters from physical considerations. If the above first-order model were for the liquid tank presented in Chap. 3, the time constant could be estimated from $\tau = R\rho A$. The resistance to flow R could come from a measurement of the flow F and the liquid height Y_0 and $R = Y_0 / F$. At least, you should use this kind of approach to check the numbers coming from the least squares fit.

9-16-2 Third-Order Models

Before the reader gets too excited, consider the third-order extension of this example where the starting model is

$$\frac{\tilde{Y}(s)}{\tilde{U}(s)} = \frac{g}{(\tau_1 s+1)(\tau_2 s+1)(\tau_3 s+1)} \qquad \tau_1 = 1 \qquad \tau_2 = 2 \qquad \tau_3 = 3 \tag{9-72}$$

This model might have been constructed for a three-tank process similar to that in Chap. 5 where $\tau_i = R_i \rho A_i, i = 1, 2, 3$ and $g = R_3$. This model has s-plane poles at $-1/\tau_1$, $-1/\tau_2$, $-1/\tau_3$ or −1, −0.5, −0.333.

The corresponding noiseless discrete time model, developed from Eq. (9-72) using Matlab model conversion routines, is

$$\frac{\widehat{Y}(z)}{\widehat{U}(z)} = \frac{0.01957z^{-1} + 0.05016z^{-2} + 0.007828z^{-3}}{1 - 1.1691z^{-1} + 0.9213z^{-2} - 0.1599z^{-3}} \tag{9-73}$$

This model has z-plane poles at 0.71653, 0.60653, and 0.36788 (the roots of the denominator). The z- and s-plane poles are related by

$$z = e^{sh} \qquad s = \frac{\ln z}{h} \qquad e^{-1} = 0.36788 \qquad e^{-0.5} = 0.6065$$

$$e^{-0.3333} = 0.7165$$

Adding the same small amount of noise to this model as was done to the first-order model and trying to fit a third-order least squares model (using `arx`) produces the results in Fig. 9-24:

As with the first-order case, the true model and empirically determined model are virtually visually indistinguishable because the added noise is so slight. The noiseless model, generating the data derived from Eq. (9-73), namely,

$$\begin{aligned} y_i = {} & 1.1691y_{i-1} - 0.9213y_{i-2} + 0.1599y_{i-3} + 0.01957U_{i-1} \\ & + 0.05016U_{i-2} + 0.007828U_{i-3} \qquad i = 0, 1, 2, \ldots \end{aligned} \tag{9-74}$$

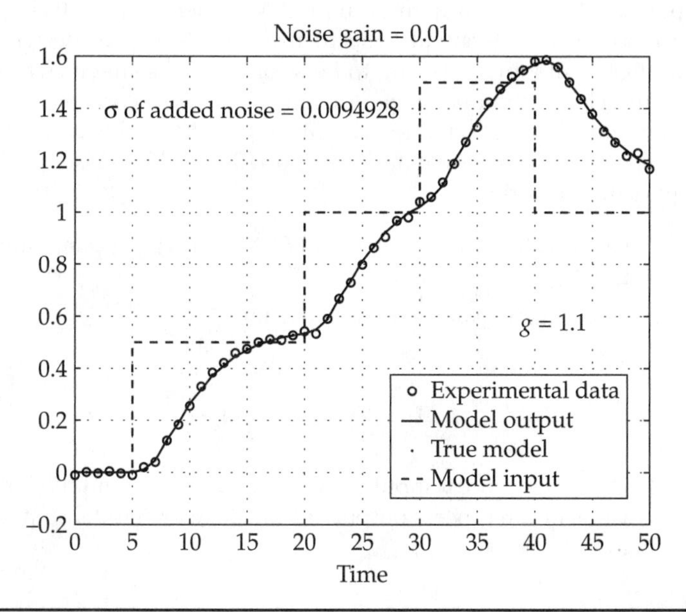

Figure 9-24 Fitting a slightly noisy third-order process with a third-order model.

has z-plane poles of 0.71653, 0.60653, and 0.36788. On the contrary, the empirically determined least squares model, derived from the noisy data in Fig. 9-24, using Matlab's `arx` routine, namely,

$$y_i = 0.7923y_{i-1} + 0.2457y_{i-2} - 0.2002y_{i-3} + 0.008279U_{i-1} + 0.06248U_{i-2} + 0.1088U_{i-3} \qquad i = 0, 1, 2, .. \tag{9-75}$$

has z-plane poles of −0.50001, 0.77656, and 0.51572. Therefore, unlike the case for the first-order model, one should not expect to be able to backsolve Eq. (9-75) for the values of the three time constants in Eq. (9-72). By the way, when I did the least squares fit for data generated from Eq. (9-74) with no noise, I got back Eq. (9-74).

Why not use Eq. (9-75) as a basis for designing a control algorithm? If you expect to operate the three-tank process only over the range of values that generated the data in Fig. 9-24 then perhaps it is OK. However, if you ever plan to operate the process under different conditions the model in Eq. (9-75) may give you problems. Since you were able to convince the process owner to let you disturb the process and generate the data in Fig. 9-24, you might ask for permission to run experiments that would yield numbers for the time constants and flow resistances. Then you might be able to generate a model that would behave well over a wider set of operating conditions.

9-16-3 A Practical Method

In practice, I have used numerical methods extensively in trying to find model parameters that I can then use for tuning or designing controllers. Without appearing to be a dinosaur, I sometimes avoid the regression techniques when dealing with noise-corrupted data (always much worse than the examples shown above). Instead, I read the step-change response data for y and U into a Matlab script that roughly has the form

1. Guess values for τ_p and g (and D if there is an apparent dead time).
2. Calculate the values of a and b using

$$a = e^{-\frac{h}{\tau_p}} \qquad b = g(1-a)$$

3. Using the process input data, U_i, $i = 1, 2, \ldots$ and the initial value of the process output y_0, calculate the model response from

$$y_i = ay_{i-1} + bU_{i-1-D} \qquad i = 1, 2, \ldots$$

4. Plot the model response and the actual response.
5. Adjust τ_p, g, and D based on visual assessment of the graph (it's not that difficult).
6. Go to step 2.

Repeat this cycle until you are satisfied. With a computer, this approach is quite quick and it will, at least, give you physically meaningful values.

In doing the parameter determination this way, I avoid the noise sensitivities of a black box regression method. As a manager, beware of the pitfalls of numerical model determination via least squares. Make sure that the engineer has visually verified the model by plotting it against the original data. In addition, make sure the verification plots are generated by feeding the process input and *just* the initial values of the process output (just enough process output data to get the model started) to the model. In effect, you are extrapolating from initial conditions. Do not use the model to predict the process output based on process input *and* process output data (one-step-ahead prediction). If this is done the model will always look great whether it is or not. Early in my career I fooled myself (and a lot of others) doing this.

9-17 Process Identification with White Noise Inputs

In App. F the Laplace transform of the impulse or Dirac delta function $I(s)$ was shown to be unity [see Eq. (F-11)].

$$L\{\delta(t)\} = I(s) = 1$$

If one were to move from the Laplace domain to the frequency domain via the $s \to j\omega$ substitution, the unity-valued transform would mean that the spectrum was constant. That is, all frequencies equally contribute power to the signal.

$$I(j\omega) = 1 \qquad -\infty < \omega < \infty$$

Alternatively, from the Fourier series point of view, one might say that to fit sinusoids to a pulse all of the frequencies are required.

Earlier, the Laplace transform of the process impulse response $I_m(s)$ was shown to be equal to the process transfer function [see Eq. (F-26) and (F-27)].

$$I_m(s) = G(s)I(s) = G(s)$$

$$L\{I_m(t)\} = G(s)$$

This suggests that if one could somehow measure the process impulse response then one could also identify the process transfer function. The enthusiasm for this idea is tempered by the difficulty in constructing a pulse in the time domain.

The spectrum of white noise is, in theory, also flat in the sense that each frequency contributes equally to the total power. In Chap. 8 the cumulative line spectrum was introduced to deal with the variability of the white noise line spectrum and to emphasize its flatness. This suggests that the impulse response is, in some way, equivalent to the white noise response. Thus, if white noise is the process input and a line spectrum is constructed from the resulting process output, the line spectrum of that output signal should have the shape of the Bode plot magnitude for the process transfer function.

Consider a first-order process with a time constant of 4.0 sec and a gain of unity (for simplicity). Its transfer function is

$$G(s) = \frac{g}{\tau s + 1}$$

The Bode plot magnitude is

$$|G(j\omega)| = \frac{g}{\sqrt{(\tau\omega)^2 + 1}}$$

which has a "corner" frequency of $\omega_c = 1/\tau$ or $f_c = 1/(2\pi\tau) =$ 0.0398 Hz.

Now, feed a discrete time white noise stream to the process and sample the response (and the input) at 1.0 Hz. The Nyquist or folding frequency is 0.5 Hz. For one run of 4096 points the spectrum of the white noise input is given in Fig. 9-25.

Note the rather strange shape because the frequency is plotted on a logarithmic scale but the actual spacing in the frequency domain is constant at

$$f_1 = 1/(Nh) = 1/(4098) = 2.4402\text{e}-004 \text{ Hz}$$

Also, note the "ratty" variation of the line spectrum. This is consistent with comments in Chap. 8 about white noise spectra.

The spectrum of the response of the first-order process to the white noise is shown in Fig. 9-26. This line spectrum does not look much like the magnitude from the Bode plot for a first-order process because of the erratic nature of the white noise. To get anything resembling the Bode plot one must repeat this exercise many

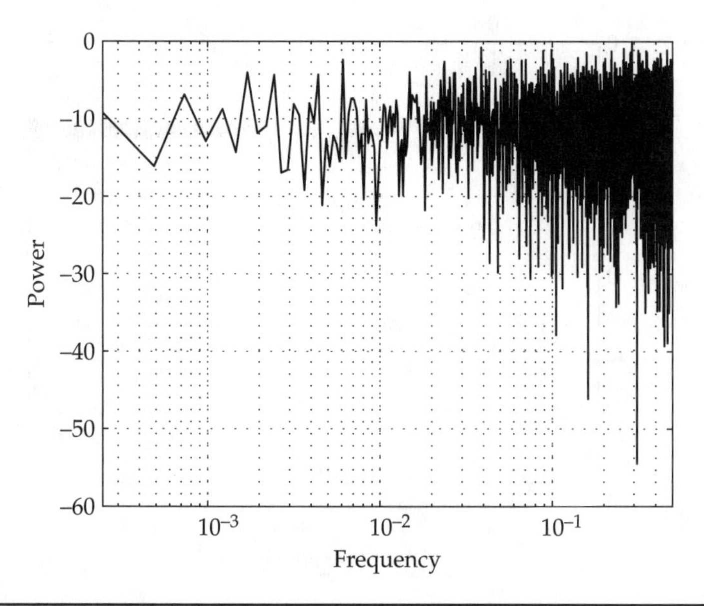

FIGURE 9-25 Spectrum of white noise input into first-order process.

times and average the results. Figure 9-27 shows the average line spectrum of the input white noise for 100 runs and Fig. 9-28 shows the average line spectrum of the process output for those same runs.

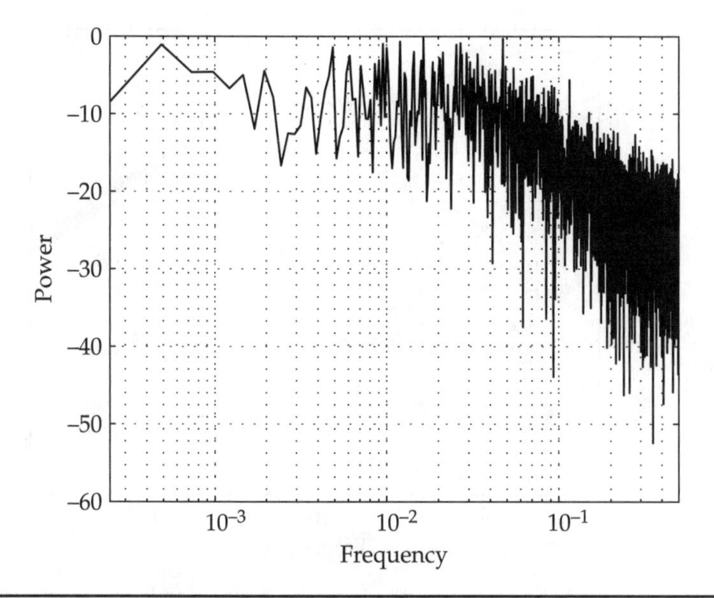

FIGURE 9-26 Spectrum of first-order response to white noise.

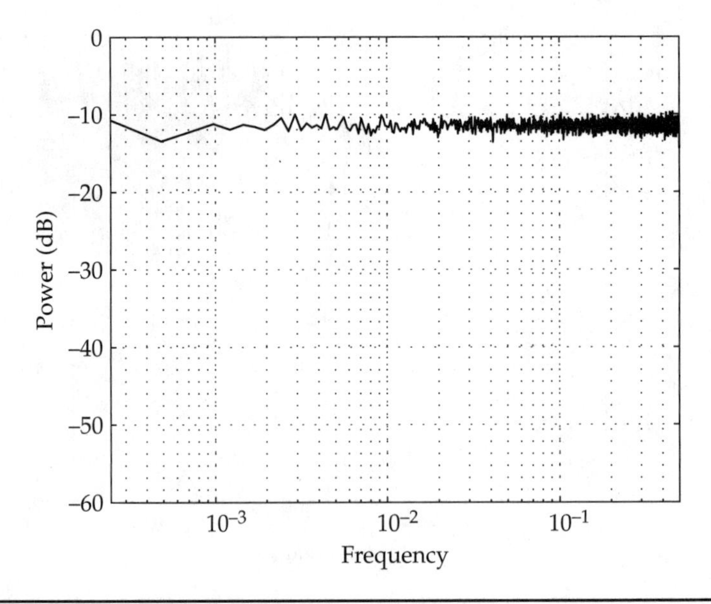

FIGURE 9-27 Spectrum of averaged white noise input.

Figure 9-28 shows the approximate shape of a first-order Bode plot magnitude where one can see the corner frequency of 0.0398 Hz. Figure 9-29 shows the theoretical Bode plot magnitude for the discrete time first-order process. The gain in Fig. 9-28 depends on the

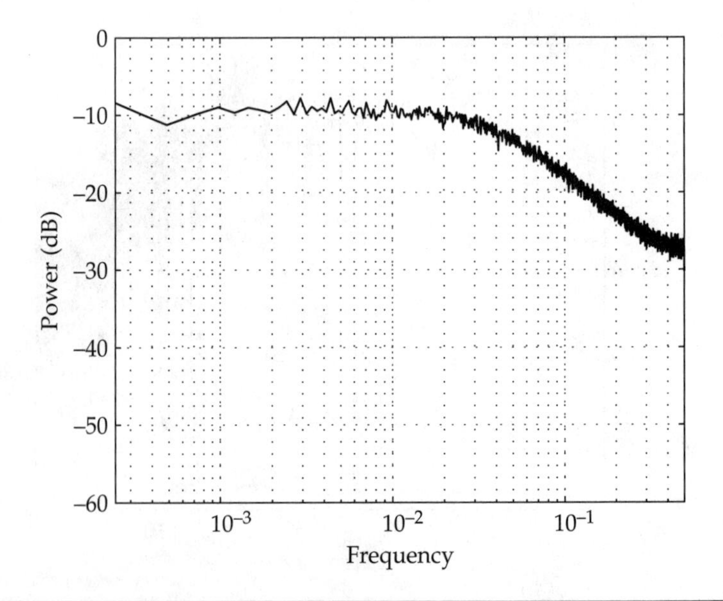

FIGURE 9-28 Spectrum of averaged white noise response.

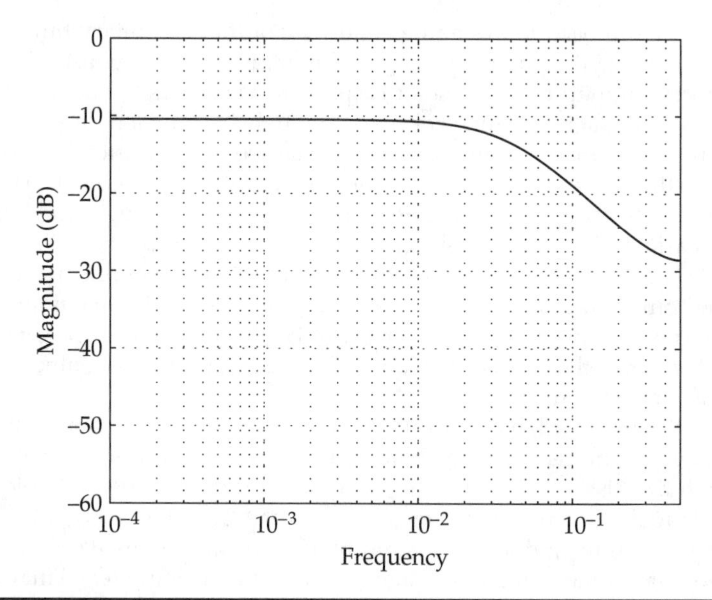

FIGURE 9-29 Bode plot magnitude for sampled first-order process ($TC = 4$, $h = 1$).

intensity of the white noise input so the graphs were adjusted vertically to make the comparison clearer.

This exercise suggests that using white noise as an input for process identification may require quite a bit of work. Alternatively, one could look at decreasing the length of the sample run which, as pointed out in Chap. 8, would decrease the number of "bins," increase the power per bin and decrease the variation while also decreasing the frequency resolution.

9-18 Summary

As warned, this has been a long chapter. We introduced the reader to the discrete time domain by discretizing the first-order model's step-change response and then breaking the input up into a series of contiguous steps. The model became an indexed algebraic equation. The backshift operator was applied to this equation and the result was called the Z-transform. A couple of mathematical subtleties associated with sampling and the zero-order hold were discussed and reference was made to App. I where a more elegant and general approach is taken. The concept of a discrete time filter was mentioned in connection with the first-order model.

This discretization operation was applied to several common equations that had been derived earlier in the book in the continuous time domain, such as the FOWDT model and the PI and PIfD control algorithms.

An approach to designing control algorithms in the continuous and discrete time domains was presented. Here the response of the process output to the set point is specified via an easily understood parameter such as a time constant. The transfer function describing this specification is combined with the closed-loop transfer function and the as-yet-undetermined control algorithm is derived. This approach was applied to the first- and second-order models, yielding an easily tunable PI control algorithm in the former case and a filtered PID control algorithm in the latter case. In the discrete time domain the control of a first-order process model yielded a discrete time PI control algorithm with another set of easily tunable control gains. When applied to the FOWDT model, a special dead-time compensation algorithm resulted.

A simple variable substitution allowed for the transition to the frequency domain where some of the special concepts associated with the discrete time domain were brought out via the Bode plot. This tool was applied to several kinds of filters: low-pass autoregressive filters, double-pass no-phase-lag autoregressive filters, moving average filters, and high-pass or differencing filters. Finally, filtering in the frequency domain was covered. The discrete time state-space system was briefly introduced. We will return to it in Chaps. 10 and 11.

The chapter closed with a brief discussion of model parameter determination. First, the method of least squares was applied to step-change response data. This exercise suggested that the presence of noise can cloud the issue when the models are of higher order. Second, spectral analysis was applied to white noise response data. This exercise suggested that extensive averaging might be required for a feasible identification.

We did not spend any time on developing time-domain solutions to problems having Z-transforms via inversion simply because there were more important and useful things to do. However, App. I does briefly touch on that topic.

Next, we tackle the problem of combining process models with noisy experiments to estimate the state of the system.

CHAPTER 10

Estimating the State and Using It for Control

In Chap. 5, matrices and the concept of the state were introduced. In Chap. 6 an underdamped process was studied where the state consisted of the position and speed (or derivative of the position) of the mass in the mass/spring/dashpot process. We showed that regular proportional-integral (PI) control, which uses only the position, did not do a good job for this process. However, the proportional-integral-derivative (PID) algorithm which uses the position and the speed, that is, the state, performed significantly better. Another method fed back the state to create a new process that had better dynamic characteristics. It appears that an estimate of the state can play a crucial part in the successful control of a process.

This chapter will present a method that combines a model of the actual process with process measurement(s) to produce an estimate of the state. It will be applied to the control problem posed in Chap. 6. The method, called the Kalman filter, was developed in the late 1950s. To use the Kalman filter, one must find values of a vector called the Kalman gain. Two ways to find this gain will be presented. The first is based on choosing variances associated with the process model and with the process noise. The second is based on placing the eigenvalues of the system.

The Kalman filter will also be applied to the three-tank problem presented in Chap. 5. However, the variables to be controlled will be extended to include all the three tank levels and the variables to be manipulated will be extended to include all the input flow rates. The resulting multidimensional control algorithm will contain integral control and will be tuned by placing the eigenvalues of the controlled system. For the sake of comparison, the same three-tank process will be controlled by three separate PI controllers. Finally, the state-space control approach will be applied to a lumped approximation of the tubular energy exchanger process presented in Chap. 7.

10-1 An Elementary Presentation of the Kalman Filter

The Kalman filter combines the predicted value of the state from a model with suitably adjusted process measurements to provide an estimate of the state. The first component of the Kalman filter is the process model.

10-1-1 The Process Model

Consider the continuous time case where the process is described by

$$\frac{d}{dt}x = Ax + BU \\ Z = Hx \tag{10-1}$$

where x is a $(n, 1)$ vector, A is a (n, n) matrix, B is a (n, m) matrix, U is a $(m, 1)$ vector, and H is the (p, n) "measurement" matrix. The quantity Z, a $(p, 1)$ vector, is the measured quantity. If all the elements of the state are measurable, then $p = n$ and the H matrix is square. If some of the states are not measurable, then $p < n$. In the case of the underdamped process, it might be the position that is the only part of the state available for measurement and therefore, $n = 2, p = 1$.

The discrete time version of Eq. (10-1) is developed in App. H as

$$x_{i+1} = \Phi x_i + \Gamma U_i \\ Z_i = H x_i \tag{10-2}$$

This discrete time model is augmented by two sources of noise, as follows:

$$x_{i+1} = \Phi x_i + \Gamma U_i + w_i \\ Z_i = H x_i + v_i \tag{10-3}$$

where w is sometimes called *process noise* and can represent the error between the model and the actual process. The symbol v is sometimes called *measurement noise*. Both of these stochastic processes are considered to be white, have zero mean, have a normal distribution, and have covariances (with zero lag), symbolized by matrices Q and R, respectively. The covariance matrix was introduced in Chap. 8. In the scalar case we will use σ_w^2 and σ_v^2. The covariance matrix Q is a measure of the model uncertainty and the covariance matrix R is a measure of the measurement noise.

10-1-2 The Premeasurement and Postmeasurement Equations

In many texts the derivation of the Kalman filter appears, in my humble opinion, to be one of the most convoluted exercises in control engineering theory. I will not attempt to derive it here. If the reader thinks, after the presentation in this section, that she needs to delve into the derivation for a better understanding, I recommend *Applied Optimal Estimation*, edited by Arthur Gelb. This book was first published in 1974 and is still probably one of the most readable books around. Do not attempt to read Rudolf Kalman's original paper! As an interesting alternative, one might visit the Internet and see what the Wikipedia has to say about the Kalman filter.

There are two stages in the estimation: Before the measurement and after the measurement. A quantity estimated before the measurement is taken (using the model) will have (–) appended to its symbol. Quantities estimated after the measurement is taken will have the (+) appendage.

Before a measurement is taken at the *k*th sample time, the model can be used to generate an estimate at time t_k, as in

$$\hat{X}_k(-) = \Phi \hat{X}_{k-1}(+) + \Gamma U_{k-1} \tag{10-4}$$

Equation (10-4) gives the premeasurement state estimate $\hat{X}_k(-)$ at time t_k based on knowledge of the process input U_{k-1} and the postmeasurement estimate of the state from time t_{k-1} which is $\hat{X}_{k-1}(+)$. Using the model to predict a value at time t_k based on information at time t_{k-1} is sometimes referred to *one-step extrapolation*. Note that the tilde symbolizes that the quantity is an estimate of the true value X_k.

The postmeasurement estimate at time t_k is calculated from

$$\hat{X}_k(+) = K_k Z_k + (I - K_k H)\hat{X}_k(-) \tag{10-5}$$

where Z_k is the measurement at time t_k and K_k is the (n, p) Kalman gain vector at time t_k. Equation (10-5) suggests that the postmeasurement is a weighted sum of the measurement Z_k and the premeasurement model-based estimate $\hat{X}_k(-)$. Equation (10-5) can also be written as

$$\hat{X}_k(+) = \hat{X}_k(-) + K_k[Z_k - H\hat{X}_k(-)] \tag{10-6}$$

which shows that the postmeasurement estimate $\hat{X}_k(+)$ is equal to the premeasurement model-based estimate $\hat{X}_k(-)$ plus a correction term proportional to the difference between the measurement Z_k and the model-based estimated state $\hat{X}_k(-)$ modified by the measurement matrix H, namely, $Z_k - H\hat{X}_k(-)$. The reader should take a moment and try to figure out the dimensions of the matrices and vectors in Eqs. (10-5) and (10-6). The measurement vector Z_k has a dimension of $(p, 1)$. The state vector $\hat{X}_k(-)$ has a dimension of $(n, 1)$, where $n \geq p$. The measurement matrix H has dimension (p, n) and the Kalman filter gain K_k has dimension (n, p). Equation (10-6) shows how the constraint of not being able to measure the full n-dimensional state does not mean it cannot be estimated (oops, double negatives again).

I would hope that the presentation of the Kalman filter, so far, does not conflict with the reader's common sense. Equation (10-4) is used to generate a premeasurement estimate via a process model and Eq. (10-5) or (10-6) is used to improve that estimate using measured values.

10-1-3 The Scalar Case

Temporarily consider the scalar first-order case where $n = 1$, $p = 1$, and $m = 1$. In this case the state is one-dimensional and the measurement of the state is available but may be noisy. The premeasurement state estimate, via extrapolation, is

$$\hat{X}_k(-) = e^{-\frac{h}{\tau}}\hat{X}_{k-1}(+) + g\left(1 - e^{-\frac{h}{\tau}}\right)U_{k-1} \tag{10-7}$$

and the postmeasurement correction is

$$\hat{X}_k(+) = \hat{X}_k(-) + K_k[Z_k - \hat{X}_k(-)] \tag{10-8}$$

where all of the quantities are scalars. If the model is quite accurate, and the measurement is noisy, then $\sigma_w \approx 0$ and σ_v would be relatively large and you might expect that K_k would be small. Conversely, if the model is only approximate but the measurement is quite good, then $\sigma_v \approx 0$ and σ_w would be relatively large and you would expect K_k to be significant.

10-1-4 A Two-Dimensional Example

For the two-dimensional case, namely, the underdamped process covered in Chap. 6, we have $n = 2$, $p = 1$, and $m = 1$, where the only measurement available is the position of the mass. Equation (10-4) would look like

$$X_{i+1} = e^{Ah}X_i + A^{-1}(e^{Ah} - I)BU_i$$

$$X_{i+1} = \Phi X_i + \Gamma U_i$$

$$\Phi = e^{Ah} \qquad \Gamma = A^{-1}(e^{Ah} - I)B$$

where

$$A = \begin{pmatrix} 0 & 1 \\ -\omega_n^2 & -2\zeta\omega_n \end{pmatrix} \qquad B = \begin{pmatrix} 0 \\ g\omega_n^2 \end{pmatrix}$$

$$X = \begin{pmatrix} \text{position} \\ \text{speed} \end{pmatrix}$$

$$H = (1 \quad 0) \tag{10-9}$$

The numerical values for the e^{Ah}, Φ, and Γ matrices are developed in App. H. The last element of Eq. (10-9), $H = (1 \quad 0)$, says that only the position is measurable. You are urged to pause here and study these equations and perhaps take a look at App. H.

The premeasurement prediction based on the model is

$$\hat{X}_k(-) = \Phi\hat{X}_{k-1}(+) + \Gamma U_{k-1}$$

and the postmeasurement correction is

$$\hat{X}_k(+) = \hat{X}_k(-) + K_k[Z_k - H\hat{X}_k(-)]$$

$$\begin{pmatrix} \hat{x}_{1k} \\ \hat{x}_{2k} \end{pmatrix}(+) = \begin{pmatrix} \hat{x}_{1k} \\ \hat{x}_{2k} \end{pmatrix}(-) + \begin{pmatrix} k_{1k} \\ k_{2k} \end{pmatrix}\left[z_k - (1 \quad 0)\begin{pmatrix} \hat{x}_{1k} \\ \hat{x}_{2k} \end{pmatrix}(-)\right] \tag{10-10}$$

where z_k is the scalar measurement of the position.

10-1-5 The Propagation of the Covariances

In addition to the pre- and postmeasurement state estimates, there are pre- and postmeasurement covariances associated with these state estimates. The premeasurement covariance matrix is denoted by $P_k(-)$ and is defined as

$$P_k(-) = E\left\{X_k^e(-)X_k^e(-)^T\right\} \tag{10-11}$$

where $X_k^e(-)$ is the state estimation error at time t_k, as in

$$X_k^e(-) = \hat{X}_k(-) - X_k \tag{10-12}$$

In the scalar case, $P_k(-)$ would be a standard deviation that gives an indication of the quality of the model. In general, $P_k(-)$ is calculated from

$$P_k(-) = \Phi P_{k-1}(+)\Phi^T + Q$$
$$P_0(-) = Q \tag{10-13}$$

where Q is the covariance matrix associated with the model uncertainty. Equation (10-13) shows that the model uncertainty propagates via extrapolation. Neither the measurement nor the measurement error appears in this equation. For the one-dimensional scalar case, Eq. (10-13) becomes

$$P_k(-) = e^{-\frac{2h}{\tau}} P_{k-1}(+) + \sigma_w^2$$

The postmeasurement covariance matrix, denoted by $P_k(+)$ and defined similarly to $P_k(-)$, is calculated from

$$P_k(+) = (I - K_k H)P_k(-) \tag{10-14}$$

This covariance matrix is a measure of how much the state estimate uncertainty is changed by making the measurement.

10-1-6 The Kalman Filter Gain

The derivation of the equation for the Kalman filter gain K_k is the point where the amazons are separated from the girls and, as mentioned in Sec. 10-1-2, I will not attempt to present it here. Instead, I will show you the results of the derivation. There are many horribly intricate derivations of this equation in the literature but one of the easiest to follow is in the book by Gelb, cited in Sec. 10-1-2.

The Kalman filter gain is calculated from

$$K_k = P_k(-)H^T (H P_k(-)H^T + R)^{-1} \tag{10-15}$$

In the scalar case, the full set of equations would be as follows.

1. The premeasurement prediction based on the model

$$\hat{X}_k(-) = \Phi \hat{X}_{k-1}(+) + \Gamma U_{k-1}$$

$$\Leftrightarrow \quad \hat{X}_k(-) = e^{-\frac{h}{\tau}} \hat{X}_{k-1}(+) + g\left(1 - e^{-\frac{h}{\tau}}\right) U_{k-1}$$

2. The propagation of the premeasurement covariance.

$$P_k(-) = \Phi P_{k-1}(+)\Phi^T + Q \quad \Leftrightarrow \quad P_k(-) = e^{-\frac{2h}{\tau}} P_{k-1}(+) + Q$$

$$P_0(-) = \mathrm{Q} \qquad P_0(-) = \sigma_w^2$$

3. The calculation of the Kalman filter gain

$$K_k = P_k(-)H^T[H P_k(-)H^T + R]^{-1} \quad \Leftrightarrow \quad K_k = \frac{P_k(-)}{P_k(-) + \sigma_v^2}$$

Note that if the model is quite good relative to the measurement then σ_v would be small and $K_k \cong 1$. On the other hand, if the measurement is quite good relative to the model then σ_w would be small and σ_v would be large and K_k would be small.

4. The correction of the premeasurement prediction with the measurement

$$\hat{X}_k(+) = \hat{X}_k(-) + K_k[Z_k - H\hat{X}_k(-)]$$

$$\Leftrightarrow \quad \hat{X}_k(+) = \hat{X}_k(-) + K_k[Z_k - \hat{X}_k(-)]$$

5. The updating of the postmeasurement covariance.

$$P_k(+) = (I - K_k H)P_k(-) \quad \Leftrightarrow \quad P_k(+) = (1 - K_k)P_k(-)$$

10-2 Estimating the Underdamped Process State

For the case of $\zeta = 0.1$ and $h = 0.5$ the discrete time process model is shown in App. H to be

$$\hat{X}_k = \Phi\hat{X}_{k-1} + \Gamma U_{k-1}$$

$$\Phi = e^{Ah} = \begin{pmatrix} 0.694054 & 0.455438 \\ -0.455438 & 0.05643988 \end{pmatrix}$$

$$\Gamma = A^{-1}(I - e^{Ah})B = \begin{pmatrix} 0.3059456 \\ 0.455438 \end{pmatrix} \tag{10-16}$$

$$z_k = (1 \quad 0)\begin{pmatrix} x_{1k} \\ x_{k2} \end{pmatrix}$$

For this example, the following covariance components were chosen:

$$Q = \begin{pmatrix} 1 & 0 \\ 0 & 1 \end{pmatrix} \sigma_w^2 \qquad \sigma_w = 0.1 \qquad \text{and} \qquad \sigma_w = 0.6$$

$$R = \sigma_v^2 \qquad \sigma_v = 0.4 \qquad \text{and} \qquad \sigma_v = 0.4$$

There are two sets of standard deviations: the first for the case where the model is better.

Figure 10-1 shows how the two elements of the Kalman gain settle out to steady-state values. The solid line represents the "poorer" model case and the magnitude of the Kalman gains are relatively large. Note that by about 30 steps the gains have reached steady values.

For the "good" model case, the estimated and "true" (from the model) states are shown in Fig. 10-2. In addition, for the good model case, the measured value, the "true" value from the model, and the estimated value of the position is shown in Fig. 10-3. The "true" values were calculated from the model, sans noise.

Note that the estimate is relatively close to the model and puts less weight on the measurements.

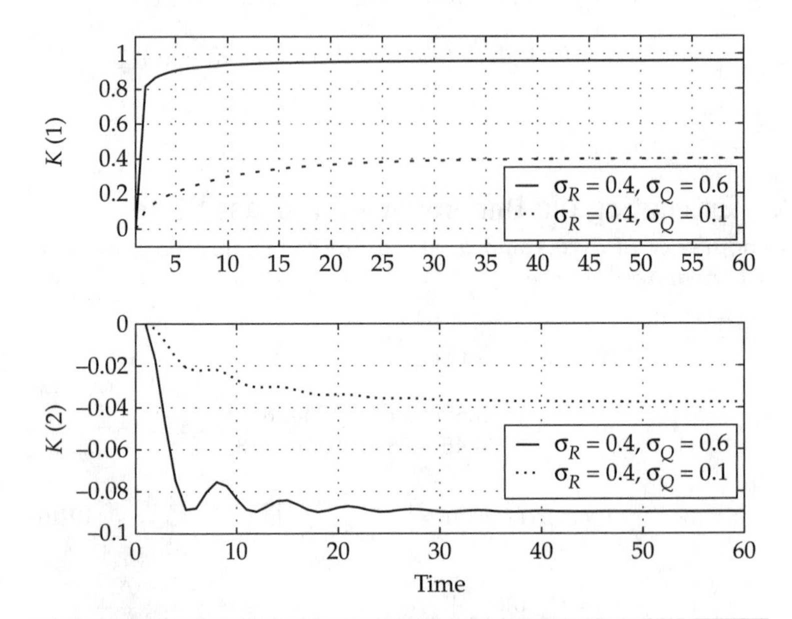

FIGURE 10-1 Kalman filter gains for underdamped process.

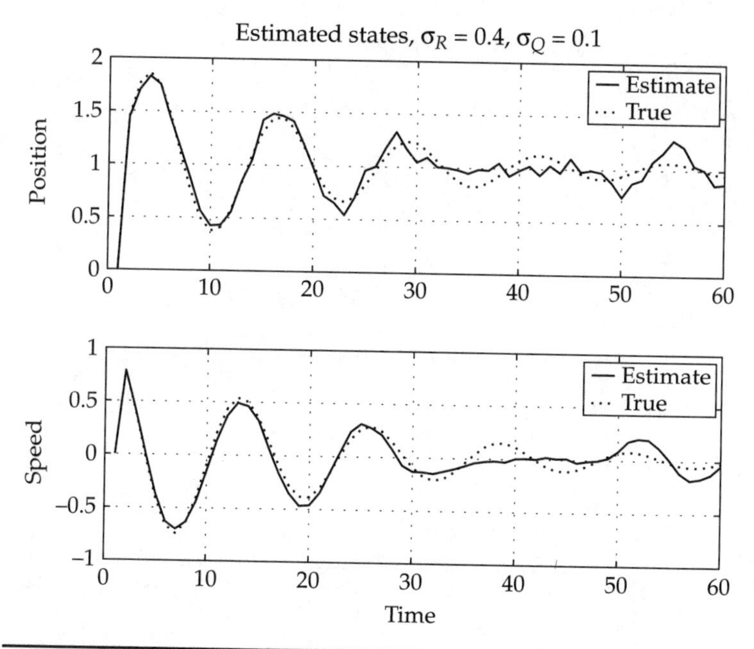

FIGURE 10-2 "Good" model case: estimated and true states.

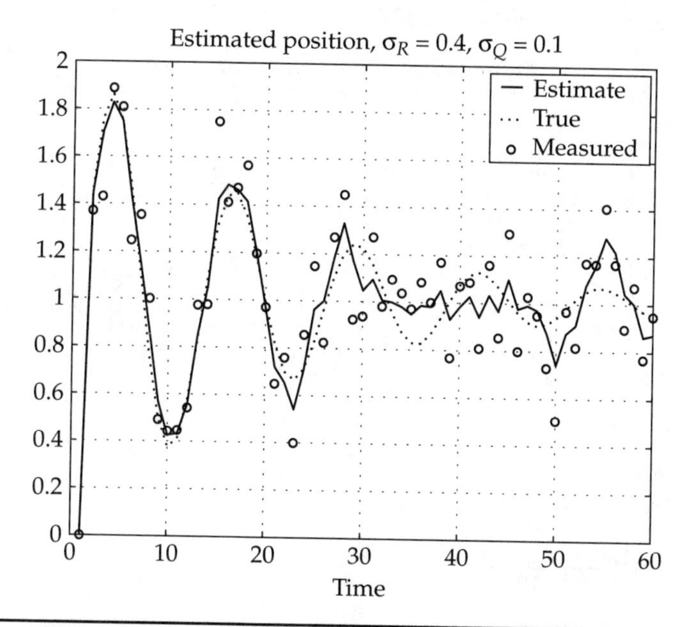

FIGURE 10-3 "Good" model case: estimated, true, and measured position.

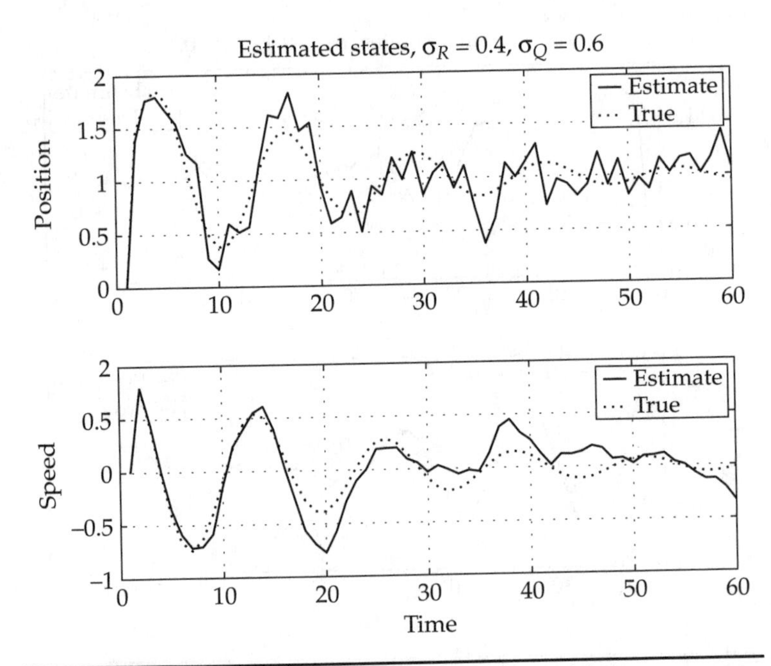

FIGURE 10-4 "Poorer" model case: estimated and true states.

For the "poorer" model case, the estimated and "true" states are shown in Fig. 10-4. Finally, for the "poorer"model case, the measured value, the "true" value from the model, and the estimated value of the position is shown in Fig. 10-5.

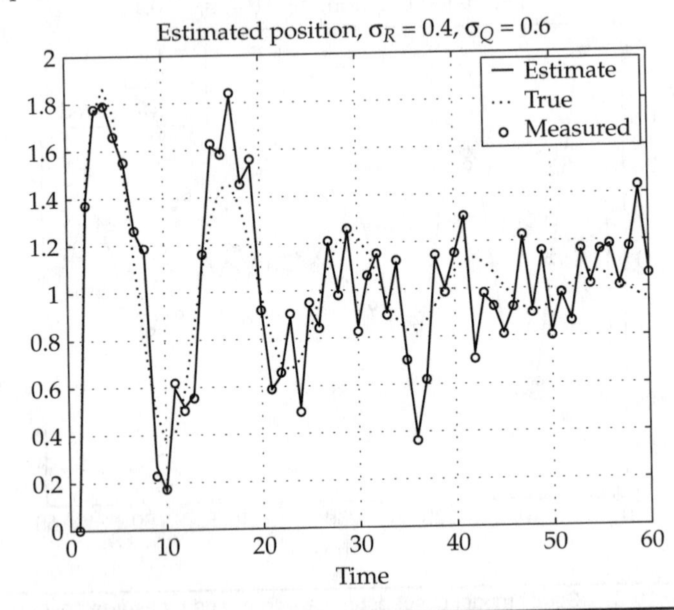

FIGURE 10-5 "Poorer" model case: estimated, true, and measured position.

Compared with Fig. 10-3, Fig. 10-5 shows that more weight is placed on the measurements.

The element-by-element calculations come from the following equations (sometimes it is not obvious to a novice how these vectors and matrices go together so I will show you the gory details):

1. The premeasurement prediction based on the model.

$$\hat{X}_k(-) = \Phi \hat{X}_{k-1}(+) + \Gamma U_{k-1} \quad \Rightarrow$$

$$\begin{pmatrix} \hat{x}_{1k} \\ \hat{x}_{2k} \end{pmatrix}(-) = \begin{pmatrix} 0.694054 & 0.455438 \\ -0.455438 & 0.05643988 \end{pmatrix} \begin{pmatrix} \hat{x}_{1k-1} \\ \hat{x}_{2k-1} \end{pmatrix}(+) + \begin{pmatrix} 0.3059456 \\ 0.455438 \end{pmatrix} U_{k-1}$$

2. The propagation of the premeasurement covariance.

$$P_k(-) = \Phi P_{k-1}(+) \Phi^T + Q \quad \Rightarrow$$

$$\begin{pmatrix} p_{11k} & p_{12k} \\ p_{21k} & p_{22k} \end{pmatrix}(-) = \begin{pmatrix} 0.694054 & 0.455438 \\ -0.455438 & 0.05643988 \end{pmatrix} \begin{pmatrix} p_{11k} & p_{12k} \\ p_{21k} & p_{22k} \end{pmatrix} (+) \begin{pmatrix} 0.694054 & -0.455438 \\ 0.455438 & 0.05643988 \end{pmatrix} + \begin{pmatrix} 1 & 0 \\ 0 & 1 \end{pmatrix} \sigma_w^2$$

$$\begin{pmatrix} p_{110} & p_{120} \\ p_{210} & p_{220} \end{pmatrix}(-) = \begin{pmatrix} 1 & 0 \\ 0 & 1 \end{pmatrix} \sigma_w^2$$

3. The calculation of the Kalman filter gain

$$K_k = P_k(-) H^T [H P_k(-) H^T + R]^{-1} \quad \Rightarrow$$

$$\begin{pmatrix} k_{1k} \\ k_{2k} \end{pmatrix} = \begin{pmatrix} p_{11k} & p_{12k} \\ p_{21k} & p_{22k} \end{pmatrix}(-) \begin{pmatrix} 1 \\ 0 \end{pmatrix} \left[\begin{pmatrix} 1 & 0 \end{pmatrix} \begin{pmatrix} p_{11k} & p_{12k} \\ p_{21k} & p_{22k} \end{pmatrix}(-) \begin{pmatrix} 1 \\ 0 \end{pmatrix} + \frac{1}{\sigma_v^2} \right]^{-1}$$

After multiplying out the matrices, the element in the square brackets is a scalar so taking the inverse is trivial.

4. The correction of the premeasurement prediction with the measurement

$$\hat{X}_k(+) = \hat{X}_k(-) + K_k [Z_k - H \hat{X}_k(-)] \quad \Rightarrow$$

$$\begin{pmatrix} \hat{x}_{1k} \\ \hat{x}_{2k} \end{pmatrix}(+) = \begin{pmatrix} \hat{x}_{1k} \\ \hat{x}_{2k} \end{pmatrix}(-) + \begin{pmatrix} k_{1k} \\ k_{2k} \end{pmatrix} \left[z_k - \begin{pmatrix} 1 & 0 \end{pmatrix} \begin{pmatrix} \hat{x}_{1k} \\ \hat{x}_{2k} \end{pmatrix}(-) \right]$$

After the matrix multiplication, the element in the square brackets is a scalar.

5. The updating of the postmeasurement covariance.

$$P_k(+) = (I - K_k H)P_k(-) \quad \Rightarrow$$

$$\begin{pmatrix} p_{11k} & p_{12k} \\ p_{21k} & p_{22k} \end{pmatrix}(+) = \left[\begin{pmatrix} 1 & 0 \\ 0 & 1 \end{pmatrix} - \begin{pmatrix} k_{1k} \\ k_{2k} \end{pmatrix}(1 \quad 0)\right]\begin{pmatrix} p_{11k} & p_{12k} \\ p_{21k} & p_{22k} \end{pmatrix}(-)$$

In this example the "true" values came from the process model. A more correct approach would have used a separate model for the actual process to generate the measured values.

Before getting too excited about the results of this idealized simulation, remember that we have assumed that the process is perfectly represented by the state-space model. In reality, this will not be true and, if the model is really poor, the Kalman filter may do more harm than good. Also, choosing the elements of the covariance matrices Q and R can sometimes be more of an art than a science.

The graphs show that the values of the Kalman gain tend toward a steady-state value which we will denote simply as K. Similarly, the covariance matrices, $P_k(+)$ and $P_k(-)$ converge to steady values. The equations of the Kalman filter show that the calculation of these three quantities is independent of the measurements and that they can be calculated separately. Often, the steady-state values are calculated initially and stored before they are used in the estimation calculations.

10-3 The Dynamics of the Kalman Filter and an Alternative Way to Find the Gain

The dynamics of the Kalman filter can be studied by combining the model with the measurement, when the steady-state value of the Kalman gain K is used. First, the extrapolation equation using information at time t_{k-1}

$$\hat{X}_k(-) = \Phi \hat{X}_{k-1}(+) + \Gamma U_{k-1}$$

is combined with the measurement update equation using information at time t_k

$$\hat{X}_k(+) = \hat{X}_k(-) + K[Z_k - H\hat{X}_k(-)]$$

to eliminate $\hat{X}_k(-)$, resulting in

$$\hat{X}_k(+) = \Phi \hat{X}_{k-1}(+) + \Gamma U_{k-1} + K[Z_k - H\Phi \hat{X}_{k-1}(+) + H\Gamma U_{k-1}]$$

Collecting coefficients of $\hat{X}_{k-1}(+)$ yields

$$\hat{X}_k(+) = (\Phi - KH\Phi)\hat{X}_{k-1}(+) + (I + KH)\Gamma U_{k-1} + KZ_k \tag{10-17}$$

This indexed discrete time equation, in a manner analogous to differential equations, has a homogeneous part and a nonhomogeneous part, as in

$$\hat{X}_k(+) = \underbrace{(\Phi - KH\Phi)\hat{X}_{k-1}(+)}_{\text{homogeneous}} + \underbrace{(I + KH)\Gamma U_{k-1} + KZ_k}_{\text{nonhomogeneous}}$$

To solve the homogeneous part,

$$\hat{X}_k^h(+) = [\Phi - KH\Phi]\hat{X}_{k-1}^h(+) \tag{10-18}$$

one can try

$$\hat{X}_k^h(+) = C\lambda^k \tag{10-19}$$

where C is a vector and λ is a scalar. This is similar to trying $Ce^{\lambda t}$ for a continuous time differential equation.

When Eq. (10-19) is tried, Eq. (10-18) becomes

$$C\lambda^k = [\Phi - KH\Phi]C\lambda^{k-1}$$

or, after dividing both sides by the scalar quantity λ^k,

$$[\Phi - KH\Phi - \lambda I]C = 0 \tag{10-20}$$

As shown in App. H, λ is an eigenvalue of the matrix $\Phi - KH\Phi$. If the size of this matrix is (n, n), then the solution will have the form

$$\hat{X}_k^h(+) = \sum_{i=1}^{n} C_i(\lambda_i)^k \tag{10-21}$$

Common sense suggests that for the Kalman filter to behave in a stable manner, all of the eigenvalues of $\Phi - KH\Phi$ should lie inside (or on) the unit circle in the complex z-plane just as we required the eigenvalues of the differential equations to lie in the left-hand side of the complex s-plane. In general, the eigenvalues will be complex, as in $\lambda_i = a_i + jb_i$, therefore, stability requires that

$$|\lambda_i| = |a_i + j_i b| \leq 1 \tag{10-22}$$

Furthermore, as the location of the eigenvalues (or poles) moves toward the origin of the z-plane, the transient or homogeneous component will die away more quickly (and the Kalman filter will be more aggressive).

The requirement represented by Eq. (10-22) and the discussion about the dependence of the dynamics on the placement of the eigenvalues inside the unit circle suggests an alternative approach to designing the Kalman filter. That is, instead of playing number games with the covariance matrices, Q and R, one specifies where they want the eigenvalues of $\Phi - KH\Phi$ to lie inside the unit circle and then determines the value of the Kalman gain K from the resulting equations. In fact, Matlab has a built-in function, `place`, that can be used to do this for you (and me) transparently.

10-3-1 The Dynamics of a Predictor Estimator

A slight modification to the preceding Kalman filter equations uses information, including the measurement, at time t_{k-1}, rather than at time t_k, as in

$$\hat{X}_k(-) = \Phi \hat{X}_{k-1}(+) + \Gamma U_{k-1}$$

$$\hat{X}_{k-1}(+) = \hat{X}_{k-1}(-) + K_k[Z_{k-1} - H\hat{X}_{k-1}(-)]$$

Depending on how you configure your estimation/control problem, this structure may pop up rather than that in Eq. (10-17). The reader should take a moment and compare this predictor estimator with the equations presented above.

Combining these two equations to eliminate $\hat{X}_{k-1}(+)$ and collecting coefficients of $\hat{X}_{k-1}(-)$, gives

$$\hat{X}_k(-) = (\Phi - K_k H)\hat{X}_{k-1}(-) + K_k Z_{k-1} + \Gamma U_{k-1} \tag{10-23}$$

When the dynamics of this indexed equation are studied in the same manner as earlier in this section, one obtains the following eigenvalue problem:

$$[\Phi - KH - \lambda I]C = 0 \tag{10-24}$$

where the eigenvalues of the matrix $\Phi - KH$ are to be found. Equation (10-24) will be referred to in Sec. 10-5.

10-4 Using the Kalman Filter for Control

In Sec. 6-5 the state of the underdamped process was fed back to make the compensated system behave differently, namely, without the ripples. We then applied integral-only control to the compensated system with reasonable success. The state was constructed from the measured position and the estimated filtered derivative.

In this section, the Kalman filter will be used to estimate the two components of the state, which will be fed back just as in Sec. 6-5. In addition, the estimated position will be used in an integral-only control loop. Figure 10-6 shows a condensed version of a Matlab Simulink model of the controlled system. If you are not familiar with Simulink, treat the figure as a block diagram. Box 1 contains the compensation gain K_u, which is applied to the controller output. Box 2 contains the two compensation gains, which are applied to the state

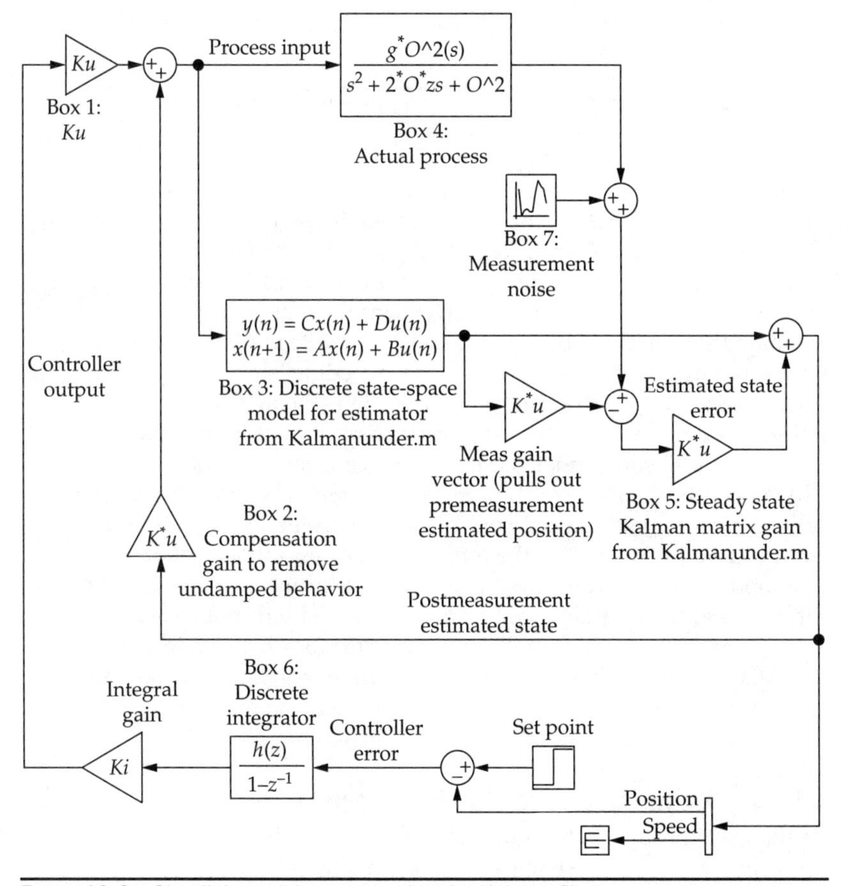

FIGURE 10-6 Simulink model: control using the Kalman filter.

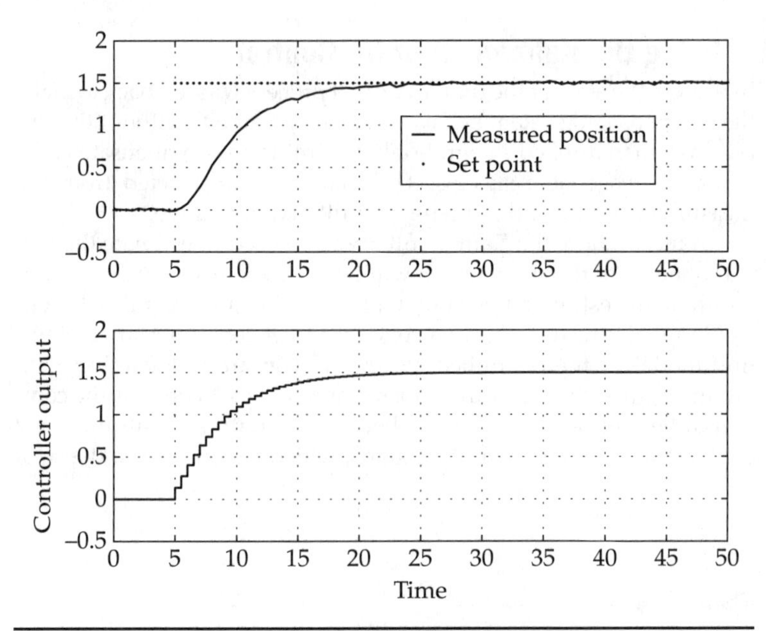

FIGURE 10-7 Closed-loop control of compensated underdamped process.

before it is fed back. These three compensation gains are the same as those applied in Sec. 6-5. The state is estimated in box 3 while box 4 simulates the actual process. The steady-state Kalman filter, the same as that used in the so-called "good" model example of Sec. 10-3, is applied in box 5. Box 6 contains an integrator that is applied to the estimated state and fed back as the controller output. Therefore, after state estimation and the compensation via feeding back the estimated state, integral-only control is applied to the estimated position. The noise added in box 7 has the same variance as that used in Sec. 6-5.

Figure 10-7 shows that the performance is similar to that when PIfD was used. Since the estimated states are fed back for compensation and also used in the control, there may be less noise in the controller output. However, the continuous process model in the Simulink model uses the same parameters as those in the discrete Kalman filter. In reality, the Kalman filter process model will not match the actual process and there will be error introduced. In spite of the potential problems that can arise from differences between the process and the model, I have used this control scheme with success for a similarly underdamped process.

10-4-1 A Little Detour to Find the Integral Gain

By the way, the logic behind tuning the integral-only controller is the following. Feeding the state back creates a new process that has different dynamics and unity gain, especially at the new natural frequency

ω_D (review Sec. 6-4 for a discussion of ω_D and ω_n). In the frequency domain this means

$$|G(j\omega_D)| \cong 1$$

(This might require a little thought and a short review of Bode plot analysis.) When integral-only control is applied the open-loop transfer function is

$$\frac{K_I}{s}G(s)$$

After making the substitution $s = j\omega$ and deriving the magnitude, we get (when $\omega = \omega_D$)

$$\frac{K_I}{\omega_D}|G(j\omega_D)|$$

A rule of thumb, not too widely known, that often provides a satisfactory phase margin (see Chap. 4 for a discussion of phase margin) specifies that the open-loop gain at the desired frequency be 1/6. Therefore,

$$\frac{K_I}{\omega_D}|G(j\omega_D)| = \frac{1}{6}$$

which yields $K_I = \omega_D / 6$ because $|G(j\omega_D)| \cong 1$. Note that other approaches to finding K_I could have been used and that PI could also have been used.

10-5 Feeding Back the State for Control

In previous chapters one-dimensional proportional-only control was accomplished by feeding back the process variable and subtracting it from the set point. In state-space one could apply this approach as follows.

$$\begin{aligned} X_k &= \Phi X_{k-1} + \Gamma U_{k-1} \\ U_k &= K_c(S_k - X_k) \end{aligned} \tag{10-25}$$

where K_c is a feedback gain. The state X_k can be available through measurements or through estimation via the Kalman filter.

When these two equations are combined we get

$$\begin{aligned} X_k &= \Phi X_{k-1} + \Gamma K_c(S_{k-1} - X_{k-1}) \\ &= (\Phi - \Gamma K_c)X_{k-1} + BK_c S_{k-1} \end{aligned} \tag{10-26}$$

which describes the dynamics of a closed-loop system. As with the Kalman filter equations in Sec. 10-1-2, this is an indexed equation that has a homogeneous and nonhomogeneous part. The homogeneous part, namely,

$$X_k^h = (\Phi - \Gamma K_c)X_{k-1}^h \tag{10-27}$$

has a solution of the form

$$X_k^h = C\lambda^k$$

which, when applied to Eq. (10-27), gives

$$C\lambda^k = (\Phi - \Gamma K_c)C\lambda^{k-1} \tag{10-28}$$

Equation (10-28) can be rearranged to give

$$(\Phi - \Gamma K_c - \lambda I)C = 0$$

where, for a solution to exist, λ must satisfy the eigenvalue-yielding equation of

$$\left|\Phi - \Gamma K_c - \lambda I\right| = 0 \tag{10-29}$$

Therefore, the dynamics of the controlled system are dictated by the eigenvalues of the matrix $\Phi - \Gamma K_c$ which, in turn, depend on the control gain K_c.

As mentioned in Sec. 10-3, Matlab has a built-in function that will calculate K_c if the eigenvalues of $\Phi - \Gamma K_c$ are specified. This provides an alternative method of finding control parameters.

10-5-1 Integral Control?

In the case of the underdamped process, the state consists of the position and the speed. For there to be integral control one must augment the state with the integral of the position, or preferably, the integral of the difference between the position and the set point. This will be done in Sec. 10-6.

10-5-2 Duals

When the reader compares Eq. (10-23) for the prediction estimator with Eq. (10-26), they notice that the solution of the estimator problem is similar to that of the control problem. The estimation problem requires finding the eigenvalues of $\Phi - KH$ while the control problem requires the same thing of $\Phi - \Gamma K_c$. If the reader were to take the transpose of $\Phi - KH$ and get $\Phi^T - H^T K^T$ they would see that both eigenvalue problems have the same structure and the same Matlab `place` algorithm can be used to place the eigenvalues. In this case, the control literature calls the control problem the "dual" of the estimation problem and vice versa.

10-6 Integral and Multidimensional Control

Feeding back the state for control purposes is often equivalent to proportional-only control and therefore exhibits offset. To be more effective, integral control should be added. This can be accomplished by adding the integral of the state and constructing an augmented state-space model. When the augmented state is fed back, the resulting control will contain an integral component and there should be no offset.

10-6-1 Setting Up the Example Process and Posing the Control Problem

It would be attractive if the state-space approach could be used to address multidimensional problems where there is more than one variable to be driven to set point and more than one variable to be adjusted. The three-tank process with backflow will be chosen to illustrate the approach. Figure 10-8 shows the three-tank process with two additional process inputs that will be used to drive the three-tank levels to their set points.

Since you now have a familiarity with estimating the state of a process using the Kalman filter, I will specify that in this problem the state is measurable (or estimable). Should the state not be measurable, I will assume that you can construct a Kalman filter that would supply estimates of the state that could be used for control.

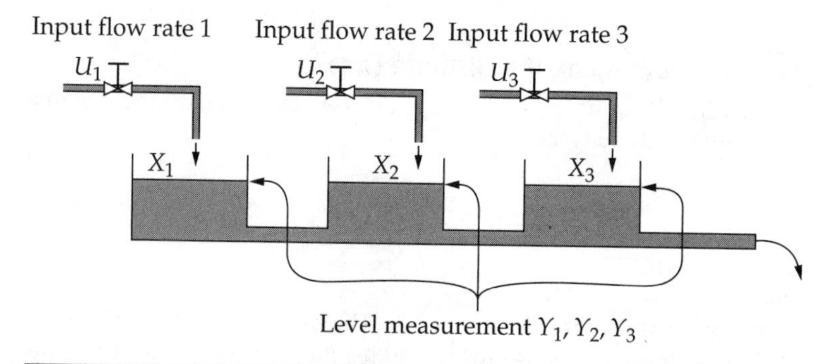

FIGURE 10-8 A three-tank system with backflow.

In the continuous time domain the process is described by

$$\frac{d}{dt}\begin{pmatrix} x_1 \\ x_2 \\ x_3 \end{pmatrix} = \begin{pmatrix} -\frac{1}{\rho A_1 R_{12}} & \frac{1}{\rho A_1 R_{12}} & 0 \\ \frac{1}{\rho A_2 R_{12}} & -\left(\frac{1}{\rho A_2 R_{12}} + \frac{1}{\rho A_2 R_{23}}\right) & \frac{1}{\rho A_2 R_{23}} \\ 0 & \frac{1}{\rho A_3 R_{23}} & -\left(\frac{1}{\rho A_3 R_{23}} + \frac{1}{\rho A_3 R_3}\right) \end{pmatrix} \begin{pmatrix} x_1 \\ x_2 \\ x_3 \end{pmatrix}$$

$$+ \begin{pmatrix} \frac{1}{\rho A_1} & 0 & 0 \\ 0 & \frac{1}{\rho A_2} & 0 \\ 0 & 0 & \frac{1}{\rho A_3} \end{pmatrix} \begin{pmatrix} U_1 \\ U_2 \\ U_3 \end{pmatrix}$$

$$\begin{pmatrix} y_1 \\ y_2 \\ y_3 \end{pmatrix} = \begin{pmatrix} 1 & 0 & 0 \\ 0 & 1 & 0 \\ 0 & 0 & 1 \end{pmatrix} \begin{pmatrix} x_1 \\ x_2 \\ x_3 \end{pmatrix} \tag{10-30}$$

or more compactly as

$$\frac{d}{dt}X = AX + BU$$

$$Y = CX$$

We use that same parameter values as those in Chap. 5, namely,

$$\rho = 1, A_1 = 0.1, A_2 = 0.1, A_3 = 0.1, R_{12} = 10, R_{23} = 10, R_3 = 10$$

10-6-2 Developing the Discrete Time Version

Using Matlab routines cited in App. H, one can develop the discrete time version directly, as in

$$X_{i+1} = \Phi X_i + \Gamma U_i$$

$$Z_i = H X_i$$

where H, Φ, and Γ are (3, 3) matrices. The measurement matrix H is a (3, 3) unit matrix because all the states are considered measurable, perhaps due to a Kalman filter.

The three components of the integral of the state can be introduced as the first three elements of the new augmented state:

$$(x_1^I \quad x_2^I \quad x_3^I \quad x_1 \quad x_2 \quad x_3)^T$$

where

$$\begin{pmatrix} x_1^I \\ x_2^I \\ x_3^I \end{pmatrix}_k = \begin{pmatrix} x_1^I \\ x_2^I \\ x_3^I \end{pmatrix}_{k-1} + \begin{pmatrix} x_1 \\ x_2 \\ x_3 \end{pmatrix}_{k-1}$$

That is, x_i^I, $i = 1, 2, 3$ are the sums (integrals) of the three states, x_i, $i = 1, 2, 3$.

The augmented system now looks like

$$\begin{pmatrix} x_1^I \\ x_2^I \\ x_3^I \\ x_1 \\ x_2 \\ x_3 \end{pmatrix}_k = \begin{pmatrix} 1 & 0 & 0 & 1 & 0 & 0 \\ 0 & 1 & 0 & 0 & 1 & 0 \\ 0 & 0 & 1 & 0 & 0 & 1 \\ 0 & 0 & 0 & \Phi_{11} & \Phi_{12} & \Phi_{13} \\ 0 & 0 & 0 & \Phi_{21} & \Phi_{22} & \Phi_{23} \\ 0 & 0 & 0 & \Phi_{31} & \Phi_{32} & \Phi_{33} \end{pmatrix} \begin{pmatrix} x_1^I \\ x_2^I \\ x_3^I \\ x_1 \\ x_2 \\ x_3 \end{pmatrix}_{k-1} + \begin{pmatrix} 0 & 0 & 0 \\ 0 & 0 & 0 \\ 0 & 0 & 0 \\ \Gamma_{11} & \Gamma_{12} & \Gamma_{13} \\ \Gamma_{21} & \Gamma_{22} & \Gamma_{23} \\ \Gamma_{31} & \Gamma_{32} & \Gamma_{33} \end{pmatrix} \begin{pmatrix} U_1 \\ U_2 \\ U_3 \end{pmatrix}_{k-1}$$

or

$$X_k = \Theta X_{k-1} + \Psi U_{k-1} \tag{10-31}$$

or

$$X_k = \begin{pmatrix} I & I \\ 0 & \Phi \end{pmatrix} X_{k-1} + \begin{pmatrix} 0 \\ \Gamma \end{pmatrix} U_{k-1}$$

where

$$X_k = \begin{pmatrix} x_1^I \\ x_2^I \\ x_3^I \\ x_1 \\ x_2 \\ x_3 \end{pmatrix}_k \qquad \Theta = \begin{pmatrix} I & I \\ 0 & \Phi \end{pmatrix} \qquad \Psi = \begin{pmatrix} 0 \\ \Gamma \end{pmatrix}$$

This augmented system is really just two matrix equations put into one matrix equation. The two matrix equations, obtained from

Eq. (10-31) by multiplying across and down, as specified by the definition of matrix multiplication, are

$$\begin{pmatrix} x_1^I \\ x_2^I \\ x_3^I \end{pmatrix}_k = \begin{pmatrix} 1 & 0 & 0 \\ 0 & 1 & 0 \\ 0 & 0 & 1 \end{pmatrix} \begin{pmatrix} x_1^I \\ x_2^I \\ x_3^I \end{pmatrix}_{k-1} + \begin{pmatrix} 1 & 0 & 0 \\ 0 & 1 & 0 \\ 0 & 0 & 1 \end{pmatrix} \begin{pmatrix} x_1 \\ x_2 \\ x_3 \end{pmatrix}_{k-1}$$

and

$$\begin{pmatrix} x_1 \\ x_2 \\ x_3 \end{pmatrix}_k = \begin{pmatrix} \Phi_{11} & \Phi_{12} & \Phi_{13} \\ \Phi_{21} & \Phi_{22} & \Phi_{23} \\ \Phi_{31} & \Phi_{32} & \Phi_{33} \end{pmatrix} \begin{pmatrix} x_1 \\ x_2 \\ x_3 \end{pmatrix}_{k-1} + \begin{pmatrix} \Gamma_{11} & \Gamma_{12} & \Gamma_{13} \\ \Gamma_{21} & \Gamma_{22} & \Gamma_{23} \\ \Gamma_{31} & \Gamma_{32} & \Gamma_{33} \end{pmatrix} \begin{pmatrix} U_1 \\ U_2 \\ U_3 \end{pmatrix}$$

The first is the integration (summation) of the state and the second is the original discrete time state-space model. Make sure that you understand where these last two equations came from.

10-6-3 Finding the Open-Loop Eigenvalues and Placing the Closed-Loop Eigenvalues

The *open*-loop dynamics are described by the six eigenvalues of the augmented matrix Θ. If the augmented state is fed back then the *closed*-loop dynamics will be described by the eigenvalues of the matrix $\Theta - \Psi K_{aug}$ where K_{aug} is the control gain matrix of size (3, 6), applied to the augmented state. (You may want to go back to the development of Eq. 10-29 to review why this is so.) The matrix K_{aug} is of size (3, 6) because $K_{aug} X$ is fed back to be the controller output U which is size (3, 1). That is,

$$(3,1) \sim (3,6)\ (6,1)$$

$$U = K_{aug} X \tag{10-32}$$

To find the value of K_{aug} we first find the six eigenvalues of the *open*-loop system by solving the equation

$$|\Theta - \lambda I| = 0$$

We can use these eigenvalues for the *closed*-loop system or we can modify them. For this example, the open-loop eigenvalues are 1.0, 1.0, 1.0, 0.8203, 0.2112, and 0.0389, where the first three are from the integration equation and the last are from the process model. All of these eigenvalues lie on or inside the unit circle. For the closed loop, after a couple of trials, we decided to multiply all six of the eigenvalues by 0.7 and use the following Matlab statement

```
Kcaug = place(ADaug, BDaug, eigaugspec);
```

to determine the control gain. Here Kcaug is K_{aug}, ADaug is Θ, BDaug is Ψ, and eigaugspec is a vector of the six modified eigenvalues: 0.7, 0.7, 0.7, 0.5743, 0.1478, and 0.0272. The multiplication by a factor less than 1.0 tends to move the location of the eigenvalues closer to the origin in the z-plane, thereby decreasing the transients and making the control more aggressive.

10-6-4 Implementing the Control Algorithm

The application of the control vector K_{aug} is a bit tricky and perhaps it is best explained via the Matlab Simulink diagram in Fig. 10-9.

Boxes 1 and 2 show how the basic model was constructed using Φ (or AD) and Γ (or BD). Box 3 contains the three set points. Box 4 is a multiplexer that combines the three-tank level errors with the integrals of the tank level error into a six-dimensional state vector. Box 5 contains the control gain K_{aug} that multiplies the state vector to generate the three-dimensional controller output. Boxes 6 and 7 are one-sample delays. The control outputs are clipped to keep them nonnegative.

Note that the control gain multiplies not the three-tank levels and their integrals but rather the three-tank level *errors* and the integrals of those *errors*. Therefore, Eq. (10-32), the control output equation, becomes

$$(3,1) \sim (3,6)\ (6,1)$$

$$U = (K_{aug})\begin{pmatrix} \left(\sum e\right)_1 \\ \left(\sum e\right)_2 \\ \left(\sum e\right)_3 \\ e_1 \\ e_2 \\ e_3 \end{pmatrix} \tag{10-33}$$

The control equation from Eq. 10-33, with the numerical values in place, and after partitioning is

$$\begin{pmatrix} U_1 \\ U_2 \\ U_3 \end{pmatrix} = \left[\begin{pmatrix} 0.0480 & -0.0177 & 0.005 \\ -0.0294 & 0.060 & -0.0057 \\ 0.0017 & -0.0154 & 0.0295 \end{pmatrix}\begin{pmatrix} 0.1095 & 0.0194 & 0.0002 \\ 0.0103 & 0.0729 & 0.0574 \\ 0.0074 & 0.0251 & -0.307 \end{pmatrix}\right]\left[\begin{pmatrix} \left(\sum e\right)_1 \\ \left(\sum e\right)_2 \\ \left(\sum e\right)_3 \end{pmatrix} \\ \begin{pmatrix} e_1 \\ e_2 \\ e_3 \end{pmatrix}\right] \tag{10-34}$$

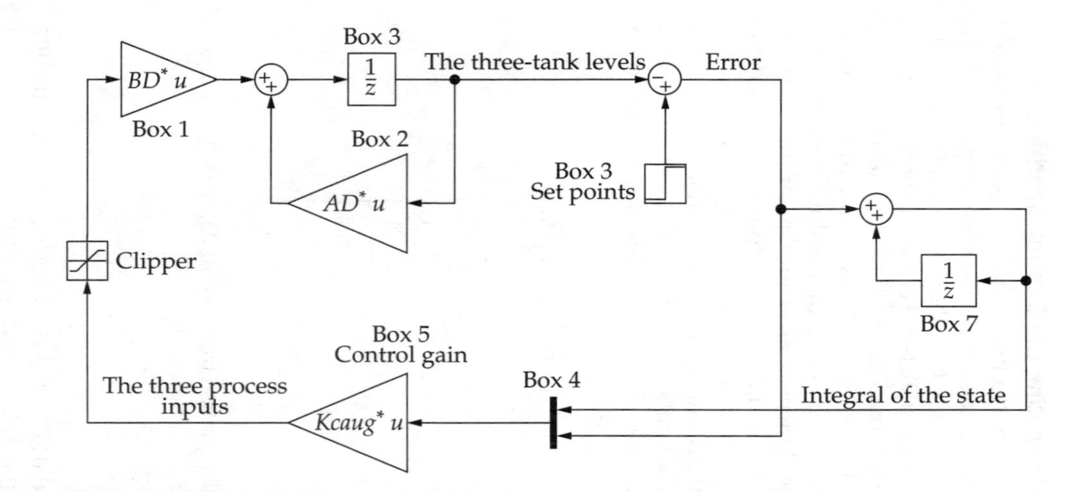

FIGURE 10-9 Simulink model: augmented state control of three-tank system with back flow.

The diagonal elements of the left hand (3, 3) matrix in the partitioned (3, 6) augmented gain matrix relate the sums of the *i*th state variable to the *i*th control output and generate the integral components of the controller output. Note that those diagonal elements are all positive and are a little larger than the nondiagonal elements in that (3, 3) part of the (3, 6) matrix. That the nondiagonal elements are not zero or negligible means the interaction between the tanks is being used in the control algorithm. The diagonal elements of the right-hand (3, 3) matrix in the partitioned augmented gain matrix relate the error in the *i*th state to the *i*th control output. Again, they are a little larger in magnitude from the nondiagonal elements in that (3, 3) part of the (3, 6) matrix. However, note that $K_{aug_{36}}$ element is negative.

Figure 10-10 shows the performance of this control algorithm. The first of the two plots in Fig. 10-10 contains the process outputs and their respective set points.

One must choose the set points carefully. For example, it would not make sense to have a set point of 100 for the second tank while the first and third tanks had set points of 50. Were this the case both the flow rates into tanks one and three would be driven to zero and the levels would significantly miss their targets. As an exercise, the reader should verify this by modifying the Matlab code that generates Fig. 10-10.

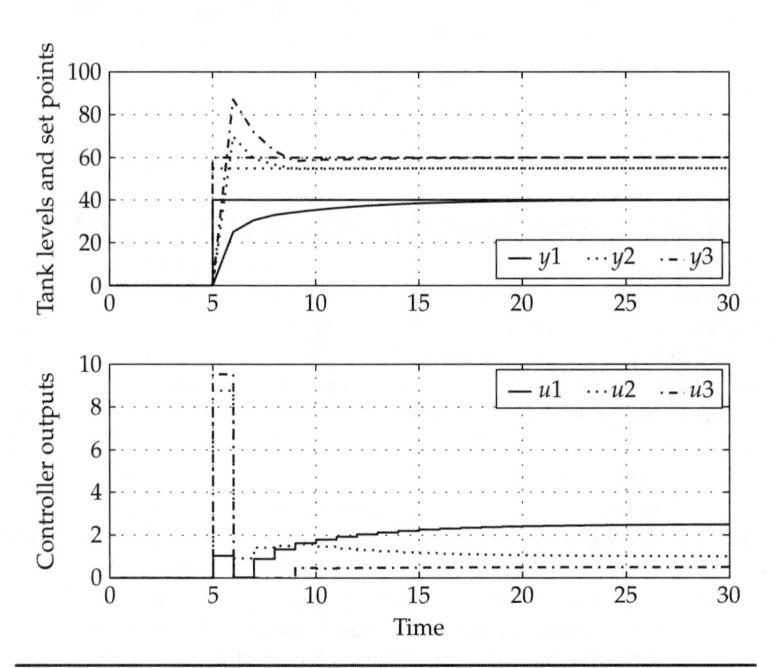

FIGURE 10-10 Augmented state-space control of three-tank system with back flow.

10-7 Proportional-Integral Control Applied to the Three-Tank Process

For the sake of comparison, three PI controllers were applied to the control of the same three-tank process. As with the previous example, we are assuming that all three elements of the state are measurable. If this is not the case then the Kalman filter could be used to estimate them. Figure 10-11 shows the Matlab Simulink diagram. After some trial and error tuning of the six controller gains, the following performance shown in Fig. 10-12 was obtained.

Comparing Fig. 10-10 with Fig. 10-12, it appears that both approaches perform about the same. Given the Matlab routines to aid in the control design and the task of tuning the three PI controllers, I think I would prefer the approach of Sec. 10-6.

As another exercise, take a look at Eq. (10-34) and see if you could reformulate it to fit the three separate PI control loop approach. The noninteraction of the three loops means that the nondiagonal elements in the two 3×3 matrices would be zero, as in

$$\begin{pmatrix} U_1 \\ U_2 \\ U_3 \end{pmatrix} = \begin{pmatrix} I_1 & 0 & 0 & P_1 & 0 & 0 \\ 0 & I_2 & 0 & 0 & P_2 & 0 \\ 0 & 0 & I_3 & 0 & 0 & P_3 \end{pmatrix} \begin{pmatrix} \left(\sum e\right)_1 \\ \left(\sum e\right)_2 \\ \left(\sum e\right)_3 \\ e_1 \\ e_2 \\ e_3 \end{pmatrix}$$

10-8 Control of the Lumped Tubular Energy Exchanger

In Secs. 7-7 through 7-9, the tubular energy exchanger was approximated by a series of continuous stirred tanks (CSTs) or lumps. The equations describing the lumping can be written in state-space form where each CST has an adjustable steam jacket temperature. This suggests a control problem, challenging but perhaps of academic interest only, where one has the means to adjust the steam temperatures of the individual lumps. The method of Sec. 10-6 could be brought to bear on this problem.

We choose to approximate the process with 20 lumps so the state will consist of the temperatures of these lumps. With only a few detail modifications, one can develop the discrete time model equations from Eq. (7-27) using the Matlab routines mentioned in App. H.

Because integral control will be used, the open-loop dynamics are described by the eigenvalues of the augmented matrix Θ and there will be 40 of them. If the augmented state is fed back then the closed

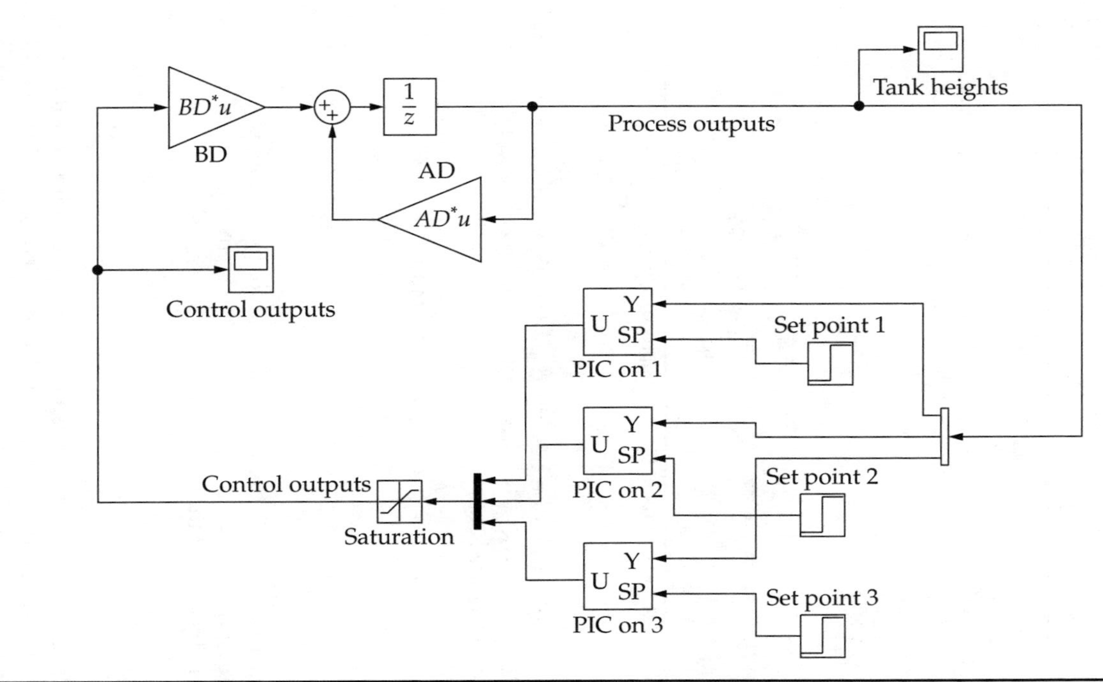

FIGURE 10-11 Simulink model: PI control of three-tank process.

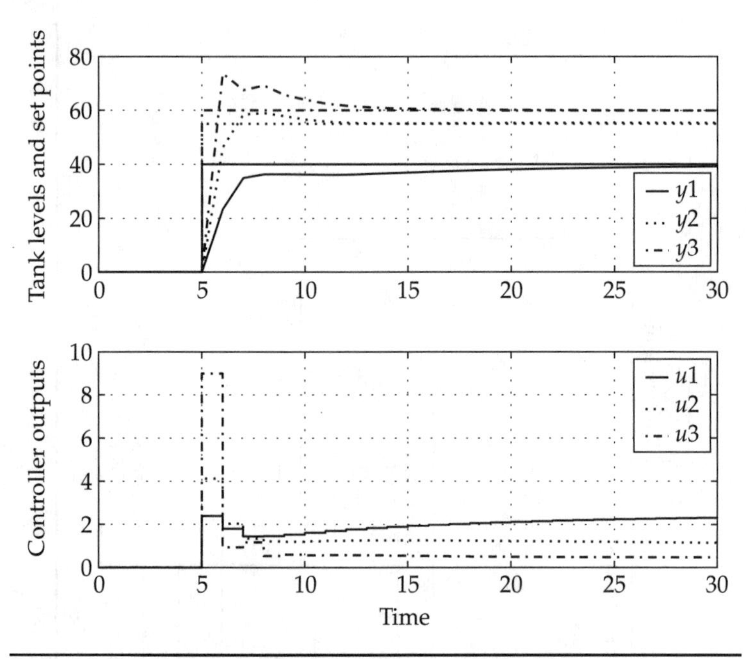

FIGURE 10-12 PI control of three-tank process with backflow.

loop dynamics will be described by the eigenvalues of the matrix $\Theta - \Psi K_{aug}$ where K_{aug} is the control gain matrix of size 20×40 applied to the augmented state.

$$(20,1) \sim (20,40)\ (40,1)$$

$$U = K_{aug} X \tag{10-35}$$

(Note that the preceding verbiage is paraphrased from Sec. 10-6.)

To find the value of K_{aug} we find the eigenvalues of the *open*-loop system by solving the equation

$$|\Theta - \lambda I| = 0$$

for the 40 eigenvalues. We can use these eigenvalues for the closed-loop system or we can modify them. As in Sec. 10-6, I chose to attenuate them by a factor of 0.7. The results are shown in Figs. 10-13 through 10-16. Note that the control outputs (the steam jacket temperatures) were clipped at 200°.

For the large-diameter tube, Fig. 10-13 suggests that the steam jacket temperatures of lumps 1 through 10 stay maxed-out at 200°

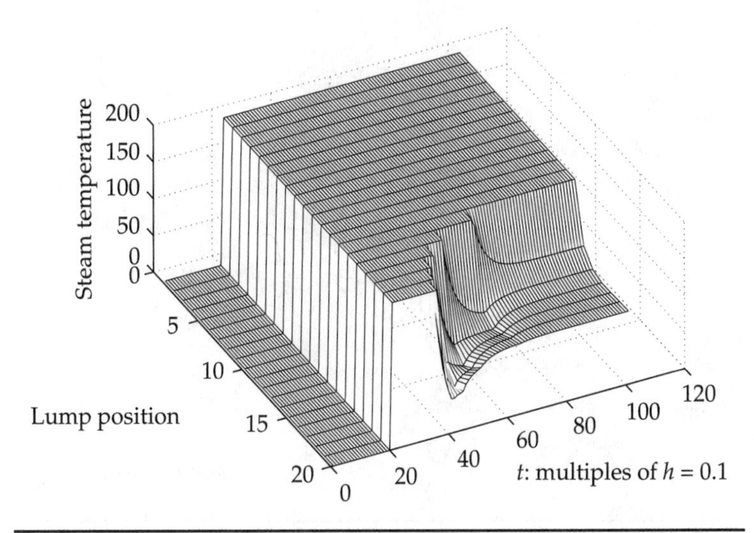

FIGURE 10-13 Control of the large-diameter tube energy exchanger, steam temperatures along the tube.

and that only the jacket temperatures near the end of the tube settle down with time. Figure 10-14 shows that only the lumps near the outlet actually reach the set point of 100°.

The behavior of the small-diameter tube energy exchanger is a little different. Figure 10-15 shows that most of the lump's jacket temperatures come off the maximum allowed values.

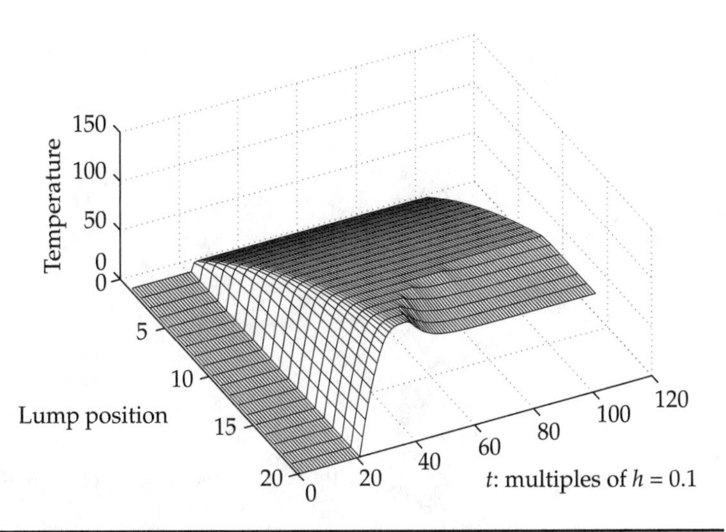

FIGURE 10-14 Control of the large-diameter tube energy exchanger, internal temperature along the tube.

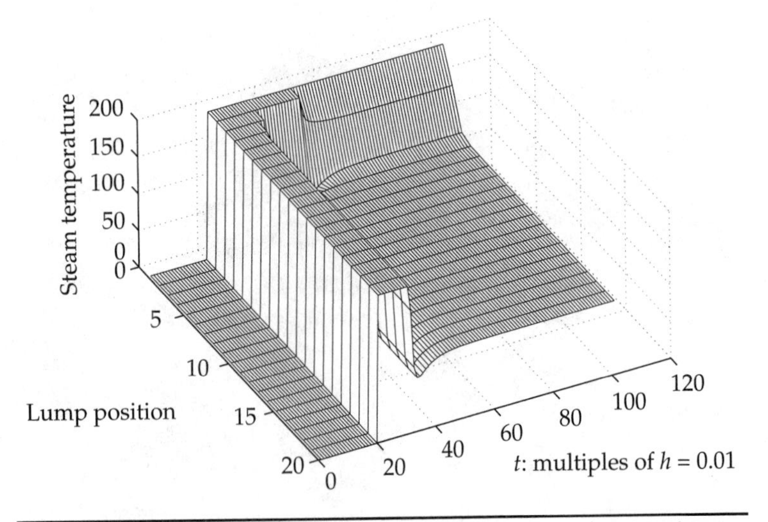

FIGURE 10-15 Control of the small-diameter tube energy exchanger, steam temperatures along the tube.

Finally, Fig. 10-16 shows that the most of the lumps reach the set point. Admittedly, this may not be a frequently occurring problem but I am including it to show how the general state-space approach can be applied to a variety of problems with minor changes in the dimensions and in the elements of the matrices involved. Note that many of the sophisticated mathematical operations like eigenvalue determination and control gain calculation are carried out transparently by Matlab routines.

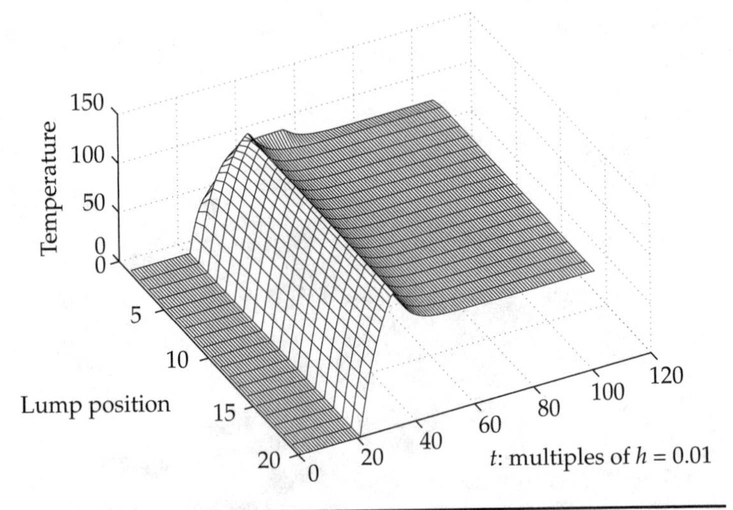

FIGURE 10-16 Control of the small-diameter tube energy exchanger, internal temperatures along the tube.

10-9 Miscellaneous Issues

This section is a kind of grab bag of material that I do not think is worth presenting in detail but which you might be exposed to.

10-9-1 Optimal Control

The gains in Eq. (10-25) or (10-33) can also be found by minimizing the objective function J in the following equation

$$J = \frac{1}{2}\sum_{k=0}^{N}(X_k^T Q_1 X_k + U_k^T Q_2 U_k) \tag{10-36}$$

subject to the constraint that the state-space model be satisfied, as in

$$-X_k + \Phi X_{k-1} + \Gamma U_{k-1} = 0 \qquad k = 1,2,\ldots,N \tag{10-37}$$

The object is to drive the state to zero over the N instants of time such that the weighted sums of the state and the control output are minimized.

Minimizing J in Eq. (10-36) subject to the constraint in Eq. (10-37) is a reasonably standard problem, solvable by the method of Lagrange multipliers. The matrices Q_1 and Q_2 are adjustable by the user. If, for example, Q_2 were small or zero, then the minimal (or optimal) solution would probably have X_k quickly going to zero with massively (and probably unacceptably) large movement in the controller output U_k. Conversely, a small Q_1 would have relatively small control output movement and large deviations of X_k from zero. There is an art to choosing the weighting matrices and many books have been written on this approach to control. However, I would rank optimal control a little bit below the other techniques presented in this book (consider the source).

In passing, it should be mentioned that the Kalman filter problem can be presented in a similar manner, that is, as a minimization problem, although I will not attempt it here. This may not be too surprising since it was mentioned earlier that the estimation problem and the control problem are duals of each other.

10-9-2 Continuous Time Domain Kalman Filter

All of the Kalman filter equations have an analog in the continuous time domain. I will not spend much time on them simply because the implementation of the Kalman filter is usually done in the discrete time domain with microprocessors. However, some authors prefer to develop all the equations and solutions in the continuous time domain and then convert the solutions to the discrete time domain using a variety of tools provided by Matlab or other computer-aided design software packages.

The continuous time domain state-space description is

$$\frac{d}{dt}x = Ax + BU + w$$
$$Z = Hx + v \tag{10-38}$$

A continuous time Kalman filter would look like

$$\frac{d}{dt}\hat{x} = A\hat{x} + BU + K(Z - H\hat{x})$$
$$\frac{d}{dt}P = AP + PA^T + BQB^T - KRK^T \tag{10-39}$$
$$K = PH^T R^{-1}$$

where Q and R are the covariance matrices associated with the stochastic sequences w and v, respectively. If the reader looks closely, he will hopefully see parallels between the discrete time and continuous time formulations. There are many ways of developing these continuous time equations and perhaps the least painful is the somewhat formal derivation in the book by Gelb where the discrete time equations are shown to morph into the continuous time equations as the time interval is shrunk to infinitesimal size.

10-10 Summary

The Kalman filter has been presented, without derivation, as a method of using a process model to augment the process measurements available such that the state is estimated. Two methods of using the Kalman filter were presented. In the first, the user chooses the two covariances Q and R and calculates the state estimates accordingly. The second, allows the user to specify the location of certain eigenvalues (poles) and to apply this specification to a Matlab routine that can generate the steady-state values of the Kalman filter gain. The literature sometimes refers to this approach as "pole placement."

With this state estimation tool in hand, we represented a control approach that fed the state (probably estimated by a Kalman filter) back. Via augmentation, integral control was added to this state-space approach that also required the user to specify the eigenvalues of a certain matrix.

This augmented state-space control approach was applied to the three-tank process originally presented in Chap. 5. Although this control approach was found to be satisfactory, similar performance was obtained from three PI control loops applied to the same process. The same augmented state-space control approach was applied to a 20-lump approximation of the tubular energy exchanger.

CHAPTER 11

A Review of Control Algorithms

If you have read through the first 10 chapters I am pleased and amazed at your effort. This chapter will be much easier. As the title suggests, it is indeed a review of some of the control algorithms that we have covered (or uncovered). It also looks at a couple of extensions like cascade control. Just in case the reader has some exposure to statisticians, this chapter spends some time dealing with an often misunderstood subject: statistical process control (SPC) and controlling processes in the face of white noise disturbances (again).

To get started, we visit the Strange Motel Shower Stall (at the Bates Motel?).

11-1 The Strange Motel Shower Stall Control Problem

You are on a business trip for your company, visiting a far-flung plant in the hinterlands. You check into the local motel/hotel and decide to take a shower (Fig. 11-1).

Not being familiar with the plumbing in this motel you have to develop a strategy for adjusting the shower water temperature before getting into the shower. I suggest that it would be something like that shown in Fig. 11-2.

Let's try to quantify the algorithm outlined in Fig. 11-2. The stick figure (you) is standing outside the stall and sampling the shower head spray. Once you have turned the valve you might carry out the following steps:

1. Sample the water temperature at time t_i, $i = 1$, with your finger (the start of "digital" control). You will not have a numerical value but we will still denote the temperature by $T(t_i)$.
2. Adjust the valve to an amount that is proportional to the perceived error $E(t_i)$

$$E(t_i) = S - T(t_i)$$

FIGURE 11-1 The Strange Motel shower stall control algorithm.

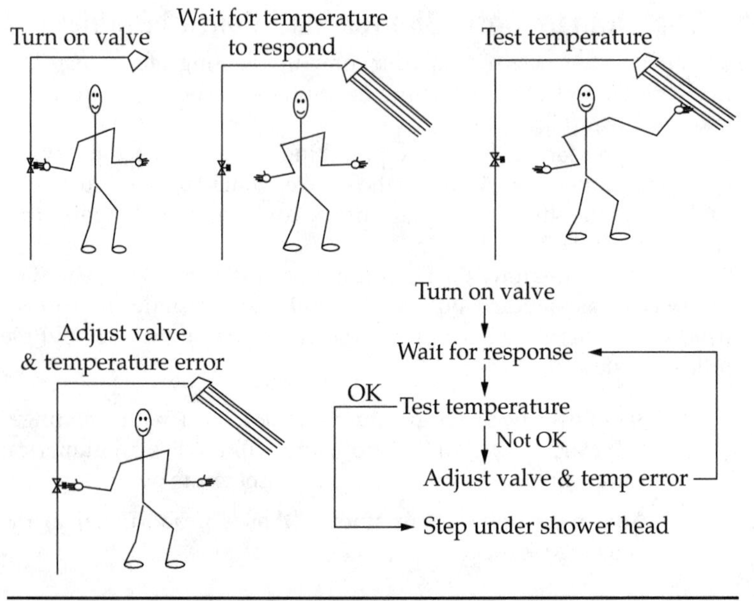

FIGURE 11-2 The Strange Motel shower stall control strategy.

As with the temperature, you will not have a numerical value for the error but you probably will have a feeling for the deviation as to sign and approximate amount. The adjustment will be

$$\Delta U(t_i) = K E(t_i) = K[S - T(t_i)]$$

$$U(t_i) = U(t_{i-1}) + \Delta U(t_i)$$

where K is a proportionality constant that is a measure of your patience and aggressiveness and ΔU is the change in the valve position. Note that the second equation simply says that you added the increment (positive or negative) to the previous valve position.

3. Wait a period of time h for the water temperature to respond to your adjustment. This wait time will probably include any dead time and at least one time constant. The time is now $t_i = t_{i-1} + h,\ i = 2$. Note that you have implicitly incremented the time index i.
4. Sample the water at time t_i with your fingers and go to step 2.

You would continue this loop until the error is perceived to be acceptable and you would then step into the shower stall.

Figure 11-3 shows how the temperature and adjustments might proceed if you were a patient and conservative person. Note that after each adjustment there is a small dead time, probably associated with the transport of the water through the piping, followed by a first-order-like response. The wait time h is long enough for the expiration of the dead time and 99% of the time constant response.

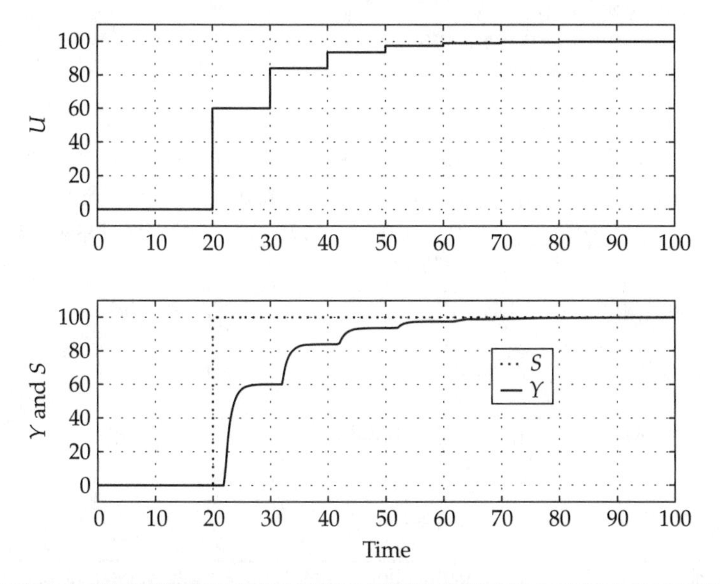

FIGURE 11-3 Conservative Strange Motel shower stall control.

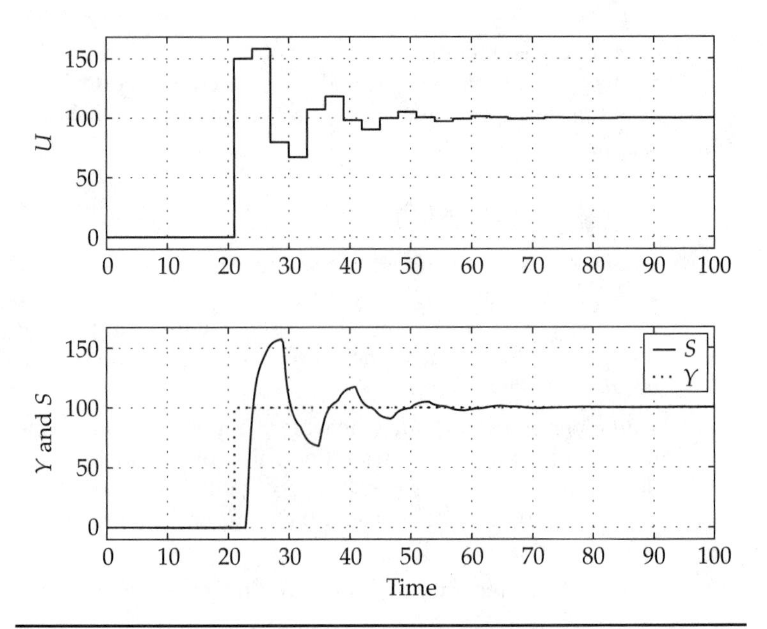

FIGURE 11-4 Aggressive Strange Motel shower stall control.

If you used the same strategy but had less patience and felt more aggressive, the results might be like those shown in Fig. 11-4.

In this case, you did not wait for the full response of the temperature and your adjustment sizes were greater for the same perceived size of the error. As a result, there was overshoot, although the desired temperature may have been arrived at earlier than with the more conservative strategy of Fig. 11-3.

The control strategy fits the closed-loop structure that we have been using in the rest of the book as shown in Fig. 11-5.

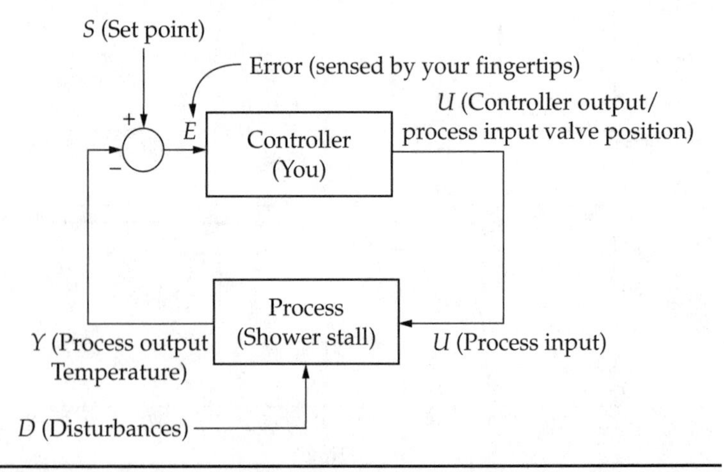

FIGURE 11-5 The Strange Motel shower stall control strategy, block diagram.

11-2 Identifying the Strange Motel Shower Stall Control Approach as Integral Only

The basic control algorithm is

$$\begin{aligned} \Delta U(t_i) &= IhE(t_i) = Ih[S - T(t_i)] \\ U(t_i) &= U(t_{i-1}) + \Delta U(t_i) \end{aligned} \tag{11-1}$$

where the proportionality constant has been replace with Ih so as to include the control interval in the algorithm.

Combine the control moves, as in

$$\begin{aligned} \Delta U(t_i) &= IhE(t_i) \\ U(t_i) &= U(t_{i-1}) + IhE(t_i) \\ U(t_{i+1}) &= U(t_{i-1}) + IhE(t_i) + IhE(t_{i+1}) \\ &\cdots \\ U(t_n) &= U(t_0) + IhE(t_1) + IhE(t_2) + \cdots + IhE(t_n) \\ U(t_n) &= U(t_0) + Ih\sum_{i=1}^{n} E(t_i) \end{aligned} \tag{11-2}$$

The reader should recognize Eq. (11-2) as a discrete time version of integral-only control algorithm. This is especially apparent if we shrink the control interval h to an infinitesimal value while increasing n such that nh is held constant at $t_f - t_0$.

$$\begin{aligned} U(t_f) &= U(t_0) + \lim_{h\to 0}\left\{Ih\sum_{i=1}^{n} E(t_i)\right\} \\ &= U(t_0) + I\int_{t_0}^{t_f} du E(u) \end{aligned} \tag{11-3}$$

Note that integral control will continue to modify the control output until the error is driven to zero. Also, based on the previous chapters, one can guess that becoming aggressive and impatient when there is a dead time can lead to overshoot.

Each person visiting the Strange Motel will have a slightly different approach to adjusting the water temperature before stepping under the shower but for the most part I suggest this integral-only control strategy approximates the initial strategy that you might apply. I often used this example in the internal company courses that I used to teach and several attendees have commented that they reflexively think of this example whenever they arrive at a motel while traveling for the company. The course was given to a wide variety of engineers and was entitled *Digital Control.* After presenting the strange shower stall algorithm one gentleman who had little formal background in engineering but a wealth of practical experience in analog instrumentation, asked me "why not just keep your right hand in the water stream all the time and continually adjust the faucet valve with the left hand until a satisfactory temperature is obtained?"—a valid and interesting question from someone with an entirely different background than mine.

Finally, note that since the control change is proportional to the error (not the *change* in the error), this algorithm is often mislabeled and misinterpreted as proportional control. This is a key distinction because, as pointed out in Chap. 4, integral control adds 90° of phase lag whereas proportional control does not. On the other hand, proportional-only control exhibits an offset between the set point and the process output.

11-3 Proportional-Integral, Proportional-Only, and Proportional-Integral-Derivative Control

11-3-1 Proportional-Integral Control

The integral-only control algorithm can be modified by the addition of a term proportional to the change in the error, as in

$$\begin{aligned} \Delta U(t_i) &= IhE(t_i) + P\Delta E(t_i) \\ U(t_i) &= U(t_{i-1}) + \Delta U(t_i) \end{aligned} \tag{11-4}$$

Figure 11-6 shows how the shower control strategy is changed with the addition of the proportional component. The first step is larger because of the one-time step in the set point. At each subsequent instant of control (every 10 time units) the proportional component bounces up and down because it is proportional to the error which is alternating in sign.

There are opportunities to improve this situation. The reader notices that the time constant of the shower stall process is about 1.0 and the dead time is about 2.0. This small time constant suggests a smaller control interval.

When the control interval is decreased from 10.0 to 0.1 and the proportional and integral gains are modified to 0.5 and 0.3, respectively, the result is shown in Fig. 11-7.

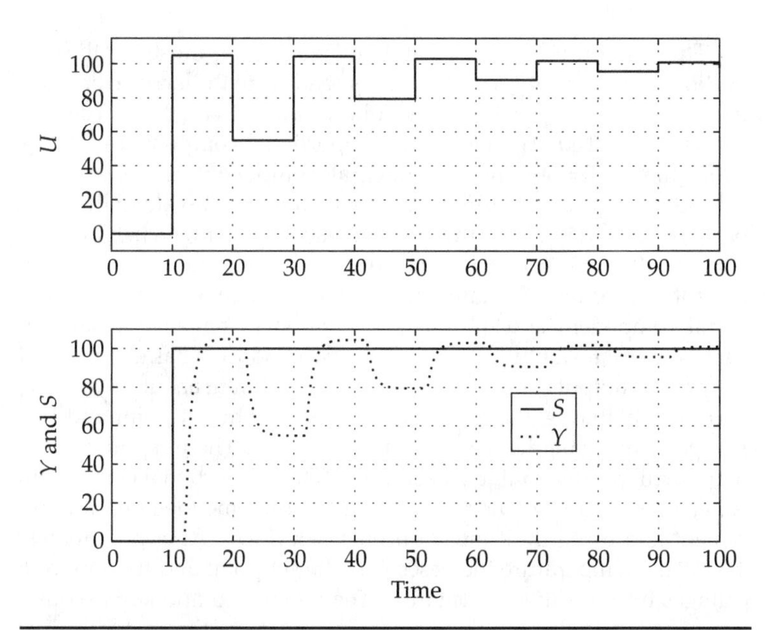

Figure 11-6 PI shower stall temperature control, large control interval.

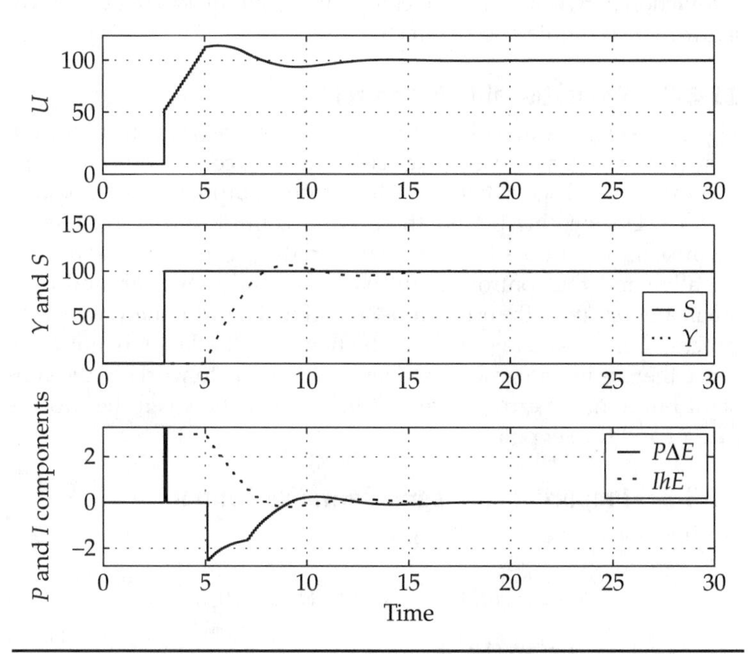

Figure 11-7 PI control of shower stall temperature, short control interval.

The first control move ΔU at time $t = 3$ consists mostly of the proportional component $P\Delta E$ which is responding to the large error when the set point is stepped. At this point in time, $E = 100 - 0 = 100$ and $\Delta E = 100 - 0 = 100$. Therefore, the proportional component of ΔU is $P\Delta E = 0.5 \times 100 = 50$ and the integral component is $IhE = 0.3 \times 0.1 \times 100 = 3$. The proportional component goes off-scale in the third part of Fig. 11-7. At the next control instant, when $t = 3.1$, $E = 100 - 0 = 100$, and $\Delta E = 100 - 100 = 0$, so, the proportional component is zero and the integral component is again 3. The proportional component and the integral component remain the same at every control instant until $t = 5$ when the dead time has elapsed and the process output starts to respond. From $t = 5$ until about $t = 8$, $E \geq 0$, and $\Delta E < 0$, that is, the temperature is below the set point and it is rising so the change in the error is decreasing. The integral component is still positive but it is decreasing. The proportional component is negative but rising and it is starting to overcome the integral component and bring the control output back down. At approximately $t = 8.0$ the temperature increases past the set point and the sign of E changes from positive to negative. The integral component becomes negative while the proportional component continues to rise. At about $t = 9$ the proportional component changes sign and becomes positive.

Therefore, the proportional and the integral components sometimes augment each other and sometimes oppose each other. It is the interaction between these two components that makes the PI control algorithm so simple and so effective.

11-3-2 Proportional-Only Control

Figure 11-8 shows the effect of removing the integral control for the same conditions as those in Fig. 11-7. Here the control output jumps to 50 at $t = 3$ and stays there until the process output starts to respond at $t = 5$. During this period there is no control output movement because the error does not change. When the process responds, ΔE is negative and the control output backs off and moves around by a small amount until the error stops changing. Unfortunately, when the process output and the error stop changing, the latter is not zero. Since there is no integral component to continue to work on the constant but nonzero error, there will be an offset between the process output and the set point.

11-3-3 Proportional-Integral-Derivative Control

Adding derivative to Eq. 11-4 gives

$$\begin{aligned} \Delta U(t_i) &= IhE(t_i) + P\Delta E(t_i) + D_g \Delta[\Delta E(t_i)] \\ U(t_i) &= U(t_{i-1}) + \Delta U(t_i) \end{aligned} \tag{11-5}$$

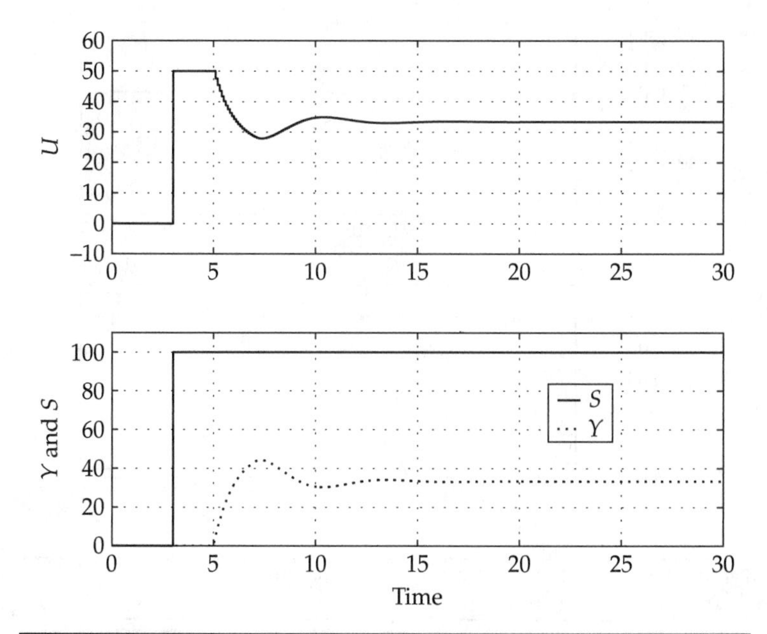

FIGURE 11-8 Proportional-Only control of shower stall temperature.

The first line contains a difference of a difference, $\Delta[\Delta E(t_i)]$, which is

$$\Delta[\Delta E(t_i)] = \Delta(E_i - E_{i-1}) = E_i - 2E_{i-1} + E_{i-2}$$

In Chap. 6 the derivative was shown to amplify noise. In the strange shower stall example, I have conveniently ignored noise so as to illustrate the basic features of the various components of the control algorithm. In the case of noise, you might consider the use of a filter applied to the derivative component as shown in Chap. 6.

Returning to the strange shower stall example, I kept the same P and I gains at 0.5 and 0.3, respectively, and added extremely small D_g values until I arrived at reasonable performance which is shown in Figs. 11-9 and 11-10.

Figure 11-9 shows that the addition of a small amount of derivative ($D_g = 0.1$) changes the nature of the control output by adding spikes at the moment of the set-point change and when the process output starts to respond. The overshoot of the process variable is decreased as a consequence of this extra-jerky activity. Figure 11-10, when compared to Fig. 11-7, shows that the presence of the derivative component causes the proportional component to change considerably from its performance in the PI case.

Adding derivative can often improve performance but there is a risk of spikes and noise amplification.

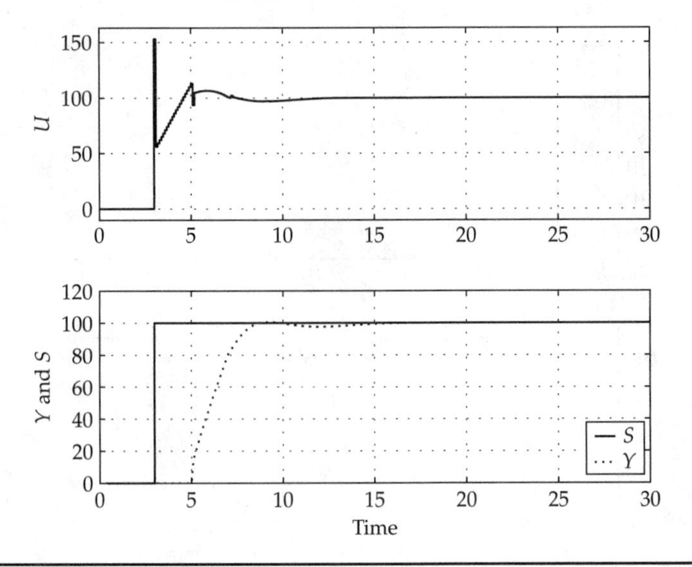

FIGURE 11-9 PID control of shower stall temperature.

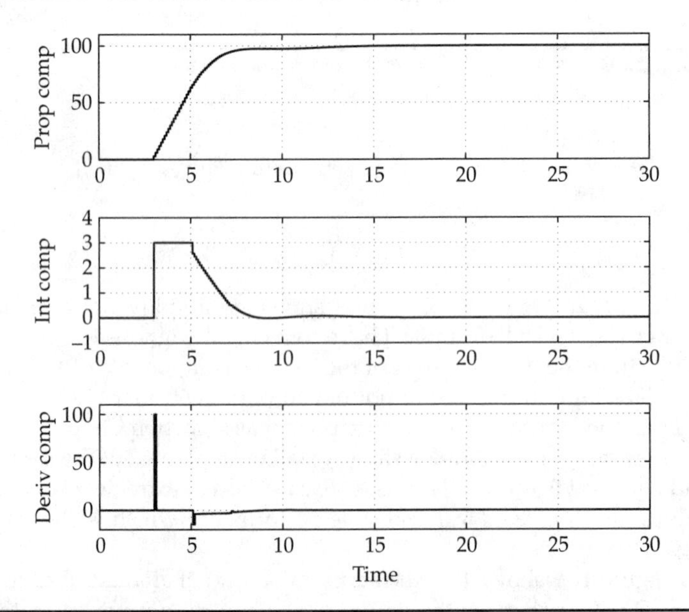

FIGURE 11-10 PID control of shower stall temperature—components of control output.

11-3-4 Modified Proportional-Integral-Derivative Control

If the set point is removed from the error term in the derivative, the algorithm is

$$\begin{aligned} \Delta U(t_i) &= IhE(t_i) + P\Delta E(t_i) - D_g \Delta[\Delta T(t_i)] \\ U(t_i) &= U(t_{i-1}) + \Delta U(t_i) \end{aligned} \tag{11-6}$$

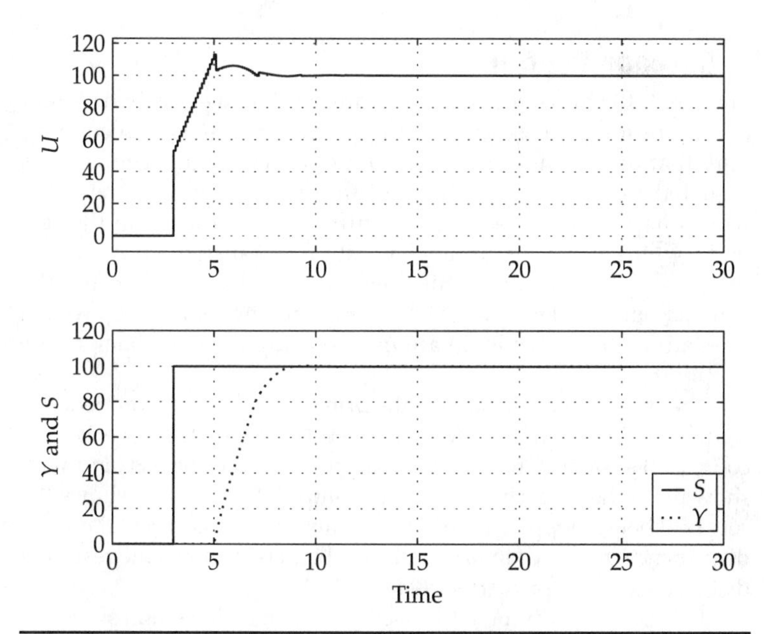

Figure 11-11 Modified PID control of shower stall.

Steps in the set point will no longer generate spikes in the controller output. The performance is about the same as is shown in Figs. 11-11 and 11-12. The derivative gain was raised slightly to 0.2.

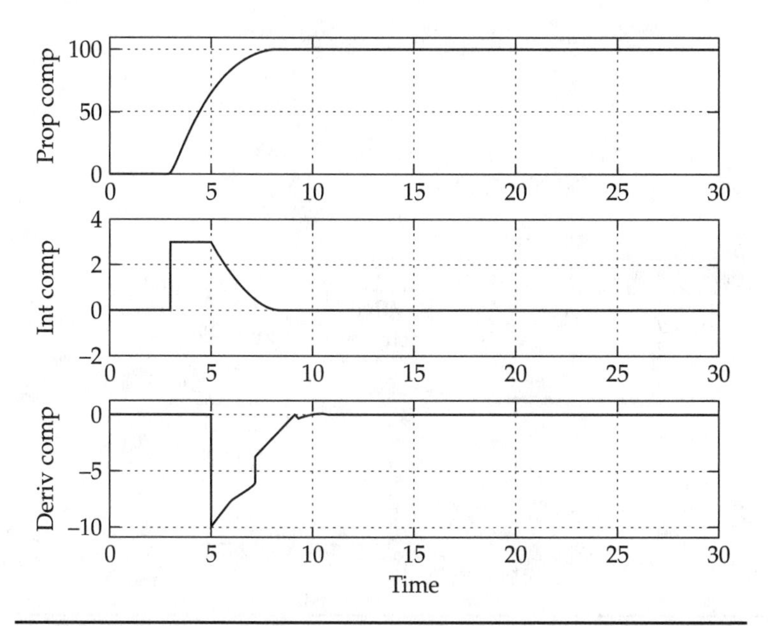

Figure 11-12 Modified PID control of shower stall—components of control output.

11-4 Cascade Control

Figure 11-13 shows the familiar water tank in a slightly different configuration. The source of the process input is a secondary tank that has an input flow rate of unknown origin. The valve is adjusted to maintain the level in the primary tank. Now, what would happen if there were a significant disturbance in the secondary tank? This disturbance would first cause the flow rate to the primary tank to vary. This flow rate variation would cause the primary tank level to deviate from set point. The control loop would then adjust the valve in an attempt to bring the level back to the set point.

The process output, namely the primary tank level, experiences a significant deviation in response to the upstream disturbance. For the controlled system to react to the disturbance, an error has (and will) show up in the primary tank process output. Figure 11-14 shows the set point being stepped at time $t = 1$. Later on, at time $t = 30$ there is a disturbance in the secondary tank and Fig. 11-14 shows the resulting disturbance in the primary tank level.

This problem can be addressed if a second flow-control loop is added, as shown in Fig. 11-15. In this case, the flow rate coming into the primary tank is controlled to a flow-rate set point generated by the level control loop. Should there be a disturbance in the secondary tank, it will be sensed by the flow-rate controller and quickly corrected such that there may be little or no variation in the primary tank level.

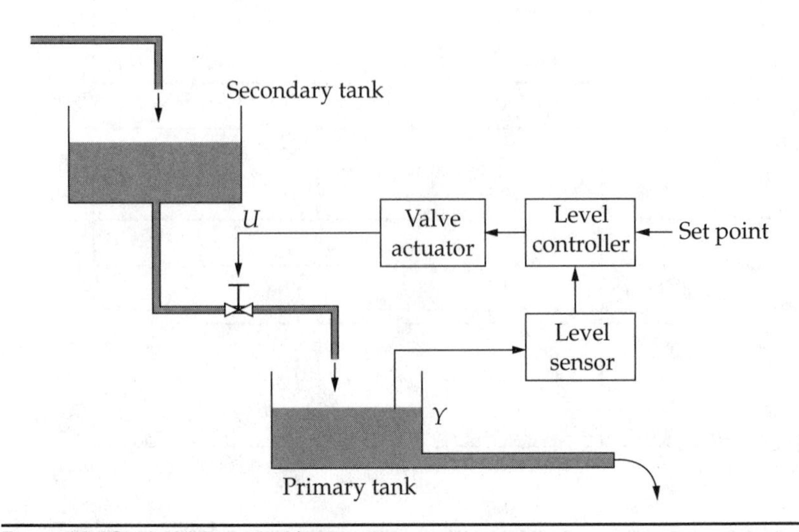

FIGURE 11-13 A single control loop.

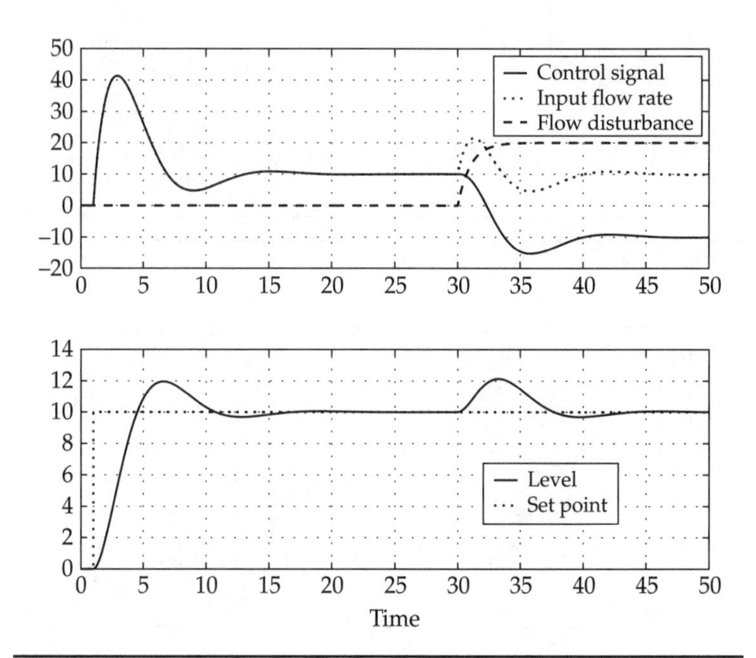

FIGURE 11-14 Single-loop control performance.

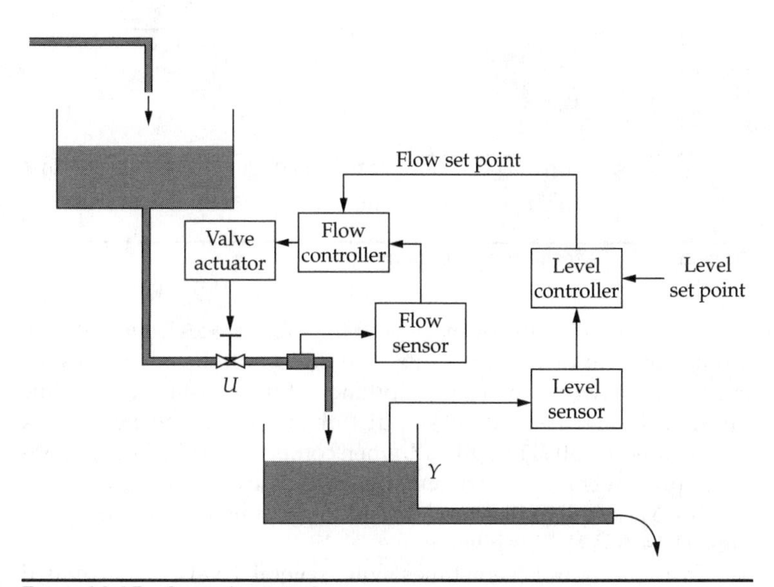

FIGURE 11-15 Cascade control.

The schematic in Fig. 11-16 shows how the master loop (level control) generates a set point for the slave loop (flow control). Refer to Fig. 11-17 where the same primary tank as in Fig. 11-13 has a process gain of unity and a time constant of 10.0 time units. The secondary

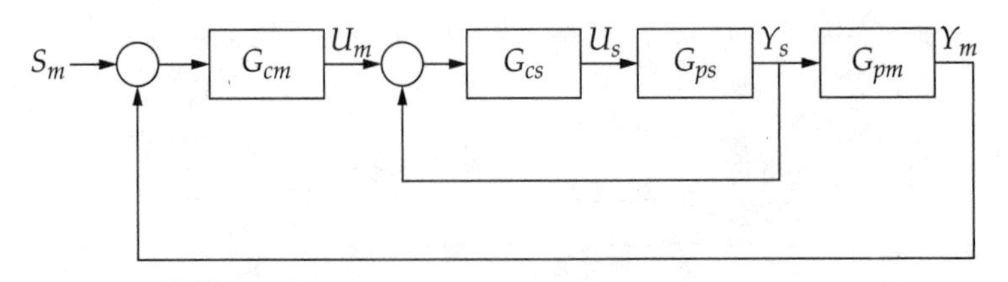

FIGURE 11-16 Cascade control schematic.

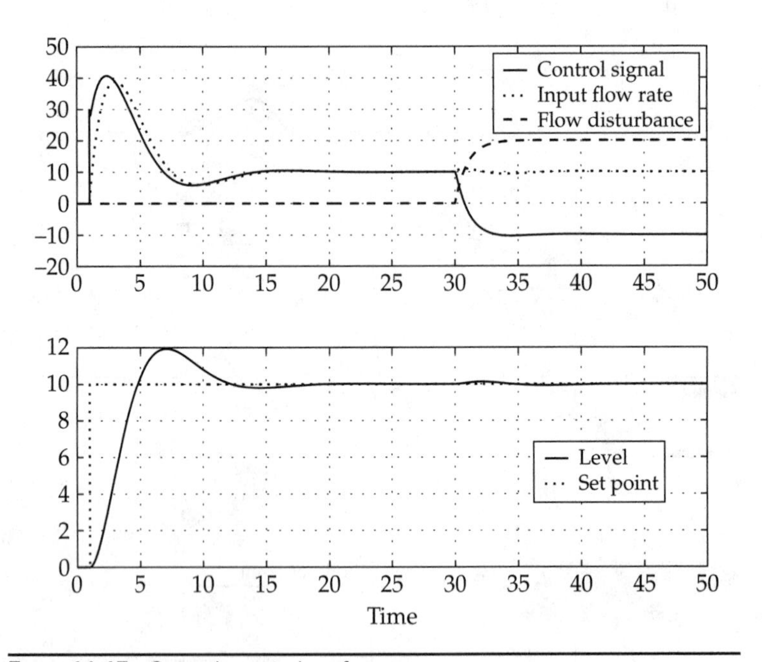

FIGURE 11-17 Cascade control performance.

tank is smaller than the primary with the same process gain but with a time constant of 1.0 time units. The flow-controller dynamics are even quicker with a gain of unity and a time constant of 0.5 time units. As in Fig. 11-14, there is a disturbance in the secondary tank level at time $t = 30$. Figure 11-17, when compared to Fig. 11-14, shows the improvement in performance by using cascade control.

The Matlab Simulink model used to generate the simulations in Figs. 11-14 and 11-17 is given in Fig. 11-18.

Cascade control, sometimes with several levels of embedded master/slave structure, is widely used in industry. It is especially effective where a secondary loop is much faster than a primary loop.

In Chap. 1, Sec. 1-7, cascade control appeared in an example process that tended to behave like a molten glass forehearth. The master control loop reads the glass temperatures via a thermocouple and sends a temperature set point to the combustion zone slave controller.

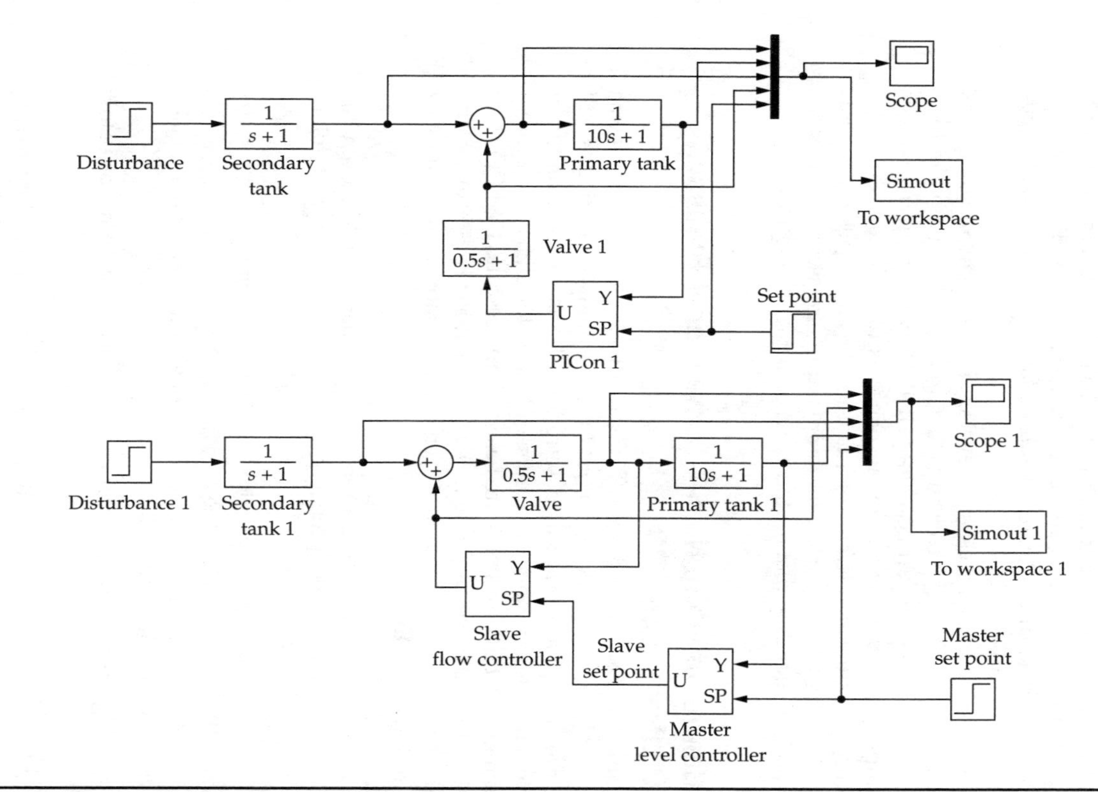

Figure 11-18 Simulink model for single and cascade control.

Although not mentioned there, the combustion zone slave control loop sends a position signal (in percent open) to another slave controller that positions a valve. So, in this example there are three levels of controllers in a cascade configuration.

In Chap. 7, the master controller reads the temperature of the liquid in the tubular energy exchanger and sent a temperature set point to a steam jacket temperature slave controller. As with the forehearth example, there would probably be another slave controller that would respond to the steam jacket temperature slave controller and manipulate a steam valve.

In these two examples each level in the hierarchy of controllers deals with effective time constants that are significantly smaller than those associated with the master loop. In the tubular energy exchanger example, the liquid temperature response would be characterized by a time constant much larger than that for the steam jacket temperature. Likewise, the steam jacket temperature effective time constant would be larger than that for the valve adjustment subprocess.

11-5 Control of White Noise—Conventional Feedback Control versus SPC

In the 1980s there was a great rush to a relatively old concept that was relabeled statistical process control (SPC). Although statisticians will go into cardiac arrest at this description, SPC is basically an alarm system that detects non–white noise riding on the signal of a process variable. Most SPC systems are based on the so-called WECO rules that were published by Western Electric in 1956. These rules claim that a process is "out of control" when one or more of the following conditions are satisfied:

1. One sample of the process output has deviated from the nominal value (probably a set point) by three standard deviations.
2. Two out of three samples have deviated from the nominal by two standard deviations.
3. Three out of four samples have deviated from the nominal by one standard deviation.
4. Eight samples in succession have occurred above or below the median line.

All the above conditions have a 1.0% probability of occurring if the process variable is behaving as a normally distributed un-autocorrelated stochastic sequence.

An important, in fact critical, part of the SPC strategy is to commit to a search for the "assignable cause" of the out-of-control condition and to solve the associated problem. During my career I have seen

SPC teams rigorously apply these rules and thereby solve many problems. The mindset of committing to find the "assignable cause" and do what is necessary to solve the problem often provides a tremendously open-minded environment.

To many control engineers, SPC is a sophisticated alarming system associated with a nearly religious commitment to "make the process right." It is, however, not a feedback control system in the sense that control engineers understand the term. In spite of this, there have been many times in my professional career when managers, in the face of a process problem, would call for the engineers to "just apply SPC." For a short period of time in the 1980s and 1990s SPC became a universal solution.

Correlated with the rise in the stature of SPC was the influx of statisticians into control engineering areas. Statisticians consistently claim that processes subject to white noise should not be controlled because the act of control amplifies the white noise riding on the process variable. The logic (which we have already touched on in earlier chapters) goes something like this. Consider the case where you are the controller and you are responsible for making control adjustments based on a stream of samples coming at you at the rate of, say, one per minute. Assume that you know that a sample is deviating from the target solely because of white noise. Therefore, the deviation of the ith sample is completely unautocorrelated with the deviation of the $i-1$th sample and will be completely unautocorrelated with the $i+1$ th sample. Consequently, it would be useless to make a control adjustment. If you did make an adjustment based on the ith sample's deviation, it would likely make subsequent deviations larger. On the other hand, if you knew the deviation of the ith sample was the result of a sudden offset that would persist if you did nothing, then you would likely make an adjustment.

I certainly agree with this logic but there are some realities on the industrial manufacturing floor where automatic feedback control of process variables subject to white noise is unfortunately necessary, especially when a load disturbance comes through the process or when there is a need to change the set point.

Furthermore, there is the question of degree. Processes with large time constants act as low-pass filters and though the feedback controllers may increase the standard deviation about the set point, the increase may be negligible.

To illustrate this idea, consider two processes. The first has a time constant of 40 time units and the second has a time constant of 0.5 time units. Both have a process gain of 2.0 and are subject to white noise. Both are initially in manual (no control adjustments). Both will be put into automatic PI control with a new set point. The standard deviation, before and after control, will be computed.

Figures 11-19 and 11-20 show the performance of the two processes. At time $t=200$ the controllers are activated with a set point

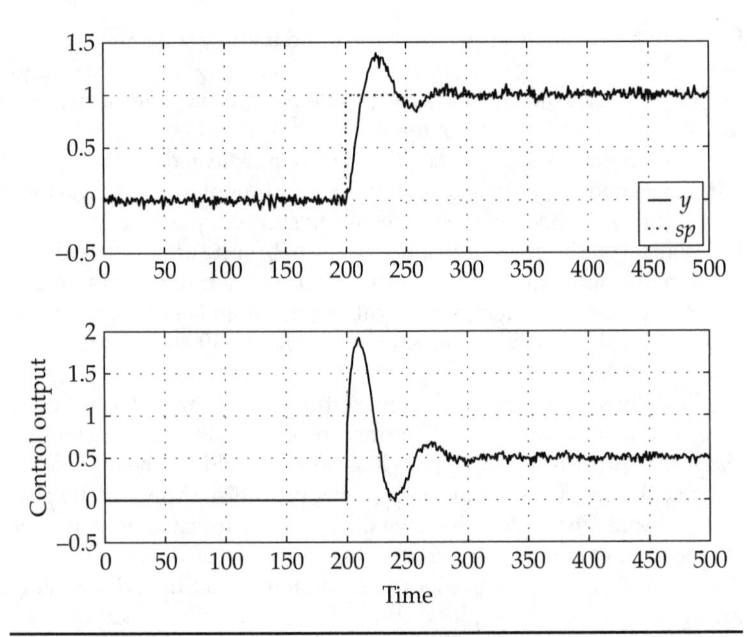

Figure 11-19 Control of a long-time constant process subject to white noise.

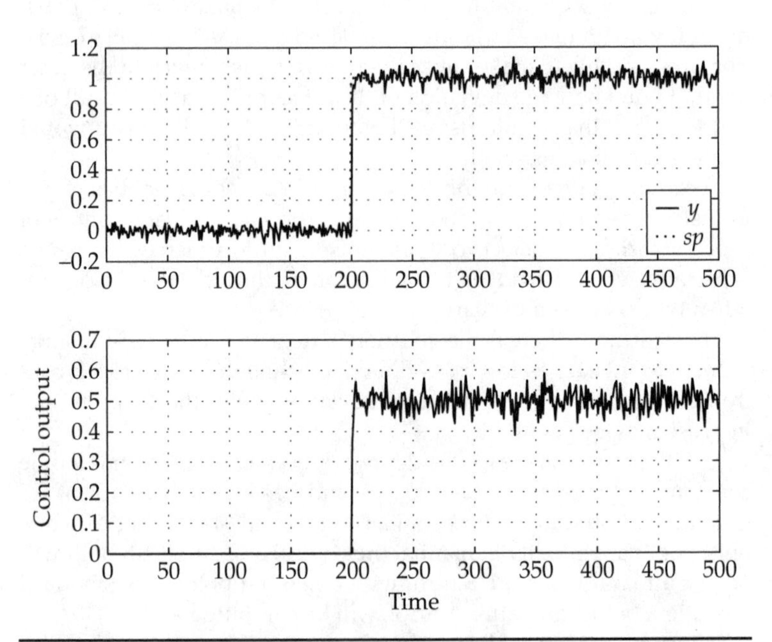

Figure 11-20 Control of a short-time constant process subject to white noise.

of 1.0. Before that time both process outputs had been bouncing about 0.0 in a unautocorrelated white noise manner. Both PI controllers were tuned such that the desired time constant was half of the actual process time constant using the tuning rules presented in Sec. 9-9.

For a long-time constant process, visual observation suggests that the intensity of the hash riding on the process variable after control has been initiated (and after the controller has settled out) is about the same as that before control. For samples one to 199, the standard deviation is 0.031811 and for samples 300 to 500 it is 0.032456—an increase of less than 2%. (I am keeping way more decimal places than I need!)

For the short time constant process, visual observation suggests that, after control is initiated, the intensity of the hash has significantly increased, especially on the control output. The standard deviation is 0.031811 before control and 0.041859 after control—an increase of almost 31%.

Processes act as low-pass filters and the long-time constant process does significantly more filtering. For the long-time constant process, the control output is not as active (because of the filtering effect of the process) and the increased activity shows up as less white noise intensity riding on the process output.

Many industrial processes are subject to white noise but they also often have large time constants, relative to the control interval, such that the application of an automatic feedback controller will do more good than harm.

11-6 Control Choices

We have stopped the deluge of different control algorithms that you or your control engineer can choose from. This does not imply that there are not more—there definitely are—however, I think we have covered the "big picture" of control algorithms.

The proportional-only control algorithm was presented first in Chap. 3 and then again in this chapter. For industrial situations it would probably not be your first choice. However, it occurs in many places. For example, your automobile engine coolant flow is regulated by a thermostatic valve. When the engine is cold, the thermostat closes the valve to restrict coolant flow and allow the engine to quickly reach a satisfactory operating temperature. As the engine heats up, the thermostat opens the valve and allows more coolant to circulate. The movement of the valve is proportional to the temperature of the coolant and there really is no set point as such. There also is no history of engine temperatures available to the thermostat so there is no integral effect that might be able to slowly work the temperature back to the desired value.

The proportional-integral control algorithm is the workhorse of the process control industry. In my opinion, it should be the first choice. Before some more sophisticated approach is taken it should be conclusively shown why PI is not acceptable.

The PI tuning rules were presented in Chap. 9

$$P = \frac{\tau_p}{g\tau_d} \qquad I = \frac{1}{g\tau_d}$$

or

$$K_c = \frac{\tau_p}{g\tau_d} \qquad \tau_I = \tau_d$$

To use these effectively the control engineer should identify the effective process time constant and gain. Actually, the identification should be part of the thorough study of the process that was presented in Chap. 2.

The PID control algorithm was shown to be effective for processes that have unusual characteristics such as underdamped behavior. The presence of noise riding on the process variable may require the control engineer to apply a low-pass filter to the derivative before using it in the algorithm. I did not present tuning rules for the PID because I really do not feel that comfortable with those in the literature. I usually tune the P and I components with a zero derivative gain using the above approach and then slowly increase the derivative gain. When I arrive at something that improves the behavior without amplifying the noise unacceptably, I iterate on the P and I gains which sometimes can be increased after the derivative has been added. However, Zeigler-Nichols PID tuning rules have been in the literature for 60 some years and you might suggest them to your control engineer.

An alternative approach of feeding back the "state" of the process to produce a modified process that has more desirable properties was presented and applied to the underdamped process.

The so-called "Q method" was presented in Chap. 9 as a means of developing control algorithms. For first-order processes the Q method yielded the PI control algorithm and associated tuning rules. For the underdamped process, it yielded a variant of the PID algorithm with a built-in low-pass filter. For processes with dead time, one could use the Q method to derive a special control algorithm that did dead time compensation in a manner similar to the famous Smith Predictor.

The idea of feeding back the state, mentioned earlier, prompted a presentation of the Kalman filter. If a good model of the process is available, the Kalman filter can provide a neat method of mixing measurements which may be noisy with model predictions to produce an estimate of the state. Two methods were presented for determining the Kalman filter gains. The first required the user to pick elements in covariance matrices associated with the process and sensor noise. The second required the user to place the eigenvalues (or poles) of the dynamical system.

The state could also be fed back for control purposes perhaps in concert with a Kalman filter that would estimate the state. In the one example presented in Chap. 10, the control gains were chosen by the eigenvalue placement method although an alternative method based on picking the covariance matrices was mentioned in passing.

I think you could conclude that there is a relatively broad spectrum of control approaches to choose from. I hope you will agree that before embarking on any of them you or your control engineer should thoroughly study the dynamics of the process.

11-7 Analysis and Design Tool Choices

We started with the simple first-order process model and used an ordinary differential equation in the continuous time domain to describe its behavior. As the models became more involved, the Laplace transform was used to move from the continuous time domain to the *s*-domain where differential equations became algebraic equations and life was often simpler. Laplace transforms were used to generate transfer functions which in turn could be used in a block diagram algebra that opened up many new methods of design and analysis. The dynamics of process models were shown to be characterized by the location of poles in the *s*-plane.

A simple substitution allowed us to move from the Laplace *s*-domain to the frequency domain where we could use concepts like phase lag, phase margin, and gain margin to develop insight into dealing with dynamics, both open loop and closed loop, often without having to solve differential or algebraic equations.

Matrices were shown to be a compact method of dealing with higher dimensional problems. The state-space approach brought us back to the time domain but presented us with an enlarged kit of tools. Eigenvalues of certain matrices were shown to be equivalent to the poles of transfer functions.

The movement from the continuous time domain to the discrete time domain was facilitated by the Z-transform where another simple substitution allowed us to move to the frequency domain to develop more insight. The state-space approach was represented in this new domain.

Finally, the Kalman filter was introduced and shown to provide a means of estimating the state from a noisy measurement if a process model was available. Several control approaches using the Kalman filter and the state-space concept were presented.

As with the control choices, I think you have been presented with a broad spectrum of analysis and design tools. Use them wisely and good luck.

APPENDIX A
Rudimentary Calculus

You probably had a passing exposure to calculus in college but never really used it during your career and the dust has gathered. Perhaps we can refresh and perhaps even enhance your understanding. However, if you were never exposed to calculus *at all* then this appendix may not get you out of the starting blocks. You might want to read this appendix completely and then go back to Chap. 2 or 3. Alternatively, you can refer to it while you read Chap. 3 and beyond.

A-1 The Automobile Trip

This section uses the metaphor of an automobile trip to introduce the concepts of integration and differentiation. Consider taking a trip with an instrumented automobile that can log the time, distance, and speed of the automobile. Figure A-1 shows a plot of the speed of the automobile as a function of time.

This shows a gradual, idealized acceleration up to 50 mi/hr, taking 10 min. Then there is a period of 100 min when the car's speed is constant at 50 mi/hr.

A-2 The Integral, Area, and Distance

How far does the automobile travel in the 110 min shown in the figure? Since distance S is related to *constant* speed v and time t as

$$S = vt \tag{A-1}$$

we can quickly estimate that the distance covered between 10.0 and 110.0 min (where the speed is constant at 50.0 mi/hr) as

$$S = vt = 50 \text{ mi/hr} \times 100 \text{ min/(60 min/hr)} = 83.33 \text{ mi}$$

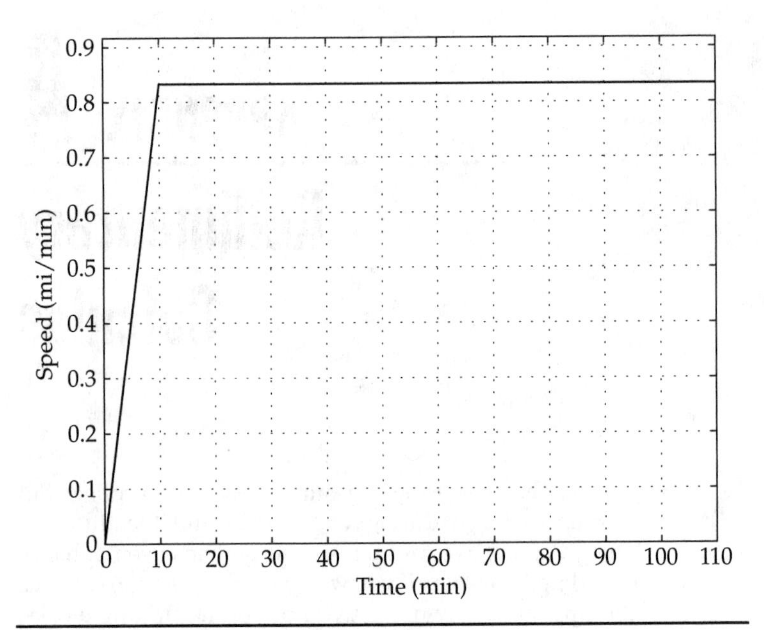

FIGURE A-1 Automobile speed during a trip.

This calculation suggests that S is the *area* under the speed curve (which is a straight horizontal line) between 10.0 and 100.0 min.

Since the speed curve, up until 10.0 min, forms the side of a right triangle, the distance traveled during this period is the area of the appropriate triangle:

$$S = \frac{1}{2} 10 \text{ min}/(60 \text{ min/hr}) \times 50 \text{ mi/hr} = 4.17 \text{ mi}$$

The total distance covered for the whole time period from 0 to 110 min is the total area under the speed curve or 83.33 + 4.17 = 87.5 mi.

In general, the distance covered is the *integral* of the speed over the time period of interest

$$S = \int_{t_1}^{t_2} v\,dt \tag{A-2}$$

which is also the area under the v curve between t_1 and t_2. For the case between 10 and 110 min the distance covered is

$$S = \int_{10}^{110} v\,dt = \frac{50}{60} t \Big|_{10}^{110} = \frac{50}{60}(110 - 10) = 83.33$$

Since v is a constant, valued at 50/60 mi/min, and since the integral of a constant is just that constant multiplied by the time interval (we will talk about this more in the next paragraph), the above integral is quite simple to evaluate.

The integral of constant, say C, with respect to the variable t, between the limits of a and b is

$$\int_a^b C\,dt = Ct\Big|_a^b = C(b-a) \tag{A-3}$$

If $C = 50/60 = 0.833$, then that would be pictured in Fig. A-2.

The area under the line representing $C = 50/60 = 0.833$ is the integral of the constant and, from the graph, has the value of $0.833 \times 100 = 83.33$.

Back to the trip. For the time period from 0 to 10.0 min, the speed (miles/minute) is increasing linearly and has the following formula:

$$v = t\frac{50}{60}\frac{1}{10}$$

The distance covered during this acceleration period is given by the integral of the speed with respect to time over the interval 0 to 10, as in

$$S = \int_0^{10} v\,dt = \int_0^{10} t\frac{50}{60}\frac{1}{10}\,dt = \frac{50}{60}\frac{1}{10}\int_0^{10} t\,dt$$

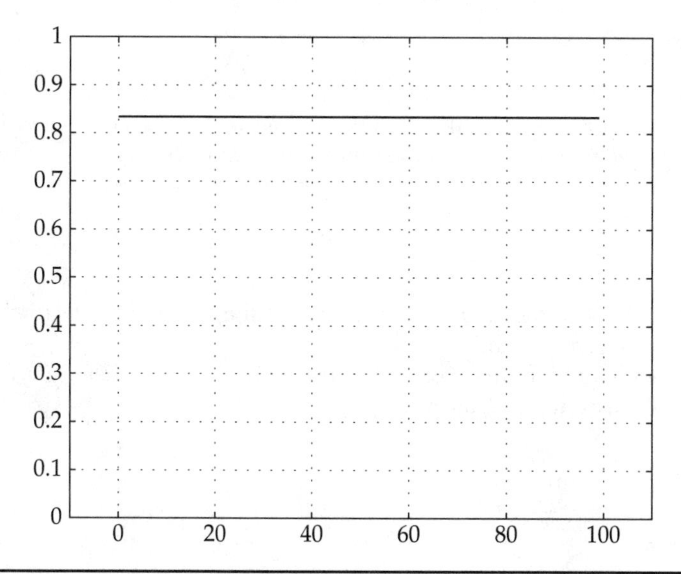

FIGURE A-2 Graph of a constant.

From the archives of your mind you might remember that the integral of t with respect to t is $t^2/2$ or

$$\int_a^b t\,dt = \frac{t^2}{2}\Big|_a^b = \frac{1}{2}(b^2 - a^2) \qquad \text{(A-4)}$$

Consequently, the distance expression becomes

$$S = \int_0^{10} t\frac{50}{60}\frac{1}{10}dt = \frac{50}{60}\frac{1}{10}\int_0^{10} t\,dt = \frac{50}{60\times 10\times 2}t^2\Big|_0^{10}$$

$$= \frac{50}{60\times 10\times 2}(10^2 - 0) = 4.17$$

During the acceleration period, the automobile covered 4.17 mi. The total distance covered for the whole time period from 0 min to 110 min is the total area under the speed curve or 83.33 + 4.17 = 87.5 miles.

$$S = \int_0^{110} v\,dt = \int_0^{10} t\frac{50}{60}\frac{1}{10}dt + \int_{10}^{110}\frac{50}{60}dt$$

$$= \frac{50}{60\times 10\times 2}t^2\Big|_0^{10} + \frac{50}{60}t\Big|_{10}^{110}$$

$$= 4.17 + 83.33 = 87.5$$

For our purposes, the integral of a variable $Y(t)$, also called the *integrand*, with respect to t over the domain of t from a to b is

$$\int_a^b Y(t)dt$$

Pictorially, the value of this integral is the *area* under the curve of $Y(t)$ between $t = a$ and $t = b$.

Sometimes in this book, the order of the integrand, here $Y(t)$, and dt will be exchanged, as in

$$\int_a^b Y(t)dt = \int_a^b dt\,Y(t)$$

This can be handy if one is looking at the integral as the *operation* of $\int_a^b dt\ldots$ on the quantity $Y(t)$.

The reader should realize that the argument of the integrand t, is a *dummy argument* and any symbol will do, as in

$$\int_a^b dtY(t) = \int_a^b duY(u)$$

Consider the speed history of another trip in an instrumented automobile shown in Fig. A-3. Here the speed changes suddenly, abruptly, and unrealistically at $t = 60$ min and again at $t = 80$ min. Temporarily ignoring the fact that an infinite braking force would be required to make the sudden changes in speed at those two times, the distance covered for the whole trip is again the area under the speed curve. In this case there are four areas. The first, from time 0 to 10.0 min, when there is acceleration, the second, from time 10.0 to 60.0 min, when the speed is constant at 50 mi/hr, the third, from time 60.0 to 80.0 min, when the speed is constant at 40 mi/hr and the fourth, from time 80.0 to 95.0 min, when the speed is constant at 30 mi/hr. The four areas can be calculated by observation. The first is the area of a right triangle and is $1/2(0.833 \times 10.0)$, the second is

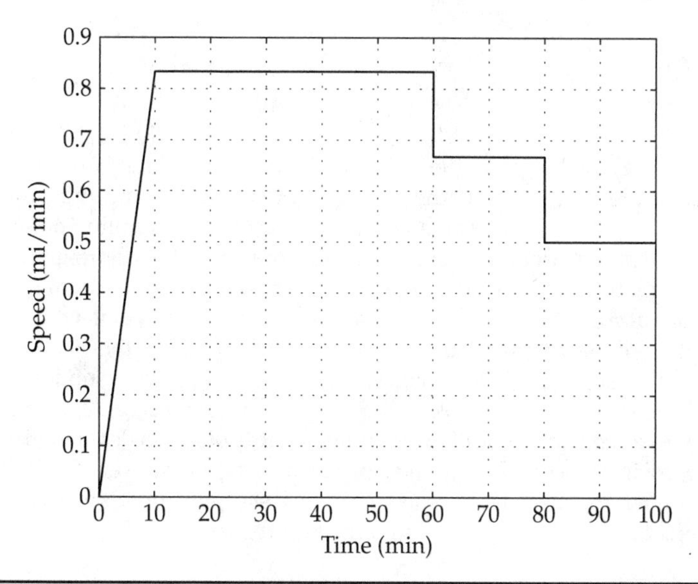

FIGURE A-3 Speed history for another trip.

0.833 × 50.0, the third is 40.0/60.0 × 20.0, and the fourth is 30.0/60.0 × 15.0 or

$$
\begin{aligned}
S &= \int_0^{95} v\,dt \\
&= \int_0^{10} v\,dt + \int_{10}^{60} v\,dt + \int_{60}^{80} v\,dt + \int_{80}^{95} v\,dt \\
&= \frac{1}{2}(0.833 \times 10) + 0.833 \times 50 + 0.666 \times 20 + 0.5 \times 15
\end{aligned}
$$

This exercise shows that the integral has been broken up into a sum of four calculations of contiguous areas.

A-3 Approximation of the Integral

In general, when the integrand is known numerically at variable sampling points of the independent variable, t in this case, as in

$$Y(t_1), Y(t_2), \ldots, Y(t_n)$$

then the integral of Y with respect to t can be *approximated* as a sum of the areas of relatively small rectangles

$$\int_{t_1}^{t_n} dt\,Y(t) \cong \sum_{i=1}^{n-1} Y(t_i)(t_{i+1} - t_i) = Y(t_1)(t_2 - t_1) + \cdots + Y(t_{n-1})(t_n - t_{n-1}) \quad \text{(A-5)}$$

Here, the height of the ith rectangle is $Y(t_i)$ and the width of the rectangle is $(t_{i+1} - t_i)$. If the spacing between the sampling points $t_1, t_2, \ldots, t_n$ can be made smaller and the number of sampling points n can be made larger, it is reasonable to expect that the approximation will get better. The reader probably can imagine that there might be more accurate ways to numerically estimate the integral when the values of the integrand are given at sample points and there certainly are. This superficial discussion of the integral suggests the notion that (1) the integral can be looked at as an area under a curve, (2) it can be approximated numerically with a sum of areas of rectangles, and (3) the approximation gets better when the rectangles get narrower.

If it happens that the spacing between sampling points is uniform, as in $t_i = t_{i-1} + h$ then Eq. (A-5) can be written as

$$\int_{t_1}^{t_n} dt\,Y(t) \cong \sum_{i=1}^{n-1} Y(t_i)(t_{i+1} - t_i) = h \sum_{i=1}^{n-1} Y(t_i) \quad \text{(A-6)}$$

A-4 Integrals of Useful Functions

First, the integral of a constant C

$$\int_a^b C\,du = Cu\Big|_a^b = C(b-a) \tag{A-7}$$

The integral of a ramp, Ct, (with a slope of C) is

$$\int_a^b Ct\,dt = C\frac{t^2}{2}\Big|_a^b = \frac{C}{2}(b^2 - a^2) \tag{A-8}$$

and the integral of an exponential function is

$$\int_a^b e^u du = e^u\Big|_a^b = e^b - e^a \tag{A-9}$$

Frequently, the exponential has an argument so the challenge is to evaluate

$$\int_a^b e^{cu} du \tag{A-10}$$

This requires a substitution to make Eq. (A-10) look like Eq. (A-9), namely,

$$v = cu \qquad \text{or} \qquad u = \frac{1}{c}v$$

$$dv = c\,du \qquad \text{or} \qquad du = \frac{1}{c}dv$$

Applying this to Eq. (A-10) gives

$$\int_a^b e^{cu} du = \int_{\frac{a}{c}}^{\frac{b}{c}} e^v \frac{1}{c} dv = \frac{1}{c}\int_{\frac{a}{c}}^{\frac{b}{c}} e^v dv = \frac{1}{c} e^v \Big|_{\frac{a}{c}}^{\frac{b}{c}} = \frac{1}{c}\left(e^{\frac{b}{c}} - e^{\frac{a}{c}}\right)$$

or

$$\boxed{\int_a^b e^{cu} du = \frac{1}{c}\left(e^{\frac{b}{c}} - e^{\frac{a}{c}}\right)} \tag{A-11}$$

The integral of x^n is useful

$$\boxed{\int_a^b x^n dx = \frac{x^{n+1}}{n+1}\Big|_a^b} \tag{A-12}$$

where n cannot be -1.

When n is -1 then

$$\int_a^b \frac{1}{x} dx = \ln(x)\Big|_a^b = \ln b - \ln a = \ln\frac{b}{a} \tag{A-13}$$

A-5 The Derivative, Rate of Change, and Acceleration

How could you estimate the acceleration during the trip represented in Fig. A-3? From physics we know that acceleration is the rate of change of velocity or speed with respect to time. In other words, acceleration is the *derivative* of speed with respect to time. Furthermore, for our trip, the acceleration is the slope of the speed curve.

A crude way to estimate the acceleration at some time $t = t'$ would be to make a ratio of the difference in velocity to that in time at time t' as in

$$a(t') \cong \frac{v(t'+h) - v(t'-h)}{(t'+h)-(t'-h)} = \frac{v(t'+h) - v(t'-h)}{2h}$$

The above is a ratio of the change in speed at time $t = t'$ to the associated change in time. The time change is $2h$ where h is a small time interval. The above expression is an approximation to the exact rate of change at time $t = t'$ and the approximation gets better as h gets smaller. In the limit, the ratio defines the *derivative* of v with respect to t which in our example is the acceleration

$$a(t') \equiv \frac{dv}{dt}\Big|_{t=t'} = \lim_{h\to 0} \frac{v(t'+h) - v(t'-h)}{2h} \tag{A-14}$$

(Note that the symbol a has occasionally been used to represent a constant but here it is the acceleration which can be a function of t.) This formula requires that the value of $v(t')$, arrived at from $v(t'+h)$ at time $t'+h$ by letting $h \to 0$, be the same as that arrived at from $v(t'-h)$ at time $t'-h$ by letting $h \to 0$. That is,

$$\lim_{h\to 0} v(t'+h) = \lim_{h\to 0} v(t'-h) = v(t')$$

In other words, one should get the same value of $v(t')$ as one approaches t' from the left as when one approaches t' from the right. That is the same thing as saying that $v(t)$ must be *continuous* at $t = t'$. This is clearly the case at $t = 5$ min.

In Fig. A-3 the acceleration at $t = 5$ min can be estimated from the ratio of differences using the formula for the speed

$$v = t\frac{50}{60}\frac{1}{10} = 0.0833t$$

as applied to Eq. (A-14)

$$\begin{aligned} a = \frac{dv}{dt} &= \lim_{h\to 0}\frac{v(t'+h)-v(t'-h)}{2h} \\ &= \lim_{h\to 0}\frac{0.0833\times(5+h)-0.0833\times(5-h)}{2h} \\ &= \lim_{h\to 0}\frac{2\times 0.0833h}{2h} \\ &= 0.0833 \end{aligned}$$

The units of the numerator are miles/minute and those of the denominator are minute so the acceleration at time $t = 5$ min is 0.0833 mi/min^2 and it stays constant at this value until time $t = 10$ min when it changes abruptly to zero and stays at zero until $t = 60$ min.

If we ask for the acceleration at $t' = 60$ min, we have a problem as can be seen by applying the above formula

$$a(60) = \frac{dv}{dt}\Big|_{t'=60} = \lim_{h\to 0}\frac{v(t'+h)-v(t'-h)}{2h}\Big|_{t'=60}$$

As pointed out above, and worth repeating here, this formula requires that the value of $v(t')$, arrived at from $v(t'+h)$ at time $t'+h$ by letting $h \to 0$, be the same as that arrived at from $v(t'-h)$ at time $t'-h$ by letting $h \to 0$. This is clearly *not* the case at $t = 60$ min. Here the limit of $v(t'-h)$ as $h \to 0$ is 0.833 but the limit of $v(t'+h)$ as $h \to 0$ is 0.666. Therefore, the speed is not uniquely defined at $t = 60$ min because it is *discontinuous*. Therefore, the acceleration, or the derivative of $v(t)$, at that time is undefined.

Question A-1 Is there a discontinuity at $t = 10$ min?

Answer Yes. The speed is continuous but acceleration is not. For $t < 10$ min the acceleration is constant and positive. At $t = 10$ min the acceleration suddenly becomes zero.

A-6 Derivatives of Some Useful Functions

The derivative of e^{at} with respect to time t is the coefficient a times the original function.

$$\boxed{\frac{d}{dt}e^{at} = ae^{at}} \tag{A-15}$$

The derivative of a constant C with respect to time is zero because it is not varying.

$$\boxed{\frac{d}{dt}\mathrm{C} = 0} \tag{A-16}$$

The derivative of the ramp Ct, where C is a constant and the rate of the ramp, with respect to t is

$$\boxed{\frac{d}{dt}\mathrm{C}t = \mathrm{C}} \tag{A-17}$$

and the derivative of Ct^2 with respect to t is

$$\boxed{\frac{d}{dt}\mathrm{C}t^2 = 2\mathrm{C}t} \tag{A-18}$$

In general, the derivative of t^n is

$$\boxed{\frac{d}{dt}t^n = nt^{n-1}} \tag{A-19}$$

Derivatives can be "chained" in the sense that the "second" derivative is the derivative of a first derivative, for example,

$$\frac{d^2}{dt^2}e^{at} = \frac{d}{dt}\left(\frac{d}{dt}e^{at}\right) = \frac{d}{dt}(ae^{at}) = a\frac{d}{dt}e^{at} = a^2e^{at}$$

The derivative of the trigonometric functions occurs frequently.

$$\begin{aligned}
&\frac{d}{dt}(\sin(at)) = a\cos(at) \\
&\frac{d}{dt}(\cos(at)) = -a\sin(at)
\end{aligned} \tag{A-20}$$

A-7 The Relation between the Derivative and the Integral

The derivative and the integral are inverses of each other. We can show this simply be using the definition of the derivative.

WARNING: *This subsection may get a little messy so you might want to breeze through it (or skip it altogether). However, it will provide a good exercise of your knowledge of calculus.*

Let

$$I(t) = \int_a^t duY(u)$$

represent the integral of $Y(t)$ where the upper limit is the independent variable t and u is a dummy variable.

The derivative of the integral of $Y(t)$ is

$$\frac{d}{dt}\int_a^t duY(u) = \frac{dI(t)}{dt} = \lim_{h\to 0}\frac{I(t+h)-I(t-h)}{2h}$$

Replace $I(t)$ by its definition

$$\frac{d}{dt}\int_a^t duY(u) = \lim_{h\to 0}\frac{\int_a^{t+h} duY(u) - \int_a^{t-h} duY(u)}{2h} \tag{A-21}$$

To make this a little more tractable, use the relation

$$\int_a^{t+h} duY(u) = \int_a^{t-h} duY(u) + \int_{t-h}^{t+h} duY(u)$$

This is simply splitting the integral from a to $t + h$ into an integral from a to $t - h$ plus an integral from $t - h$ to $t + h$. If this is hard to grasp, think in terms of a graph of $Y(t)$ versus t and remember that the integral is the area under the $Y(t)$ curve and we are just breaking one area up into two smaller contiguous ones.

Equation (A-21) becomes

$$\frac{d}{dt}\int_a^t Y(u)du = \lim_{h\to 0}\frac{\int_a^{t+h} duY(u) - \int_a^{t-h} duY(u)}{2h}$$

$$= \lim_{h\to 0}\frac{\int_a^{t-h} duY(u) + \int_{t-h}^{t+h} duY(u) - \int_a^{t-h} duY(u)}{2h}$$

$$= \lim_{h\to 0}\frac{\int_{t-h}^{t+h} duY(u)}{2h}$$

$$= \lim_{h\to 0}\frac{Y(t)2h}{2h} = Y(t)$$

The next-to-last line contains a "trick" in the sense that the integral from $t - h$ to $t + h$ is approximated by a rectangle of height $Y(t)$ and width $2h$:

$$\boxed{\int_{t-h}^{t+h} duY(u) \cong Y(t)2h} \tag{A-22}$$

This is consistent with our concept of the integral being the area under a curve. As $h \to 0$, the approximation becomes more exact and since we are letting $h \to 0$ in the definition of the derivative, all is well.

So, after all the dust is settled we have used the definitions of the integral and the derivative to show that they are *inverses* of each other.

$$\frac{d}{dt}\int_a^t Y(u)du = Y(t) \tag{A-23}$$

You will probably never have to use Eq. (A-23) but in arriving at it you have had to exercise your knowledge of calculus and perhaps a few cobwebs have been scraped away.

A-8 Some Simple Rules of Differentiation

The derivative of the product of two functions is

$$\frac{d}{dt}uv = u\frac{d}{dt}v + v\frac{d}{dt}u$$

The derivative and the integral of the product of a constant and a function is

$$\frac{d}{dt}Cu = C\frac{d}{dt}u \qquad \int_a^b dt\,Cu(t) = C\int_a^b dt\,u(t)$$

The derivative and integral of the sum of two functions is

$$\frac{d}{dt}(u+v) = \frac{d}{dt}u + \frac{d}{dt}v \qquad \int_a^b dt\,(u+v) = \int_a^b dt\,u + \int_a^b dt\,v \tag{A-24}$$

A-9 The Minimum/Maximum of a Function

A function $f(t)$ obtains localized minimum or maximum values at values of t that satisfy

$$\frac{df}{dt} = 0$$

For example, consider the function

$$y(t) = \cos(2\pi t/20)$$

The first derivative is

$$y'(t) = -\frac{\pi}{10}\sin(2\pi t/20)$$

and the second derivative is

$$y''(t) = -\frac{\pi^2}{100}\cos(2\pi t/20)$$

These three quantities are plotted in Fig A-4 where one sees that $y(t)$ takes on localized maximum values when the derivative $y'(t)$ is zero and when the second derivative $y''(t)$ is negative. On the other hand, $y(t)$ takes on localized minimum values when the derivative $y'(t)$ is zero and when the second derivative $y''(t)$ is positive.

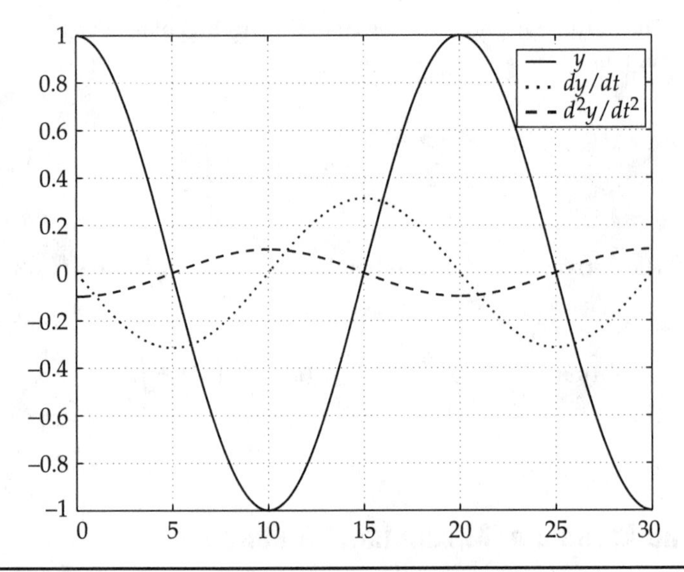

Figure A-4 First and second derivatives of a cosine.

A-10 A Useful Test Function

Consider the following function

$$Y(t) = 1 - e^{-\frac{t}{15}} \qquad t \geq 0$$
$$Y(t) = 0 \qquad t < 0 \tag{A-25}$$

The graph of this function is shown in Fig. A-5. Unlike the previous functions for the car trip examples, this function has no discontinuities. This function has the shape of "diminishing returns" in that it continually rises but the rate of change (the derivative) decreases continually.

Applying the rules of Sec. A-6 for the derivative of an exponential [Eq. (A-15)] and a constant, the derivative of the test function, where $a = -1/15$, yields

$$\frac{d}{dt}Y(t) = \frac{d}{dt}\left(1 - e^{-\frac{t}{15}}\right) = 0 + \frac{1}{15}e^{-\frac{t}{15}} = \frac{1}{15}e^{-\frac{t}{15}}$$

and Fig. A-6 shows a graph of this derivative. Note that the derivative or slope or rate of change is highest initially and then slowly, or asymptotically, decreases toward zero. Considering times before $t = 0$ where $Y(t) = 0$ and where $dY/dt = 0$, one sees that the derivative experiences a discontinuity at $t = 0$.

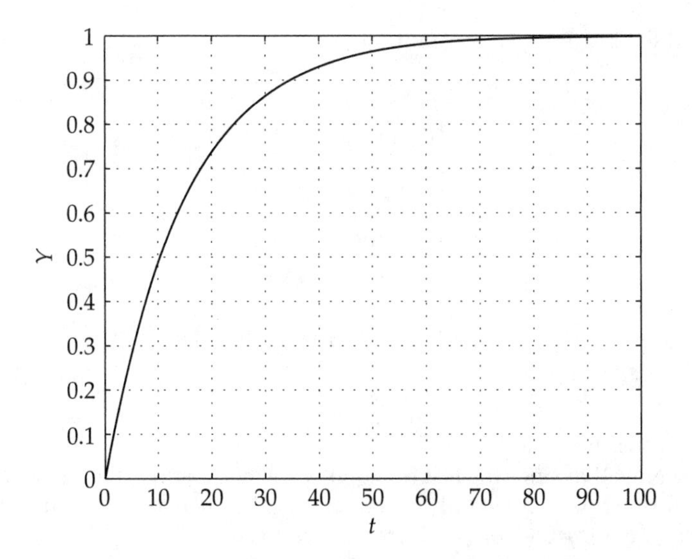

FIGURE A-5 Test function: $Y = 1 - \exp(-t/15)$.

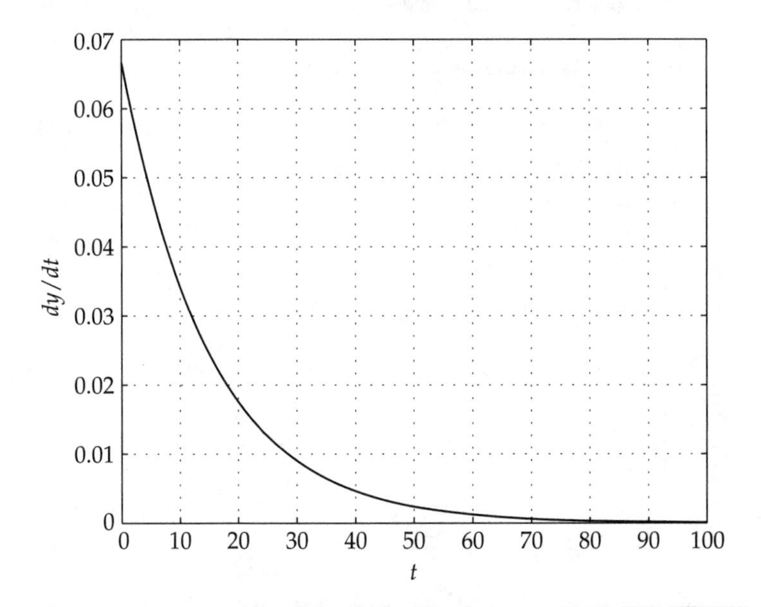

FIGURE A-6 Derivative of the test function.

The integral of exponential is also an exponential

$$\int_0^t du e^{au} = \frac{1}{a} e^{au} \Big|_0^t = \frac{1}{a}(e^{at} - 1) \tag{A-26}$$

and the integral of a constant C is a ramp

$$\int_0^t du C = Cu\Big|_0^t = Ct \tag{A-27}$$

So, the integral of our test function is

$$\int_0^t du \left(1 - e^{-\frac{u}{15}}\right) = t + 15\left(e^{-\frac{t}{15}} - 1\right)$$

which is shown in Fig. A-7.

Note that since the test function is always positive, the integral of that function is constantly, or monotonically, increasing because it is accumulating area under the test function curve.

Question A-2 Examine Figs. A-7 and A-5. Does the curve in the latter figure look like the derivative of the curve in the former?

Answer It should because the derivative of an integral is the integrand.

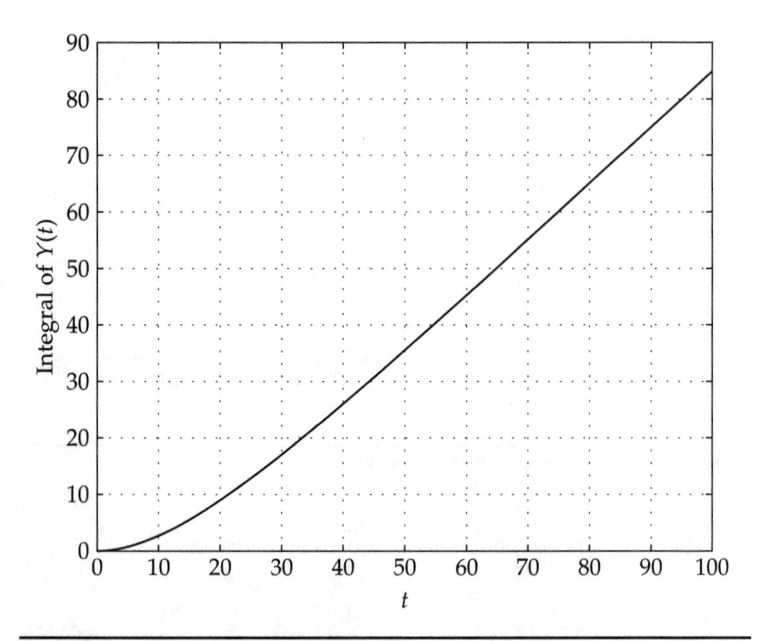

FIGURE A-7 Integral of the test function.

A-11 Summary

Using the automobile trip as an example, the concept of the integral was presented as the area under the speed curve—a way of determining how far you might travel when you have data on your speed as a function of time. The analytical integrals of a couple of common and useful functions were developed. The area under the curve was shown to be approximated numerically by a sum of rectangular areas.

The automobile trip example was used next to review the concept of the derivative. A ratio of differences was shown to evolve into the derivative when the separation in the denominator was reduced to an infinitesimal value. The derivative could provide a way of determining your acceleration throughout the trip should you have data on the speed as a function of time and should the speed be continuous. The derivative was also shown to be useful in finding locations of maxima or minima of a function. The two concepts (derivative and integral) were compared via an example that behaved in the "diminishing returns" fashion.

Summary

APPENDIX B
Complex Numbers

Consider the relatively simple challenge of solving

$$n - 1 = 0$$

for the value of the variable n that will satisfy the equation. The solution is $n = 1$ where n represents an *integer*, as in $\ldots -3, -2, -1, 0, 1, 2, 3 \ldots$.

Next, solve the following equation for x

$$2x - 3 = 0$$

The solutions to this type of equation are the *rational numbers* that are defined as ratios of integers with nonzero denominators.

The solution to an equation like

$$x^2 - 2 = 0$$

or

$$2x - 3\pi = 0$$

is an *irrational* number. That is, no ratio of integers will yield $\sqrt{2}$ or $3\pi/2$. The integers, along with the rational and irrational numbers form the *real* numbers.

Graphically, a real number can be represented as the length of a line or as a point on the abscissa. For example the quantity $x = 1.5$ could be represented as in Fig. B-1.

The solution to

$$x^2 + 2 = 0$$

is an *imaginary* number.

$$x = j\sqrt{2}$$

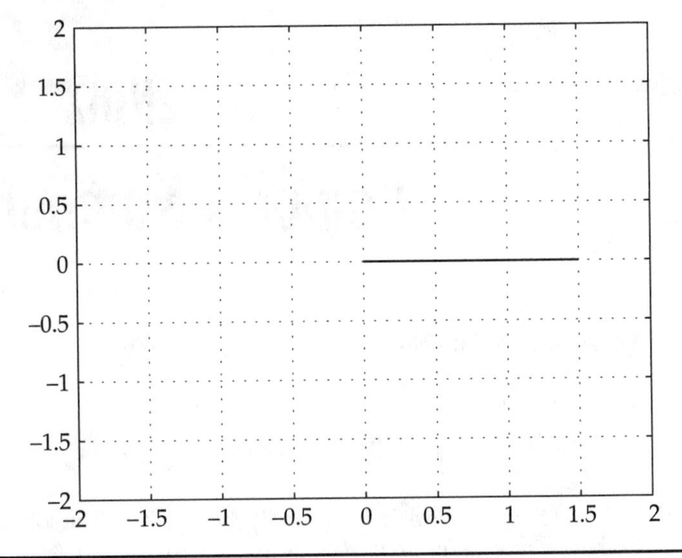

FIGURE B-1 Representation of a real number $x = 1.5$.

where the symbol j signifies that what follows is imaginary. Graphically, this number could be represented on the ordinate of a graph as in Fig. B-2

Real numbers, whether integers, rationals or irrationals, are plotted on the abscissa. Imaginary numbers are plotted on the ordinate.

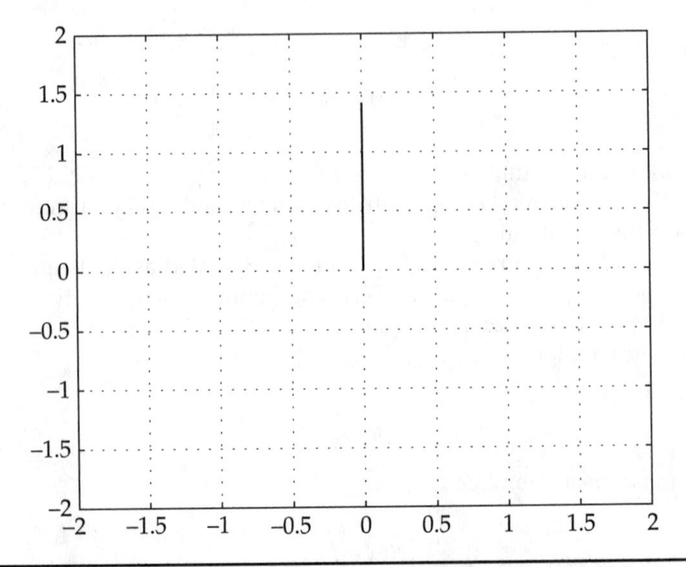

FIGURE B-2 Representation of an imaginary number $x = j1.4141414$.

Two values of x satisfy the quadratic equation

$$ax^2 + bx + c = 0 \tag{B-1}$$

and they are

$$x = -\frac{b}{2a} \pm \frac{\sqrt{b^2 - 4ac}}{2a} \tag{B-2}$$

If $b^2 < 4ac$ then the two solutions will be

$$-\frac{b}{2a} + j\frac{\sqrt{4ac - b^2}}{2a} \quad \text{and} \quad -\frac{b}{2a} - j\frac{\sqrt{4ac - b^2}}{2a} \tag{B-3}$$

which have a real *and* an imaginary part and are called *complex* numbers. While real numbers lie at a point along the abscissa and imaginary numbers lie at a point along the ordinate, complex numbers lie at a point in the so-called complex plane.

As an aside, it doesn't stop with complex numbers. In 1843, the famous Irish mathematician W. R. Hamilton developed a four-dimensional construct called *quaterions* which are basically complex numbers in four, not two, dimensions. Fortunately, there are no realistic needs for quaterions in control engineering but if your control engineer is giving you a hard time you might suggest she try using them to solve problems. Now, back to reality.

B-1 Complex Conjugates

Note that the two complex numbers appearing in Eq. (B-3) are *complex conjugates* of each other (the imaginary part of one is the negative of the imaginary part of the other) and that their product is real, as in

$$(a + jb)(a - jb) = a^2 + b^2$$

For the above expression to be true the variable j must satisfy

$$j^2 = -1$$

The basis for this equation will be developed below. In general, an asterisk indicates the complex conjugate as in

$$z = a + jb \qquad z^* = a - jb$$

B-2 Complex Numbers as Vectors or Phasors

The complex number $x = 1.5 + j\sqrt{2}$ has a real part, 1.5, that can be plotted on the abscissa and an imaginary part, $\sqrt{2}$, that can be plotted on the ordinate as in Fig. B-3. This suggests that a complex number $z = a + jb$ can be considered a *vector* (also sometimes called a *phasor* in electrical engineering) with length or magnitude M

$$M = |z| = \sqrt{a^2 + b^2} \tag{B-4}$$

and an angle (or phase) θ with the abscissa of

$$\theta = \arctan\left(\frac{b}{a}\right) = \tan^{-1}\left(\frac{b}{a}\right) \tag{B-5}$$

which, in words, says that the vector representing the complex number z has the angle in radians whose tangent is given by the imaginary part divided by the real part. Figure B-3 shows this vector with the angle θ.

Therefore, positive real numbers have an angle of 0° and positive imaginary numbers have an angle of 90° or $\pi / 2$ radians. A negative real number like –1.0 would have an angle of 180° or π radians. A negative imaginary number like $-j$ would have an angle of 270° or $3\pi / 2$ radians. Note that the magnitude of a complex number is the

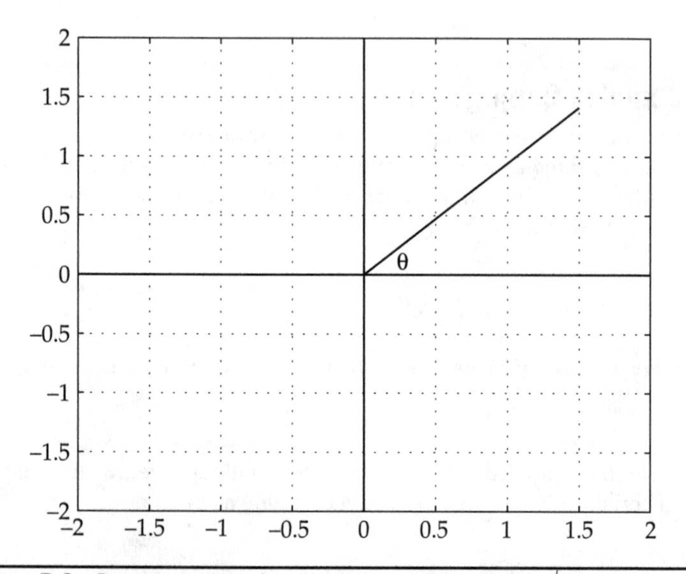

FIGURE B-3 Representation of a complex number $z = 1.5 + j\sqrt{2}$.

square root of the sum of the squares of the real and imaginary parts. Further, note that the magnitude of a complex number $z = a + jb$ can be obtained by taking the square root of the product of the number and its complex conjugate, as in,

$$M = |z| = \sqrt{z \cdot z^*} = \sqrt{(a + jb)(a - jb)} = \sqrt{a^2 + b^2}$$

Turning the magnitude and angle concept around, one could use the projections of the phasor on the real and imaginary axes and say that the real part of a complex number is $M\cos(\theta)$ and the imaginary part is $M\sin(\theta)$, as in

$$z = M\cos(\theta) + jM\sin(\theta) \tag{B-6}$$

Symbolically this is represented by

$$z = a + jb = M\cos(\theta) + jM\sin(\theta)$$

$$R\{z\} = M\cos(\theta)$$

$$I\{z\} = M\sin(\theta)$$

B-3 Euler's Equation

We will frequently use an extension of famous Euler equation for an exponential with a complex argument:

$$e^{c+jd} = e^c[\cos(d) + j\sin(d)] \tag{B-7}$$

Strictly speaking, Eq. (B-7) is not Euler's equation; rather the following is

$$e^{jd} = \cos(d) + j\sin(d)$$

which is simply Eq. (B-7) with a zero real component.

Euler's equation can be inverted to give expressions for the sine and cosine

$$\begin{aligned} \cos d &= \frac{e^{jd} + e^{-jd}}{2} \\ \sin d &= \frac{e^{jd} - e^{-jd}}{2j} \end{aligned} \tag{B-8}$$

Note that the Euler equation provides a convenient shorthand way of showing the magnitude and angle of a complex number z.

$$\begin{aligned} z &= a + jb \\ &= M[\cos(\theta) + j\sin(\theta)] \\ &= Me^{j\theta} \end{aligned} \tag{B-9}$$

where

$$M = e^c = \sqrt{a^2 + b^2}$$

$$\theta = d = \arctan\left(\frac{b}{a}\right)$$

So, in this book we will often write a complex number as

$$\boxed{\begin{aligned} z &= Me^{j\theta} \\ &= |z|e^{j\theta} \end{aligned}} \tag{B-10}$$

There are advantages of looking at a complex number as a vector or a phasor with magnitude M or $|z|$ and angle (or phase) θ. This suggests that the simple algebra of exponentials can be used in complex multiplication, as in

$$z_1 = M_1 e^{j\theta_1} \qquad z_2 = M_2 e^{j\theta_2}$$

$$z_1 \times z_2 = M_1 e^{j\theta_1} \times M_2 e^{j\theta_2} = M_1 M_2 e^{j(\theta_1 + \theta_2)}$$

So, in complex multiplication, magnitudes multiply and angles add. The phasor presentation is useful for conjugates. If the complex number of interest is

$$z = Me^{j\theta}$$

then the conjugate is

$$z^* = Me^{-j\theta}$$

Using this approach we can develop an alternative expression for the complex operator j, as in

$$e^{j\frac{\pi}{2}} = \cos\left(\frac{\pi}{2}\right) + j\sin\left(\frac{\pi}{2}\right) = 0 + j1 = j$$

This means that the complex operator j has a magnitude of unity and a phase of $\pi/2$.

As an aside, consider

$$e^{j\pi} = -1$$

which follows from Euler's equation. Isn't that a pretty expression? It contains five important parts of mathematics: the exponential, the imaginary operator, π, the unit number, and the minus sign. As a manager, you will likely never have any use for this expression but it might come in handy sometime to drop on a hotshot control engineer to impress her with your depth of mathematical understanding.

Question B-1 What is the exponential form of the number $-3-4j$?

Answer The magnitude is $\sqrt{3^2+4^2}=5$. The angle of the number is

$$\tan^{-1}\left(\frac{-4}{-3}\right) = 53.13^\circ + 180^\circ = 233.13^\circ$$

I added 180° to the phase because both the real and imaginary components are negative so the phasor lies in the third quadrant. The angle in radians is $233.13^\circ\pi/180^\circ = 4.07$. Therefore, the complex number can be presented as $5e^{j4.07}$.

Complex numbers can be added and multiplied using the $z = x + jy$ representation also. For example, addition of two complex numbers is

$$\begin{aligned} z_1 + z_2 &= x_1 + jy_1 + x_2 + jy_2 \\ &= (x_1 + x_2) + j(y_1 + y_2) \end{aligned}$$

Multiplication follows as in

$$z_1 z_2 = (x_1 + jy_1)(x_2 + jy_2) = x_1x_2 + jx_1y_2 + jx_2y_1 + j^2y_1y_2$$

Now, what is j^2? The exponential form for j can be used to answer that question.

$$j^2 = \left(e^{j\frac{\pi}{2}}\right)^2 = e^{j\pi} = -1$$

So, multiplication of two complex numbers is as follows

$$\begin{aligned} z_1 z_2 &= (x_1 + jy_1)(x_2 + jy_2) = x_1x_2 + jx_1y_2 + jx_2y_1 + j^2y_1y_2 \\ &= (x_1x_2 - y_1y_2) + j(x_1y_2 + x_2y_1) \end{aligned}$$

B-4 An Application to a Problem in Chapter 4

Early in Chap. 4, a complex sinusoid was chosen to be the input for the first-order process, as in

$$\tau \frac{dY}{dt} + Y = U$$
$$U(t) = A_U \cos(2\pi f t) = A_U \operatorname{Re}\{e^{j2\pi f t}\} \tag{B-11}$$

The input is replaced by a complex sinusoid with the agreement that when the solution has been obtained, if necessary, only the real part will be kept. Here, we have ignored the constant offset $U_{c.}$

$$U(t) = A_U e^{j2\pi f t}$$

The process output is assumed to be a sinusoid with an amplitude C and a phase θ relative to the input, as in

$$Y(t) = \operatorname{Re}\left\{Ce^{j(2\pi f t+\theta)}\right\}$$

In the proceeding manipulations, we will remove the Re operator and use

$$Y(t) = Ce^{j(2\pi f t+\theta)}$$

When the expressions for U and Y are inserted into Eq. (B-11), the result is

$$\tau(j2\pi f)Ce^{j(2\pi f t+\theta)} + Ce^{j(2\pi f t+\theta)} = A_U e^{j(2\pi f t)} \tag{B-12}$$

Now comes the busy part. We must (1) separate the real and imaginary parts of Eq. (B-12), (2) equate them, thereby developing two equations, and (3) solve for the two unknowns, C and θ. The following steps should be straightforward since they use the algebra of the exponentials and Euler's formula. See if you can go from line to line and figure out what change was made.

Factor the exponential:

$$\tau(j2\pi f)Ce^{j(2\pi ft)}e^{j\theta} + Ce^{j(2\pi f)}e^{j\theta} = A_U e^{j(2\pi f)}$$

Remove the common factor:

$$\tau(j2\pi f)Ce^{j\theta} + Ce^{j\theta} = A_U$$

Collect the $Ce^{j\theta}$ term:

$$[\tau(j2\pi f)+1]Ce^{j\theta} = A_U$$

Use Euler's formula:

$$[\tau(j2\pi f)+1]C(\cos\theta + j\sin\theta) = A_U$$

Expand the product into individual terms:

$$\tau(j2\pi f)C\cos\theta + C\cos\theta + \tau(j^2 2\pi f)C\sin\theta + Cj\sin\theta = A_U$$

Use $j^2 = -1$

$$j\tau(2\pi f)C\cos\theta + jC\sin\theta + C\cos\theta - \tau(2\pi f)C\sin\theta = A_U \qquad \text{(B-13)}$$

The imaginary part of Eq. (B-13) is

$$\tau(j2\pi f)C\cos\theta + C\sin\theta = 0$$

$$\tau(2\pi f)\cos\theta + \sin\theta = 0$$

$$\tan\theta = -\tau(2\pi f)$$

Therefore, the phase can be shown to be dependent on the time constant and the frequency.

$$\theta = -\tan^{-1}(2\pi f\tau) \qquad \text{(B-14)}$$

The real part of Eq. (B-13) is

$$C\cos\theta - \tau(2\pi f)C\sin\theta = A_U$$

$$C = \frac{A_U}{\cos\theta - \tau(2\pi f)\sin\theta}$$

Replacing θ with Eq. (B-14) and using $\omega = 2\pi f$ gives

$$C = \frac{A_U}{\cos[-\tan^{-1}(\omega\tau)] - \omega\tau\,\sin[-\tan^{-1}(\omega\tau)]} \qquad \text{(B-15)}$$

There are two trigonometric identities that can be useful, namely,

$$\cos[\tan^{-1}(x)] = \frac{1}{\sqrt{1+x^2}}$$

$$\sin[\tan^{-1}(x)] = \frac{x}{\sqrt{1+x^2}}$$

which, when combined with Eq. (B-15) yield

$$C = \frac{A_U}{\frac{1}{\sqrt{1+\omega\tau^2}} - \omega\tau\frac{\omega\tau}{\sqrt{1+\omega\tau^2}}}$$

or, finally

$$C = \frac{A_U}{\sqrt{1+(\omega\tau)^2}} \tag{B-16}$$

This means that the solution is

$$Y(t) = Ce^{j(2\pi ft+\theta)} = \frac{A_U}{\sqrt{1+(\omega\tau)^2}} e^{j\theta}\, e^{j2\pi ft}$$

In other words, $Y(t)$ is a sinusoid with an amplitude of $\frac{A_U}{\sqrt{1+(\omega\tau)^2}}$ and phase θ.

B-5 The Full Monty

You may wonder what the full solution to Eq. (B-11), including the transient part, looks like and you may be a bit lazy like me. If so, then you would probably apply Matlab's symbolic toolbox via the following script (called `solvetransient`):

```
syms y t w f g T Au % declare time domain quantities
syms Y s U % declare s-domain quantities
Y=1/(T*s+1); % transfer function of unity gain 1st order
process
U=Au*w/(s^2+w^2); % LT of the sinusoidal input
Y=Y*U;  % combine
disp('Laplace Transform of Y')
pretty(Y)
y=ilaplace(Y); % find the inverse LT
disp('inverse')
y=simple(y);
pretty(y)
```

which would yield the following computer output:

```
>> solvetransient
Laplace Transform of Y

                  Au w
           ------------------
                    2   2
           (T s + 1) (s  + w )
```

```
inverse

         Au (-T exp(- t/T) w + w T cos(w t) - sin(w t))
       - ----------------------------------------------
                        2 2
                        w T + 1

>>
```

This Matlab script uses the Laplace transform (see App. F) to solve Eq. (B-11) in full. You can see that the transient component does in fact die away at a rate dictated by the time constant. Unfortunately, I had to do some Web searching to find a trigonometric identity

$$a\sin(x) + b\cos(x) = \sqrt{a^2 + b^2}\sin(x + \vartheta) \qquad \vartheta = \tan^{-1}(b/a)$$

that would give me the final result of

$$\frac{A_U \omega\tau}{\tau^2\omega^2 + 1} e^{-\frac{t}{\tau}} + \frac{A_U}{\sqrt{\tau^2\omega^2 + 1}} \sin(\omega t + \vartheta)$$
$$\vartheta = -\tan^{-1}(\omega\tau) \qquad \text{(B-17)}$$

So, we have used two computer-based tools to show that Eqs. (B-16) and (B-17) are related.

B-6 Summary

I can remember being introduced to the infamous evil imaginary *i* in high school (to become *j* in engineering school). It was something to be feared—even the amazing convention of using the word "imaginary" is kind of scary. It should have been an exciting experience to learn that the numbers we had been using actually had another dimension that could be quite useful in solving problems. Numbers were really locations in a plane rather than just along a line!

APPENDIX C

Spectral Analysis

In Chap. 2, spectral analysis of process data was discussed briefly. Now that complex numbers have been introduced in App. B, a more detailed look at spectral analysis can be taken.

Spectral analysis can be considered as

1. A transformation of a data stream from the time domain to the frequency domain using the Fast Fourier Transform (FFT).
2. A least squares fit of a series of sines and cosines with a fixed frequency grid to a data stream. Furthermore, the least squares fit can take advantage of the orthogonality of the sines and cosines to speed up the calculations.
3. A cross-correlation of the data stream with a selected set of sinusoids.
4. A special fit of sines and cosines with a specialized grid of frequencies that might require time-consuming calculations.

Each of these viewpoints has its advantages and it is important for the manager and engineer to be aware of the differences. This appendix will examine the first two viewpoints in some detail but the reader should spend a few moments considering the second two.

Using the FFT to transform the data stream into the frequency domain is the most popular because it is the easiest and quickest. Matlab (see App. J) has a built-in function that carries out the FFT as does the widely used program, Excel. On the other hand, the least squares approach is perhaps more easily digested from a mathematical point of view.

C-1 An Elementary Discussion of the Fourier Transform as a Data-Fitting Problem

In the least squares approach to fitting data, the problem is usually stated as, "given the data, $y_i = y(t_i)$, $i = 1,\ldots,N$, where the sample points are $t_i, i = 1,\ldots,N$, find the parameters in the fitting function $F(t_i)$ such that the error at each point, namely, $e_i = y(t_i) - F(t_i)$, is minimized in the least squares sense—that is, such that $\sum_{i=1}^{N} e_i^2$ is minimized."

For the line spectrum and for the Fourier transform, the data-fitting function is

$$F(t_i) = a_0 + a_1 \cos(2\pi f_1 t_i) + b_1 \sin(2\pi f_1 t_i) + a_2 \cos(2\pi f_2 t_i) + b_2 \sin(2\pi f_2 t_i) + \ldots \tag{C-1}$$

and the sampling points are equally spaced, as in $t_i = ih, i = 1,\ldots,N$. The data fitting problem is therefore, "find the coefficients, $a_0, a_1, a_2, \ldots, b_1, b_2, \ldots$, in Eq. (C-1) such that $F(t_i)$ fits y_i for $i = 1,\ldots,N$." For each one of the N data points there will be an equation like

$$y_i = a_0 + a_1 \cos(2\pi f_1 t_i) + b_1 \sin(2\pi f_1 t_i) + a_2 \cos(2\pi f_2 t_i) + b_2 \sin(2\pi f_2 t_i) + \cdots + e_i \tag{C-2}$$

Note that in Eqs. (C-1) and (C-2), the zero frequency term is a_0 because $\cos(0) = 1$ and $\sin(0) = 0$. Since all of the sinusoids in Eq. (C-1) vary about zero, a_0 is the average value of the $y_i = y(t_i)$, $i = 1,\ldots,N$ data stream. If the average is removed from the data, then one would expect $a_0 = 0$.

Each sine and cosine term in Eq. (C-1) has a frequency f_k that is a multiple of the fundamental frequency f_1, which has a period equal to $L = nh$, the length of the data set. That is, the fundamental frequency is given by

$$f_1 = \frac{1}{L} = \frac{1}{Nh} \tag{C-3}$$

The *sampling frequency* f_s is the reciprocal of the sampling interval h.

$$f_s = \frac{1}{h} \tag{C-4}$$

If the size of the data stream is 1000 samples and the sampling interval is 2 sec then $N = 1000$, $L = 2000$, $f_1 = 1/2000 = 0.0005\text{Hz}$, $f_s = 1/2 = 0.5$ Hz, $h = 2$ sec. The other frequencies, $N/2$ of them, appearing in Eq. (C-1), are multiples of the fundamental frequency and are called harmonics of the fundamental frequency.

$$f_2 = 2f_1$$

$$f_3 = 3f_1$$

$$\ldots$$

$$f_{\frac{N}{2}} = \frac{N}{2} f_1$$

or, since
$$f_1 = \frac{1}{Nh} \tag{C-5}$$

$$f_k = \frac{k}{Nh} \qquad k = 1,\ldots,\frac{N}{2}$$

and, since
$$f_s = \frac{1}{h}$$

$$f_k = \frac{k}{N} f_s \qquad k = 1,\ldots,\frac{N}{2}$$

Thus, the spacing between frequencies is f_1 and Eq. (C-5) shows that one can get a tighter grid of frequencies by increasing the number of sample points with the same sampling interval. Increasing the sampling frequency (or decreasing the sampling interval in the time domain) actually widens the spacing of the frequencies for the same number of samples. The last and largest frequency in the Eq. (C-5) sequence is the so-called *folding frequency* or *Nyquist* frequency which can also be written as

$$f_{N/2} = f_{Ny} = \frac{N}{2} f_1 = \frac{N}{2}\frac{1}{N} f_s = \frac{f_s}{2}$$

The frequency interval between zero and the folding frequency is sometimes called the *Nyquist interval.*

Since $t_i = ih$, Eq. (C-2) can be written as

$$\begin{aligned} y_i &= a_0 + a_1\cos(2\pi f_1 ih) + b_1\sin(2\pi f_1 ih) + a_2\cos(2\pi f_2 ih) \\ &\quad + b_2\sin(2\pi f_2 ih) + \cdots + a_{N/2}\cos(2\pi f_{N/2} ih) \\ &\quad + b_{N/2}\sin(2\pi f_{N/2} ih) + e_i \qquad i = 1,\ldots,N \end{aligned} \tag{C-6}$$

But, $f_k = k\frac{1}{Nh}$, so Eq. (C-6) becomes

$$\begin{aligned} y_i &= a_0 + a_1\cos\left(\frac{2\pi i}{N}\right) + b_1\sin\left(\frac{2\pi i}{N}\right) + a_2\cos\left(\frac{4\pi i}{N}\right) + b_2\sin\left(\frac{4\pi i}{N}\right) \\ &\quad + \cdots + a_{N/2}\cos(\pi i) + b_{N/2}\sin(\pi i) + e_i \end{aligned} \tag{C-7}$$

The last sine term in Eq. (C-7) is zero because $\sin(\pi i) = 0$ for $i = 1,\ldots,N$.

Using the summation operator, Eq. (C-7) can be rewritten as

$$y_i = a_0 + \sum_{k=1}^{N/2}\left[a_k \cos\left(\frac{2\pi ki}{N}\right) + b_k \sin\left(\frac{2\pi ki}{N}\right)\right] + e_i \qquad \text{(C-8)}$$

There are N unknown coefficients: $a_0, a_1, \ldots, a_{N/2}, b_1, b_2, \ldots, b_{N/2-1}$ and there are N data points. Since the number of data points equals the number of unknown coefficients in the data-fitting function, we can expect that the data-fitting function will indeed go through every data point and that the least squares errors will be zero.

One could go through the least squares exercise of finding the coefficients such that sum of the squares of the fitting errors, $e_i, i = 1,\ldots,N$, would be minimized, where the fitting error is defined as

$$e_i = y_i - F(t_i) \qquad i = 1,\ldots,N \qquad \text{(C-9)}$$

To minimize the sum of the squares of the errors with respect to the coefficients, one would generate the following N equations:

$$\frac{\partial}{\partial a_k}\sum_{i=1}^{N} e_i^2 = 0 \qquad k = 0,1,\ldots,N/2$$

and

$$\frac{\partial}{\partial b_k}\sum_{i=1}^{N} e_i^2 = 0 \qquad k = 1,\ldots,N/2-1 \qquad \text{(C-10)}$$

This should be familiar from calculus (App. A) where it is demonstrated that the minimum value of a function occurs when that function's derivative is zero. Using Eq. (C-9) to replace e_i in Eq. (C-10) and doing some straightforward calculus and algebra would yield the so-called "normal equations."

However, it is simpler to multiply Eq. (C-8) by each sinusoid in turn and sum over the data points. For example, multiplying Eq. (C-8) by the mth cosine, $\cos(2\pi mi/N)$, and summing over the data points would yield

$$\sum_{i=1}^{N} y_i \cos\left(\frac{2\pi mi}{N}\right) = a_0 \sum_{i=1}^{N} \cos\left(\frac{2\pi mi}{N}\right)$$
$$+\sum_{i=1}^{N}\sum_{k=1}^{N/2}\left[a_k \cos\left(\frac{2\pi mi}{N}\right)\cos\left(\frac{2\pi ki}{N}\right) + b_k \cos\left(\frac{2\pi mi}{N}\right)\sin\left(\frac{2\pi ki}{N}\right)\right] \qquad \text{(C-11)}$$

Because of orthogonality, the sinusoidal products, when summed over the equally spaced data, satisfy

$$\sum_{k=1}^{n} \cos\left(\frac{2\pi mi}{N}\right)\cos\left(\frac{2\pi ki}{N}\right) = \begin{Bmatrix} 0 & k \neq m \\ N/2 & k = m \end{Bmatrix}$$

Orthogonality is an important property of equally spaced sinusoids and provides the cornerstone of spectral analysis.

Therefore, in Eq. (C-11) all the cross-products, where $m \neq k$, drop out, leaving only

$$\sum_{i=1}^{n} y_i \cos\left(\frac{2\pi mi}{N}\right) = a_m \sum_{i=1}^{n} \cos\left(\frac{2\pi mi}{N}\right)\cos\left(\frac{2\pi ki}{N}\right) = a_m \frac{N}{2}$$

which can be solved for a_m:

$$a_m = \frac{2}{N}\sum_{i=1}^{N} y_i \cos(2\pi mi) \qquad m = 0, 1, \ldots, N/2 \qquad \text{(C-12)}$$

Note that for $m = 0$, Eq. (C-12) shows that a_0 is indeed the average.

There is a similar set of equations for the b_m shown in Eq. (C-13).

$$b_m = \frac{2}{N}\sum_{i=1}^{N} y_i \sin(2\pi mi) \qquad m = 0, 1, \ldots, N/2 \qquad \text{(C-13)}$$

This orthogonality will occur only if the data is equally spaced in the time domain and if the frequencies in the data fitting equation are chosen as multiples of the fundamental frequency.

Neither Eq. (C-12) nor (C-13) should be used to calculate the coefficients. Instead, one would use the Fast Fourier Transform (FFT) algorithm which makes extensive use of trigonometric identities to shorten the calculation time. Fortunately, you do not have to understand the FFT algorithm to use it because of the algorithms in Matlab and Excel mentioned earlier and illustrated in App. J.

C-2 Partial Summary

The foregoing has been quite involved. First, the least squares concept was applied to a data-fitting problem. This approach was not continued to its bitter end because the orthogonality of sines and cosines, defined on an equally spaced grid provided a simpler path to the equations for the coefficients. In effect, Eqs. (C-12) and (C-13) transform the problem from

$$y_i, \; i = 1, \ldots, N$$

$$t_i = h, \; 2h, \ldots, \; N$$

in the time domain to

$$a_k, \; b_k, \; k = 1, \ldots, \frac{N}{2}$$

$$f_k = 0, \frac{1}{Nh}, \frac{2}{Nh}, \ldots, \frac{1}{2h}$$

in the frequency domain.

C-3 Detecting Periodic Components

If there is a periodic component lurking in a noisy data stream, the above data fit will yield relatively large values for the coefficients a_i and b_i associated with the sinusoidal term in the least squares fit having a frequency f_i near that of the periodic component. This periodic component lying in concealment must have a frequency that is less than the folding frequency. If the periodic component has a frequency that is higher than the folding frequency, it will show up as an *alias* and the analysis will yield a large value for coefficients associated with another frequency that does lie in the so-called Nyquist interval. We will discuss aliasing in more detail later in this appendix.

If the spectral analysis reveals a periodic component that you suspect is an alias of a higher frequency then the data collection should be repeated with a different sampling interval. If the subsequent spectral analysis shows that the periodic component has moved then you can safely conclude that it is an alias.

Question C-1 Why is this?

Answer Wait until we talk about aliasing later on in this appendix.

C-4 The Line Spectrum

The *line spectrum* or *power spectrum* is a plot of the magnitudes of the data fit coefficients $a_k^2 + b_k^2$ or $\sqrt{a_k^2 + b_k^2}$, versus the frequencies, f_k. In the example data stream shown in Fig. 2-3 (and in Fig. C-1 in this appendix) the coefficients associated with sinusoidal terms having a

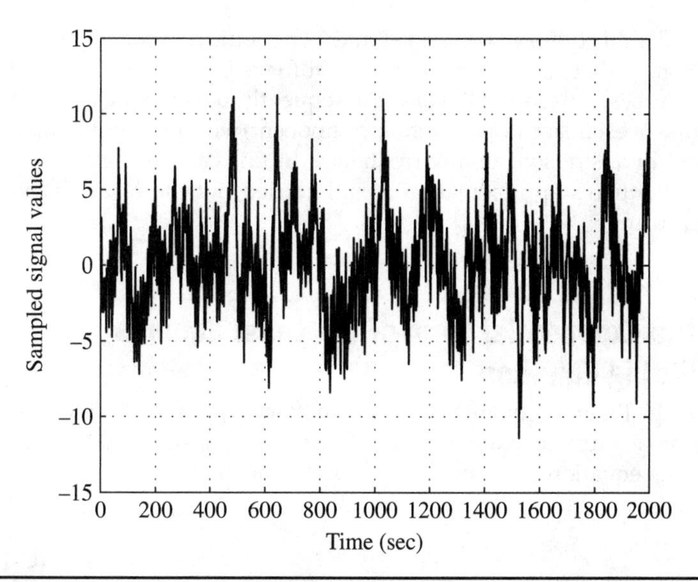

FIGURE C-1 A noisy signal containing two sinusoids.

frequency near 0.091 Hz and 0.1119 Hz were relatively large (see Fig. C-2). In effect, fitting the sinusoids to the time domain data allows one to estimate how much spectral or harmonic power is present at various frequencies.

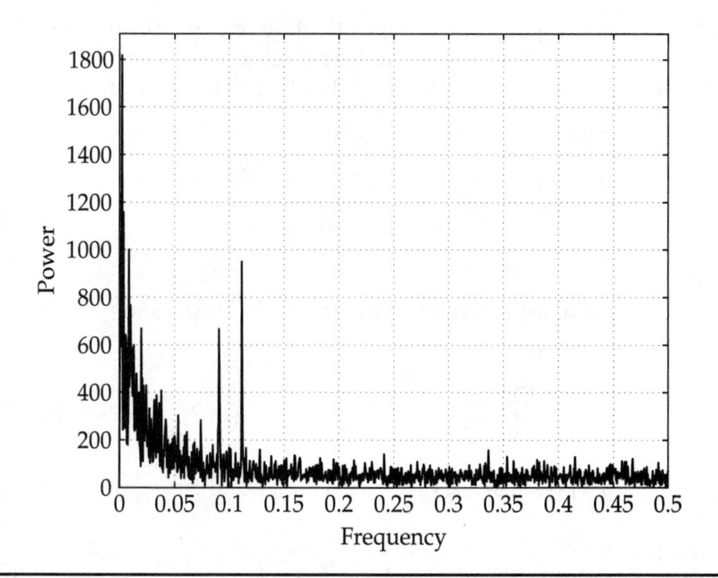

FIGURE C-2 Spectrum of a noisy signal with periodic components.

For this type of analysis to "find" a periodic component buried in a signal, that component must repeat (or really be periodic) so that the least squares fit will work. Consequently, data having an isolated pulse or excursion (a nonperiodic component) will not be approximated well by a sinusoid of any frequency. In this case, if the sinusoidal least squares approximation is still carried out the results may be confusing.

C-5 The Exponential Form of the Least Squares Fitting Equation

Using Euler's equation, an alternative equivalent (and easier) approach can be taken. Instead of a sum of real sinusoids, the data-fitting equation is a sum of complex exponentials, as in

$$y_i = \sum_{k=1}^{N} c_k e^{j\frac{2\pi ki}{N}} \tag{C-14}$$

For the sake of convenience, the average has been subtracted from the data stream so there is no constant term in Eq. (C-14).

Euler's equation

$$e^{j\frac{2\pi ki}{N}} = \cos\left(\frac{2\pi ki}{N}\right) + j\sin\left(\frac{2\pi ki}{N}\right)$$

suggests that Eq. (C-14) is equivalent to Eq. (C-8) and that the coefficients c_k in Eq. (C-14) will be complex. An expression for the coefficient c_m can be obtained relatively quickly by multiplying Eq. (C-14) by $e^{-j\frac{2\pi mi}{N}}$ and summing over the data points, as in

$$\sum_{i=1}^{N} e^{-j\frac{2\pi mi}{N}} y_i = \sum_{i=1}^{N}\sum_{k=1}^{N} c_k e^{-j\frac{2\pi mi}{N}} e^{j\frac{2\pi ki}{N}} \tag{C-15}$$

Once again, orthogonality rears its lovely head, as in

$$\sum_{i=1}^{N} e^{-j\frac{2\pi mi}{N}} e^{j\frac{2\pi ki}{N}} = \begin{Bmatrix} 0 & k \neq m \\ N & k = m \end{Bmatrix} \tag{C-16}$$

so the Eq. (C-15) collapses to

$$c_m = \frac{1}{N}\sum_{i=1}^{N} y_i e^{-j\frac{2\pi mi}{N}} \tag{C-17}$$

The *line spectrum* or *power spectrum* can be constructed from the magnitudes of the coefficients, as in $|c_m|$ or $|c_m|^2$, $m = 1,2,\ldots,N/2$. Note that only half of the coefficients are used because there is symmetry.

To illustrate this symmetry, consider a data stream of 64 equally spaced samples computed from

$$y_i = \sin\left(\frac{2\pi t_i}{14}\right) \qquad t_i = i \qquad i = 1, 2, \ldots, 64$$

The sampling interval is 1.0 sec, the sampling frequency is $f_s = 1/h = 1.0$ Hz and the folding frequency is $f_{NY} = f_s/2 = 0.5$ Hz. Figure C-3 shows the data stream and the absolute value of the coefficients, $|c_m|$ or $|Y|$, in Eq. (C-16).

Note that the absolute values are symmetrical about the folding frequency of 0.5 Hz.

Some authors combine Eqs. (C-17) and (C-14) to form a finite discrete Fourier transform pair, as in

$$Y_m = \frac{1}{N}\sum_{i=1}^{N} y_i e^{-j\frac{2\pi mi}{N}} \qquad m = 1,\ldots,N$$

$$y_i = \sum_{k=1}^{N} Y_k e^{j\frac{2\pi ki}{N}} \qquad i = 1,\ldots,N \tag{C-18}$$

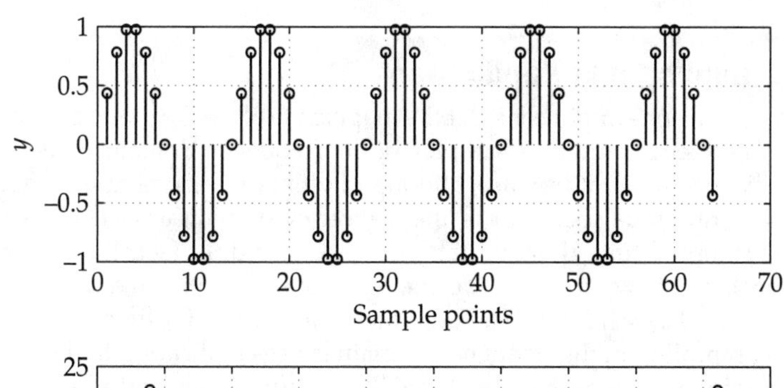

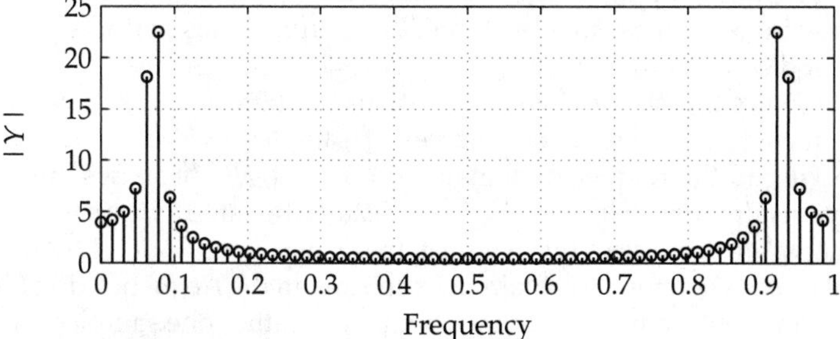

FIGURE C-3 A data stream and its Fourier transform (absolute value).

where the Y_m, $m = 1,\ldots,N$ are the elements (perhaps complex) of the discrete Fourier transform (they were the coefficients c_k, $k = 1,\ldots,N$ in the data-fitting equation) and the y_i, $i = 1,\ldots,N$ are the elements of the inverse transform (or the original data stream in the time domain). Furthermore, many authors place the $1/N$ factor in front of the inverse transform rather than the transform. Hopefully, the reader will agree that the complex exponential approach to spectral analysis is far more elegant and efficient.

C-6 Periodicity in the Time Domain

Because of the periodic nature of the sinusoids used in the data fit, whether it is in the exponential form or not, one can show that if the Fourier series fitting equation is evaluated outside the time domain interval of $[t_1, t_N]$ the series will repeat. That is,

$$\frac{1}{N}\sum_{i=1}^{N} y_i e^{-j\frac{2\pi mi}{N}} = \frac{1}{N}\sum_{i=1}^{N} y_i e^{-j\frac{2\pi m(i+N)}{N}}$$

Therefore, the act of fitting the data to a Fourier series is tantamount to specifying that the time domain data stream repeats itself with the period equal to the length of the data stream. This feature can be put another way, as in Sec. C-7.

C-7 Sampling and Replication

In general, one might consider the magnitude of the Fourier transform in Fig. C-3 as a train of samples in the frequency domain. This is a realistic viewpoint because, although I will not demonstrate it, there is a continuous spectrum in the frequency domain associated with the sampled time domain data in Fig. C-1 and, in fact, the Fourier transform shown in Fig. C-2 is the result of sampling it at a frequency interval of $f_1 = 1/L = 1/Nh$. Therefore, without proof, I suggest to you that sampling in the frequency domain causes replication in the time domain in the sense that the time domain function is periodic with a period equal to Nh.

This suggests an inverse relationship between the time and frequency domains. If the number of samples N is increased, the spacing in the frequency domain, $f_1 = 1/L = 1/Nh$ decreases but the Nyquist frequency interval $[0, 1/(2h)]$ stays the same. If the sampling interval h decreases, the Nyquist frequency interval $[0, 1/(2h)]$ is enlarged. Therefore, to obtain finer spacing in the frequency domain, one does not sample at a higher frequency; rather one samples more data. To increase the frequency range one must increase the sampling frequency.

C-8 Apparent Increased Frequency Domain Resolution via Padding

If you use the Fast Fourier Transform to analyze your time domain data stream, the spacing (or resolution) in the frequency domain will be $1/(Nh)$. This may not be enough if you are trying to determine the frequency of a periodic component with great precision. Based on the discussion so far, you might simply double the length of the time domain data stream and thereby halve the frequency domain resolution.

Alternatively, if one "pads" the original data stream with zeroes, the apparent frequency domain resolution is increased. Actually, the padding allows the analyst to *interpolate* between the frequencies in the original frequency grid but since there is no additional information the true resolution is not improved.

Figure C-4 shows the spectrum of the same noisy data stream that was used for the spectrum in Fig. C-2. However, only 512 points of the stream are used and there is no padding. This spectrum shows vague hints that there may be two periodic components lurking in the data stream. The frequency resolution is

$$\frac{1}{Nh} = \frac{1}{512} = 0.1953 \text{ Hz}$$

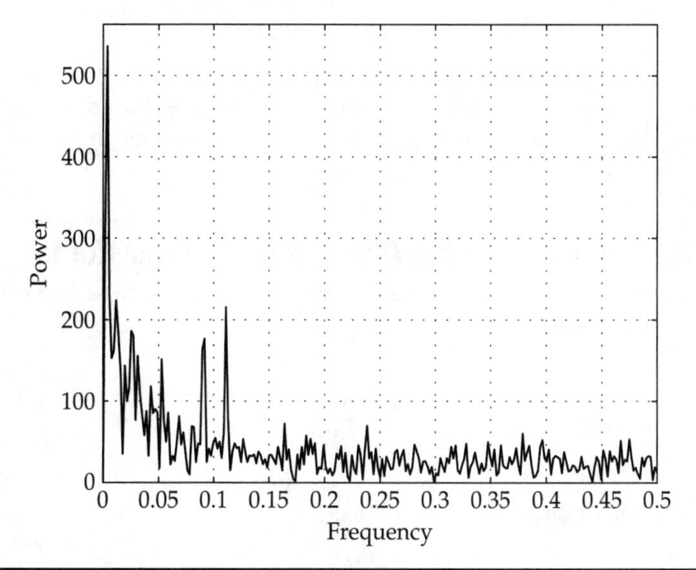

FIGURE C-4 Spectrum of a 512 point data stream sampled at 1 Hz, no padding.

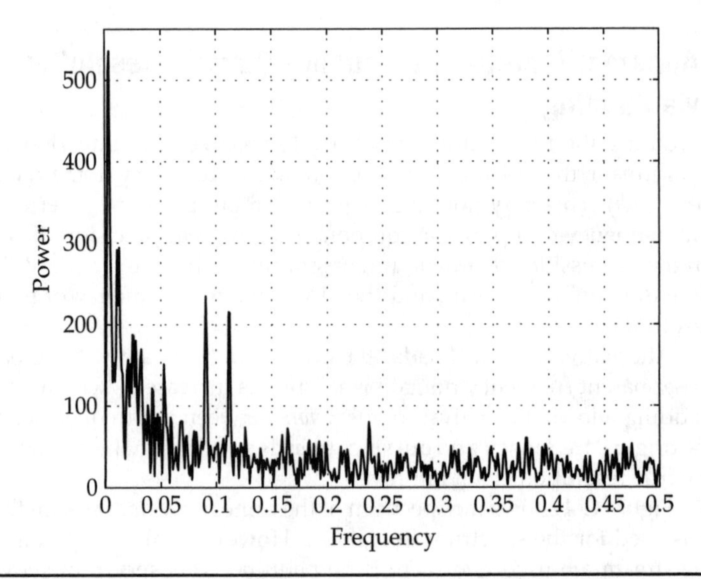

FIGURE C-5 Spectrum of a 512 sample data stream sampled at 1 Hz padded with 1024 zeroes.

Figure C-5 shows the spectrum of the same 512 samples after appending 1024 zeroes in the time domain. This padding does not add any information but the apparent resolution is now

$$\frac{1}{Nh} = \frac{1}{(512+1024)} = 0.00065104 \text{ Hz}$$

and two peaks associated with the periodic components are more clearly apparent. The spectrum in Fig. C-2 is based on 2000 samples that have been padded with 2048 zeroes in the time domain.

C-9 The Variance and the Discrete Fourier Transform

The variance of a data stream from which the average has been subtracted is

$$V = \frac{1}{N}\sum_{i=1}^{N} y_i^2 \tag{C-19}$$

In general, the data may be complex so we will modify the definition of the variance as follows

$$V = \frac{1}{N}\sum_{i=1}^{N} |y_i|^2 \tag{C-20}$$

where the absolute value can be determined from $|y_i|^2 = y\,y^*$, where y^* is the complex conjugate (see App. B).

Replacing y_i in Eq. (C-20) using Eq. (C-18) gives

$$
\begin{aligned}
V &= \frac{1}{N}\sum_{i=1}^{N}\left|\sum_{k=1}^{N} Y_k e^{j\frac{2\pi ki}{N}}\right|^2 \\
&= \frac{1}{N}\sum_{i=1}^{N}\sum_{k=1}^{N}|Y_k|^2 e^{j\frac{2\pi ki}{N}} e^{-j\frac{2\pi ki}{N}} \\
&= \frac{1}{N}\sum_{k=1}^{N}|Y_k|^2\left(\sum_{i=1}^{N} e^{j\frac{2\pi ki}{N}} e^{-j\frac{2\pi ki}{N}}\right)
\end{aligned}
\qquad \text{(C-21)}
$$

where two things have happened. In the first row of Eq. (C-21) the absolute value was replaced by the product of the quantity and its complex conjugate, as in

$$|z|^2 = z\,z^*$$

and second, the order of the summing was exchanged in the last row of Eq. (C-21). This is allowed because Y_k does not depend on the data index *i*. Equation (C-16) shows that

$$\sum_{i=1}^{N} e^{j\frac{2\pi ki}{N}} e^{-j\frac{2\pi ki}{N}} = N$$

therefore, the variance of the time domain data $y_i, i = 1,\ldots,N$, is also given in terms of a sum that is proportional to the variance of the elements of the Fourier transform which are $Y_m, m = 1,\ldots,N$. That is, Eq. (C-21) becomes

$$V = \frac{1}{N}\sum_{i=1}^{N}|y_i|^2 = \sum_{k=1}^{N}|Y_k|^2 \qquad \text{(C-22)}$$

Equation (C-22) says that the variance of the time domain data is proportional to the power of the sinusoidal components in the discrete finite Fourier transform. The sum

$$\sum_{k=1}^{N}|Y_k|^2$$

is proportional to the area under the line spectrum curve hence the comment in Chap. 2 that the variance of the time domain data stream is proportional to the area under the power spectrum.

C-10 Impact of Increased Frequency Resolution on Variability of the Power Spectrum

For each of the $N/2$ frequencies in the Nyquist interval there is a power given by $|Y_k|$ or $|Y_k|^2$, depending on your preference. Some analysts talk about the power in each of the $N/2$ "bins." As was shown in Eq. (C-22), the sum of the bin power is equal to the total power in the signal. If the number of samples, N, is increased (perhaps to increase the frequency resolution) then there are more bins and each bin has less power. Because of the variability in the time domain signal, these bin powers becomes more variable as the bins get smaller. This is somewhat similar to the increase in the variability of the histogram (see Chap. 8) as the bin sizes get smaller. To address this, some analysts will break a data stream into subsets, compute a spectrum for each subset, and then average the spectra to reduce the variability.

C-11 Aliasing

The ability to identify hidden periodic or cyclical components in a noisy data stream requires being able to sample these sinusoidal components enough times per cycle. For example, if the sampling rate is 1.0 Hz (or one sample per second) and the frequency of the suspected sinusoid is 2.0 Hz then only one sample of the suspect sinusoid will be available every two cycles. It doesn't take a rocket scientist to suspect that the sampling rate would be insufficient to identify that periodic component.

Figure C-6 shows two sine waves sampled at 1.0 Hz (once per second). The higher frequency sine wave is sampled approximately

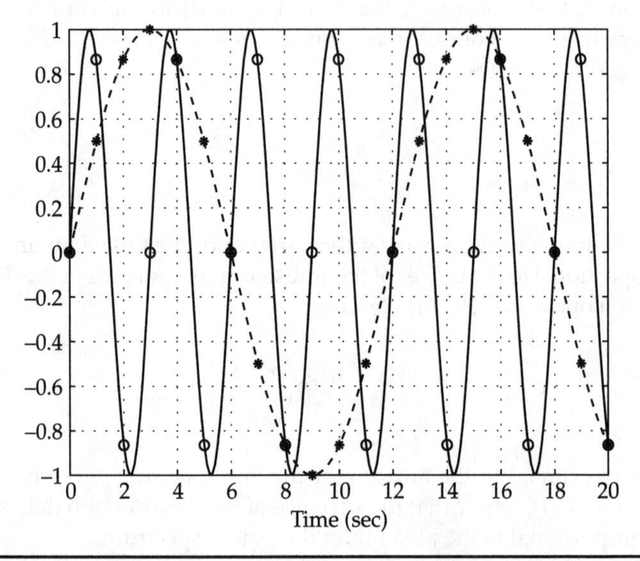

Figure C-6 Two identifiable sine waves with periods 3 and 12, sampled at 1 Hz.

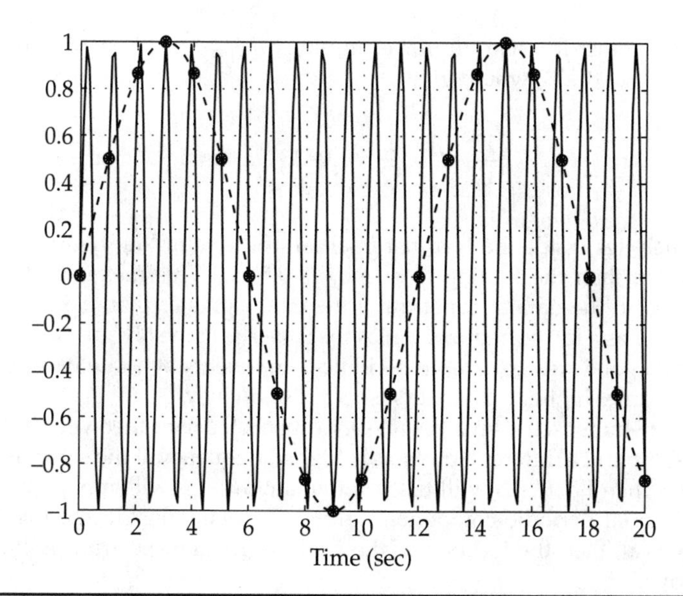

FIGURE C-7 Two sine waves that are aliases with periods 0.92308 and 12, sampled at 1 Hz.

three times per cycle and could be identified. The lower frequency wave is sampled approximately 11 times per cycle and also could be identified. Furthermore, the sampled values of the low frequency sine wave are different from those of the higher frequency sine wave.

Consider Fig. C-7 which shows two more sine waves sampled at 1 Hz. Both sine waves have identical samples. The true periodic signal may have a period of 0.92308 sec (with a frequency of 1.0833 Hz) but the samples suggest that the apparent period is 12.0 sec (with a frequency of 0.0833 Hz). The lower frequency signal, whose frequency is less that the folding frequency of 0.5 Hz and lies inside the Nyquist interval, is the *alias* of the higher frequency signal (frequency of 1.0833 Hz or 13/12 Hz). From the sampler's point of view, constrained to view life from within the Nyquist interval, these two sine waves are identical. Had this data stream been from a real-life sample set then the real signal might have had a frequency of 1.0833 Hz but it appears as one with a frequency of 0.0833 Hz.

The frequency of the alias f_{alias} can be obtained from the actual frequency f_{actual} by using the following equation:

$$f_{\text{alias}} = \left| f_{\text{actual}} - m f_s \right| \qquad m = 1, 2, 3, \ldots \tag{C-23}$$

where f_s is the sampling frequency. To use Eq. (C-23), one would try successively higher values of m until the calculated f_{alias} lies inside the Nyquist interval. In our example, $f_{\text{actual}} = 13/12$, $f_s = 1$, so for $m = 1$, Eq. (C-23) would give

$$\left|\frac{13}{12} - 1\right| = \frac{1}{12} = 0.0833 \qquad m = 1$$

which lies inside the Nyquist interval. If the sampling frequency is 1.0 Hz then the folding frequency is 0.5 Hz. This means that any signal in the data stream with a frequency greater than 0.5 Hz will appear as a lower frequency alias that does lie in the Nyquist interval. It is a good rule to sample periodic components at a rate of at least four times higher than their suspected frequency.

If your spectral analysis reveals a suspected alias then you should resample at a different, preferably higher, frequency. If the signal is an alias then Eq. (C-23) will tell you that a new alias will appear. If it is an actual periodic component with a frequency inside the folding interval, then the location of the peak in the line spectrum will not move.

C-12 Summary

The basis for spectral analysis is presented from the least squares data-fitting point of view, although other approaches that the control engineer might take are mentioned briefly. When the data is uniformly spaced, a set of sinusoids are orthogonal and they can be used to fit the data efficiently. Fast Fourier Transform packages to carry out this data fit are ubiquitous. One should keep in mind that there is a constraint on the resolution in the frequency domain. Padding can be used to increase the apparent resolution. However, if one has good information on the neighborhood of a suspected peak and wants to obtain its location precisely, they may want to set up a finer frequency grid in that neighborhood and use a least squares fit that does not take advantage of the orthogonality and therefore precludes the use of the Fast Fourier Transform. Finally, one must consider aliases when attempting to detect periodic components.

APPENDIX D

Infinite and Taylor's Series

A function can be expanded into an infinite series using the Taylor's series

$$f(x) = f(x_0) + f'(x_0)(x - x_0) + \frac{1}{2} f''(x_0)(x - x_0) + \text{h.ot.} \qquad \text{(D-1)}$$

which says that the value of a function f at x can be estimated from the known value of f at x_0 and higher derivatives of f, also evaluated at x_0, where x_0 is near x. The approximation gets better if the higher-order terms (h.o.t.) are added and if x is nearer x_0. These h.o.t.'s consist of higher-order derivatives, also evaluated at x_0. If the h.o.t.'s were removed, the second-order approximation would look like

$$f(x) \cong f(x_0) + f'(x_0)(x - x_0) + \frac{1}{2} f''(x_0)(x - x_0) \qquad \text{(D-2)}$$

We will occasionally use the first-order approximation given in Eq. (D-3).

$$f(x) \cong f(x_0) + f'(x_0)(x - x_0) \qquad \text{(D-3)}$$

For the Taylor's series to work, the derivatives of $f(x)$ at x_0 have to be available. If they are, as in the case for the exponential and trigonometric functions, the following useful expressions can be obtained:

$$\begin{aligned} e^x &= 1 + x + \frac{x^2}{2!} + \frac{x^3}{3!} + \cdots \\ \sin x &= x - \frac{x^3}{3!} + \frac{x^5}{5!} - \frac{x^7}{7!} + \cdots \\ \cos x &= 1 - \frac{x^2}{2!} + \frac{x^4}{4!} - \frac{x^6}{6!} + \cdots \end{aligned} \qquad \text{(D-4)}$$

where it's best to keep the argument x real but you could use complex arguments. The first of the infinite series in Eq. (D-4) says that a crude approximation to the exponential is simply

$$e^x \cong 1 + x$$

This is the first-order Taylor's series for $x_0 = 0$ where $f(0) = 1$ and

$$f'(0) = \frac{d}{dt} e^x \Big|_{x=0} = 1$$

In Chap. 3 a passing reference was made to Torricelli's law which relates to outlet flow rate, F, of a column of liquid that has a height of Y

$$F_o = C\sqrt{Y}$$

The flow depends in a nonlinear way on the height Y which is sometimes inconvenient for the simple math that we use in this book. A first-order Taylor's expansion about Y_0 can be useful

$$x \Rightarrow Y$$

$$f(x) \Rightarrow C\sqrt{x}$$

$$f'(x) = \frac{C}{2} x^{-\frac{1}{2}}$$

Therefore, the linearized expression for the flow rate as a function of tank height is

$$F = F_0 + \frac{C}{2\sqrt{Y_0}}(Y - Y_0) = F_0 + \frac{1}{R}(Y - Y_0)$$

$$R = \frac{2\sqrt{Y_0}}{C}$$

In Chap. 3, this equation is used with the assumption that the initial steady-state values, F_0 and Y_0, are zero (or, alternatively, that the average steady-state values are subtracted from F and Y).

Depending on the reader's energy levels, it might be interesting to use the infinite series representations in Eq. (D-4) to confirm that

$$e^{jx} = \cos x + j \sin x$$

However, if you are willing to take my word for it, don't waste time on it.

D-1 Summary

This has been the shortest appendix but the Taylor's series is an important tool and has been used often in this book.

APPENDIX E

Application of the Exponential Function to Differential Equations

E-1 First-Order Differential Equations

For the case of a constant process input U_c, Eq. (3-8) from Chap. 3 becomes

$$\tau \frac{dY}{dt} + Y = gU_c \tag{E-1}$$

There are many ways of solving this equation and we will start with the simplest. Assume that Y consists of a dynamic or *transient* part or *homogeneous* part and a *steady-state* or *nonhomogeneous* part, namely,

$$Y = Y_t + Y_{ss} \tag{E-2}$$

where Y_t, the transient part satisfies

$$\tau \frac{dY_t}{dt} + Y_t = 0 \tag{E-3}$$

which is often called the homogeneous part of the differential equation. The steady-state part Y_{ss}, sometimes called the *particular* solution, satisfies the remaining part or the nonhomogeneous part of the differential equation.

$$Y_{ss} = gU_c \qquad \text{because} \qquad \frac{dY_{ss}}{dt} = 0 \tag{E-4}$$

Since the input is constant, the steady-state solution is obtained immediately.

The transient solution to the homogeneous part of the differential equation requires a little more work. As is often the case when one is trying to solve a differential equation, one "tries" a general solution form. Experience has shown that a good form to *try* is

$$Y_t = Ce^{at}$$

where C and a are, as yet, undetermined coefficients.

Plugging this trial solution into Eq. (E-3) yields

$$\tau Cae^{at} + Ce^{at} = 0$$

or

$$\tau Cae^{at} = -Ce^{at}$$

Cancelling C and e^{at} gives

$$\tau a + 1 = 0 \qquad \text{or} \qquad \tau a = -1 \qquad \text{or} \qquad a = -\frac{1}{\tau}$$

One of the undetermined parameters a is now known and we have

$$Y_t = Ce^{-\frac{t}{\tau}}$$

By the way, the value of a that satisfies

$$\tau a + 1 = 0$$

could be considered a *root* of the above equation. We will extend this idea later on in this section.

Now that we know the transient solution, Eq. (E-2) becomes

$$Y = Ce^{-\frac{t}{\tau}} + gU_c \tag{E-5}$$

To find the coefficient C we apply the "initial" condition, which says that at time zero, that is, at $t = 0$, Y is Y_0 and Eq. (E-5) becomes

$$\begin{aligned} Y_0 &= Ce^{-\frac{0}{\tau}} + gU_c \\ &= C + gU_c \end{aligned}$$

so,

$$C = Y_0 - gU_c$$

and Eq. (E-5) becomes

$$\boxed{Y = Y_0 e^{-\frac{t}{\tau}} + gU_c(1 - e^{-\frac{t}{\tau}})} \qquad \text{(E-6)}$$

This example suggests that the exponential function has some neat properties that make it quite useful in engineering mathematics.

E-2 Partial Summary

The first-order differential equation was divided into its homogeneous part and nonhomogeneous part. A solution was constructed for each part and added together to form the total solution. This total solution contained an unknown constant which was determined by applying the initial condition. This is a procedure that will be followed for a wide variety of more complicated differential equations appearing later on.

E-3 Partial Solution of a Second-Order Differential Equation

Consider the second-order differential equation

$$\frac{d^2Y}{dt^2} + a\frac{dY}{dt} + bY = c \qquad \text{(E-7)}$$

where a, b, and c are known constants.

Following the mode of dividing and conquering, assume that the solution consists of a transient part (that will change with time t) Y_t and a constant or steady-state part Y_{ss} that will depend on the constant c:

$$Y = Y_t + Y_{ss}$$

Furthermore, assume that Y_t satisfies only the so-called "homogeneous" part of the differential equation, that is, the part on the left-hand side:

$$\frac{d^2Y_t}{dt^2}+a\frac{dY_t}{dt}+bY_t=0 \tag{E-8}$$

The following trial solution is tried for the transient part

$$Y_t=Ce^{\alpha t} \tag{E-9}$$

where C and α are as yet unknown constants. This is called a "trial" solution because once we plug it into the differential equation we may find that it is useless. Inserting the trial solution into Eq. (E-8), the homogeneous part of the differential equation, yields

$$C\alpha^2e^{\alpha t}+aC\alpha e^{\alpha t}+bCe^{\alpha t}=0$$

Now one can perhaps see that the form of the trial solution was somewhat clever because $Ce^{\alpha t}$ is in every term and it can be factored out leaving

$$\alpha^2+a\alpha+b=0$$

which is a quadratic equation for which the two values (perhaps complex) of α can be found. Let's say that the two solutions to the quadratic equation are $p+jq$ and $p-jq$. (These two solutions must be complex conjugates for the quadratic to remain real.) Each of these values for α is associated with a value for C. Since the solution has the form of Eq. (E-9), one can use Euler's equation to conclude that the solution will look like

$$\begin{aligned}Y(t) &\sim C_1e^{(p+jq)t}+C_2e^{(p-jq)t}\\ &\sim C_1e^{pt}[\cos(qt)+j\sin(qt)]+C_2e^{pt}[\cos(qt)-j\sin(qt)]\end{aligned} \tag{E-10}$$

It may be a bit of a stretch but $Y(t)$ has to be *real* to be physically acceptable as a solution so, again, take my word for it, the imaginary parts of the above solution cancel out such that $Y(t)$ is, in fact, real. However, the reader should deduce from Eq. (E-10) that, depending on the values of p and q, $Y(t)$ will have an exponential part e^{pt} that grows or dies out at a rate depending on p, and an undamped part e^{jqt} that will oscillate at a rate depending on q. One could continue with Eq. (E-10), combining it with the particular solution, applying the initial conditions, and, after many manipulations, arrive at a solution.

However, we will find there are better more insightful ways to deal with second- and higher-order differential equations—specifically, in App. F, the Laplace transform will be used with great success.

E-4 Summary

We have used the exponential form as a trial solution for a first- and second-order differential equation. Each has generated equations for the undetermined coefficients. This approach has been used widely in the book.

APPENDIX F

The Laplace Transform

Some of this section is paraphrased from Chap. 3 just in case you want to have everything you need in one place. Also, it is so important that it bears repeating.

The definition of the Laplace transform is

$$L\{Y(t)\} = \tilde{Y}(s) = \int_0^\infty dt\, e^{-st} Y(t) \qquad \text{(F-1)}$$

In words, this equation says *using a weighting factor of* e^{-st}*, integrate the time function* $Y(t)$ *from zero to infinity and generate a function depending only on* s.

With Eq. (F-1) in hand, it may be clearer why the units of $\tilde{Y}$ must be m·sec if the units of $Y(t)$ are m. By integrating over all positive time, the Laplace transform removes all dependence on time, represented by $Y(t)$ and creates a new function of s represented by $\tilde{Y}(s)$.

The Laplace transform is interested only in time after time zero so the lower limit on the integral is zero. There are exceptions which will be noted but for the most part the Laplace transform assumes that everything before time zero is zero.

The inverse operation of finding a time function for a given Laplace transform is

$$Y(t) = \frac{1}{2\pi j} \int_{c-j\infty}^{c+j\infty} \tilde{Y}(s) e^{st} ds \qquad \text{(F-2)}$$

We will not use this formula because there are less sophisticated and more effective ways of inverting Laplace transforms but it is good for you to be aware of it. If your control engineer knows how to use contour integration in the complex plane (I did, once) she may use Eq. (F-2) to invert especially complicated transforms.

F-1 Laplace Transform of a Constant (or a Step Change)

Let's do a simple example just to remove some of the awe from Eq. (F-1). Consider the step function which is zero for time less than zero but is constant at the value, say C, for $t > 0$. For this case, Eq. (F-1) becomes

$$L\{C\} = \int_0^\infty dt\, e^{-st} C = C \int_0^\infty dt\, e^{-st}$$

$$= -C \int_0^\infty \frac{1}{s} d(-st) e^{-st} = -\frac{C}{s} \int_0^{-\infty} du\, e^u$$

$$= -\frac{C}{s} e^u \Big|_0^{-\infty} = -\frac{C}{s}(0-1) = \frac{C}{s}$$

so,

$$\boxed{L\{C\} = \frac{C}{s}} \tag{F-3}$$

Another way of doing this uses the *unit step function* $\hat{U}(t)$ where

$$\begin{aligned} \hat{U}(t) &= 0 \qquad t < 0 \\ \hat{U}(t) &= 1 \qquad t \ge 0 \end{aligned} \tag{F-4}$$

It follows from Eq. (F-3) that the Laplace transform of the unit step function is

$$\boxed{L\{\hat{U}(t)\} = \int_0^\infty \hat{U}(t) e^{-st} dt = \frac{1}{s}} \tag{F-5}$$

F-2 Laplace Transform of a Step at a Time Greater than Zero

The modified unit step $\hat{U}(t-\lambda)$ is defined as

$$\hat{U}(t-\lambda) = 0 \qquad t < \tau$$

$$\hat{U}(t-\lambda) = 1 \qquad t \ge \tau$$

That is, $\hat{U}(t-\lambda)$ is a unit step that "turns on" at time $t = \lambda$.

The Laplace transform of this *delayed* unit step is quite straightforward

$$L\{\hat{U}(t-\lambda)\} = \int_0^\infty \hat{U}(t-\lambda) e^{-st} dt$$

A change of variable from t to t' will make things simple. The new variable is related to the old as follows:

$$t' = t - \lambda \qquad dt' = dt \qquad t = t' + \lambda$$

With this new variable the transform can be written as

$$L\{\hat{U}(t-\lambda)\} = \int_{-\lambda}^{\infty} \hat{U}(t')e^{-s(t'+\lambda)}dt'$$

But the term $e^{-s\lambda}$ is a constant relative to the integration, so

$$L\{\hat{U}(t-\lambda)\} = e^{-s\lambda}\int_{-\lambda}^{\infty} \hat{U}(t')e^{-st'}dt'$$

Since $\hat{U}(t)$ is zero for $t < 0$, the lower limit on the integral can be zero instead of $-\lambda$. This means that the integral is just the Laplace transform of the unit step function which is $1/s$. Therefore,

$$\boxed{L\{\hat{U}(t-\lambda)\} = \frac{e^{-s\lambda}}{s}} \tag{F-6}$$

This is a handy formula because, as will be seen in Sec. F-3, anytime there is a process dead time of length λ, one can add the factor $e^{-s\lambda}$ to the transfer function.

F-3 Laplace Transform of a Delayed Quantity

In a pure dead-time processes the process output $Y(t)$ is simply the input $U(t)$ delayed by the dead time D or

$$Y(t) = U(t-D)\hat{U}(t-D) \tag{F-7}$$

Be careful not to confuse the process input $U(t)$ with the unit step function $\hat{U}(t)$. Using Eq. (F-6) as guide, the Laplace transform of Y is seen to be

$$\tilde{Y}(s) = e^{-sD}\tilde{U}(s)$$

An alternative way to present this operation is

$$\boxed{L^{-1}\{e^{-sD}\tilde{Y}(s)\} = Y(t-D)\hat{U}(t-D)} \tag{F-8}$$

F-4 Laplace Transform of the Impulse or Dirac Delta Function

On the one hand, the Dirac delta function is a somewhat difficult concept but, as we will see, its Laplace transform is the easiest to compute. The Dirac delta function $\delta(t-\tau)$ is a "spike" at time $t=\tau$ that has indeterminate height and width but that has unit area under its curve in the time domain. At every value of time other than $t=\tau$ the Dirac delta function has a value of zero. Since we have already agreed that the integral of a function can give the area under the function's curve, the definition of Dirac delta function is simply

$$\delta(t) = 0 \qquad \text{for } t \neq 0$$

undefined for $t = 0$ (F-9)

$$\int_0^\infty dt\,\delta(t) = 1$$

Equation (F-9) says the area under the Dirac delta function is unity but the definition says nothing about the Dirac delta function's shape. An additional characteristic expands the definition of Dirac delta, namely,

$$\int_0^\infty dt f(t)\delta(t-\tau) = f(\tau) \tag{F-10}$$

Equation (F-10) says that the Dirac delta "spike" at $t=\tau$ is so high and so narrow that it "plucks" the value of $f(t)$ at $t=\tau$ and snuffs it out elsewhere. Envision a graph of $f(t)\delta(t-\tau)$. Assuming that $\tau > 0$, the graph would show a zero until $t=\tau$ at which time $f(t)\delta(t-\tau)$ would have an undetermined value. For $t > \tau$, $f(t)\delta(t-\tau)$ would again equal zero. Let's try to approximate the integral in Eq. (F-10) with a sum of the areas of small rectangles, each with a width of size Δt. The first rectangle is located at $t=0$, the second at $t=\Delta t$, and so on,

$$\int_0^\infty dt f(t)\delta(t-\tau) \cong \Delta t\, f(0)\delta(-\tau) + \Delta t\, f(\Delta t)\delta(\Delta t-\tau)$$

$$+ \Delta t\, f(2\Delta t)\delta(2\Delta t-\tau) + \cdots + \Delta t\, f(\tau)\delta(0) + \cdots$$

$$+ \Delta t\, f(\tau+\Delta t)\delta(\Delta t) + \Delta t\, f(\tau+2\Delta t)\delta(2\Delta t) + \cdots$$

All of the rectangles except the one at $t=\tau$, namely, $\Delta t\, f(\tau)\delta(0)$, would contribute zero to the sum. In the limit as $\Delta t \to 0$ and, as the rectangle approximation gets more and more accurate, this rectangle at $t=\tau$ would contribute $f(\tau)$ because, in the limit, the factor $\Delta t\delta(0)$ contributes unity, as in $\Delta t\delta(0) \to 1$.

With this in mind, look at the Laplace transform of the Dirac delta function

$$L\{\delta(t)\} = \int_0^{\infty} \delta(t) e^{-st} dt$$

The delta function plucks the value of the integrand, which is e^{-st}, at $t = 0$, so

$$\boxed{L\{\delta(t)\} = \int_0^{\infty} \delta(t) e^{-st} dt = 1} \tag{F-11}$$

F-5 Laplace Transform of the Exponential Function

$$L\{e^{-at}\} = \int_0^{\infty} dt\, e^{-st} e^{-at} = \int_0^{\infty} dt\, e^{-(s+a)t}$$

let $\quad u = (s+a)t$

so $\quad du = (s+a)dt \quad$ or $\quad dt = \dfrac{du}{s+a}$

$$= -\int_0^{\infty} \frac{1}{s+a} du e^{-u}$$

$$= -\frac{1}{s+a} \int_0^{-\infty} du e^{u}$$

$$= -\frac{1}{s+a} e^{u} \Big|_0^{-\infty} = -\frac{1}{s+a}(0-1) = \frac{1}{s+a}$$

so

$$\boxed{L\{e^{-at}\} = \frac{1}{s+a}} \tag{F-12}$$

F-6 Laplace Transform of a Sinusoid

There is but one reference to this transform in App. B so you will not be bothered with the derivation.

$$\boxed{L\{\sin(\omega t)\} = \frac{\omega}{s^2 + \omega^2}} \tag{F-13}$$

F-7 Final Value Theorem

The final value theorem is a handy way of using the Laplace transform to determine if the asymptotic value of a variable settles out in the time domain.

$$\boxed{\lim_{s\to 0} s\tilde{Y}(s) = \lim_{t\to\infty} Y(t)} \tag{F-14}$$

Question F-1 Apply the final value theorem to the transform of the constant, that is, C/s, and to the transform of the exponential, that is, $1/(s+a)$. Do the results make sense?

Answer

$$\lim_{t\to\infty} Y(t) = \lim_{s\to 0} s\frac{C}{s} = \lim_{s\to 0} C = C$$

$$\lim_{t\to\infty} Y(t) = \lim_{s\to 0} s\frac{1}{s+a} = \frac{0}{0+a} = 0$$

The results do, in fact, make sense because C/s is the transform of the constant C which has an ultimate value equal to its initial value, namely C. The expression $1/(s + a)$ is the transform of e^{-at} and its ultimate value as $t \to \infty$is zero.

F-8 Laplace Transform Tables

As the reader might imagine, there are a huge number of Laplace transform/time domain function pairs simply because there are a huge number of useful time domain functions. We really only need the transforms of the constant, the sine, the exponential, and the Dirac delta function to get off the ground. To get a better perspective the reader should browse a standard engineering mathematics textbook and find the table for Laplace transform pairs.

F-9 Laplace Transform of the Time Domain Derivative

By definition the transform of a derivative is

$$L\left\{\frac{dY}{dt}\right\} = \int_0^\infty dte^{-st}\frac{dY}{dt} = \int_0^\infty e^{-st}dY \tag{F-15}$$

To evaluate Eq. (F-15) we need to review *integration by parts*. As presented in Eq. (A-20), the differential of a product of two functions u and v is

$$d(uv) = udv + vdu$$

Likewise, the integral is given by

$$\int_a^b d(uv) = \int_a^b u\,dv + \int_a^b v\,du$$

Solving for the first integral on the right-hand side gives

$$\int_a^b u\,dv = \int_a^b d(uv) - \int_a^b v\,du = uv\,|_a^b - \int_a^b v\,du$$

$$\boxed{\int_a^b u\,dv = uv\,|_a^b - \int_a^b v\,du} \qquad \text{(F-16)}$$

Compare Eqs. (F-15) and (F-16) and match dv with dY and u with e^{-st}

$$L\left\{\frac{dY}{dt}\right\} = \int_0^\infty e^{-st}dY = e^{-st}Y\,|_0^\infty - \int_0^\infty Y d(e^{-st}) = 0 - Y(0) - \int_0^\infty Y(-se^{-st}dt)$$

$$= -Y(0) + s\int_0^\infty dt e^{-st}Y = -Y(0) + sL\{Y\}$$

so, we arrive at the desired result

$$\boxed{L\left\{\frac{dY}{dt}\right\} = s\tilde{Y} - Y(0^+)} \qquad \text{(F-17)}$$

Note that the initial value $Y(0^+)$ is the value arrived at from the right as t decreases toward zero from positive values, as in

$$Y(0^+) = \lim_{t\to 0^+} Y(t)$$

F-10 Laplace Transform of Higher Derivatives

Since the development of Eq. (F-17) was a little bit painful, I will simply give the result for the second-order derivative:

$$L\left\{\frac{d^2Y}{dt^2}\right\} = s^2L\{Y\} - sY(0^+) - \frac{dY}{dt}\,|_{0^+} \qquad \text{(F-18)}$$

$$= s^2\tilde{Y} - sY(0^+) - \dot{Y}(0^+)$$

That is, the Laplace transform of the second derivative of a quantity is s^2 times the Laplace transform of that quantity, $\tilde{Y}$, minus the initial value of that quantity times s, $sY(0^+)$, minus the initial value of that quantity's first derivative, $\dot{Y}(0^+)$.

Thus, when the initial conditions are *all* zero, the various derivatives can be transformed by replacing the derivative by Laplace transform of the quantity times the appropriate power of s.

F-11 Laplace Transform of an Integral

By definition we are dealing with

$$L\left\{\int_0^t du Y(u)\right\} = \int_0^\infty dt e^{-st} \int_0^t du Y(u) \tag{F-19}$$

Integration by parts will be used to address the evaluation of Eq. (F-19) and although its relatively straightforward you may want to skip to the result in the box.

From above, the integration by parts formula is

$$\boxed{\int_a^b u dv = uv \Big|_a^b - \int_a^b v du}$$

Match up u and v according to the following

$$u = \int_0^t dx Y(x) \quad \text{and} \quad dv = dt e^{-st} \quad \text{and} \quad a = 0 \quad b = \infty$$

so

$$du = Y(x)dx \quad \text{and} \quad v = \int_0^t dv = \int_0^t dt e^{-st} = -\frac{e^{-st}}{s}$$

This means that the integration by parts formula takes on the following form:

$$\int_0^\infty dt e^{-st} \int_0^t dx Y(x) = -\frac{e^{-st}}{s} \int_0^t dx Y(x) \Big|_0^\infty + \frac{1}{s} \int_0^\infty dt e^{-st} Y(t)$$

$$= -\frac{e^{-s\infty}}{s} \int_0^\infty dx Y(x) + \frac{e^{-s0}}{s} \int_0^0 dx Y(x) + \frac{1}{s} \int_0^\infty dt e^{-st} Y(t)$$

$$= \frac{1}{s} \int_0^\infty dt e^{-st} Y(t)$$

$$= \frac{1}{s} \tilde{Y}(s)$$

So, after all the dust settles the result is

$$\boxed{L\left\{\int_0^t du\,Y(u)\right\} = \frac{\tilde{Y}(s)}{s}} \tag{F-20}$$

Therefore, to get the Laplace transform of a *derivative* one *multiplies* by s and to get the Laplace transform of the *integral* one *divides* by s. This symmetry is consistent with the observation that the derivative and the integral are inverses of each other.

F-12 The Laplace Transform Recipe

The set of rules presented in Chap. 3 are repeated here, with a few extras, as a summary of the Laplace transforms developed earlier in this appendix.

$$\frac{d}{dt} \Rightarrow s$$

$$\frac{d^2}{dt^2} \Rightarrow s^2$$

$$Y(t) \Rightarrow \tilde{Y}(s)$$

$$U(t) \Rightarrow \tilde{U}(s)$$

$$C \Rightarrow \frac{C}{s}$$

$$\int_0^t Y(u)du \Rightarrow \frac{\tilde{Y}(s)}{s}$$

$$\lim_{s\to 0} s\tilde{Y}(s) = Y(\infty)$$

$$e^{-at} \Rightarrow \frac{1}{s+a}$$

$$\sin(\omega t) \Rightarrow \frac{\omega}{s^2+\omega^2}$$

$$\delta(t) \Rightarrow 1$$

$$\hat{U}(t-D)Y(t-D) \Rightarrow e^{-sD}\tilde{Y}(s)$$

Note that these rules are a special case of the rules derived in this appendix in that the initial value of Y is assumed to be zero.

F-13 Applying the Laplace Transform to the First-Order Model: The Transfer Function

Return to Eq. (3-22) in Chap. 3 where we studied the tank of liquid that had an input flow rate of U and a tank height of Y. The behavior of the tank was described by the following differential equation

$$\tau \frac{dY}{dt} + Y = gU \tag{F-21}$$

After following the rule of replacing the derivative with the operator s, and assuming that initial values were all zero, Eq. (F-20) would become

$$s\tau\tilde{Y} + \tilde{Y} = g\tilde{U}$$

$$\tilde{Y} = \frac{g}{\tau s + 1}\tilde{U}$$

$$\tilde{Y} = G(s)\tilde{U}$$

This suggests that the response $Y(t)$ can be obtained from the inversion of the product of the quantity $G(s)$, which is the transfer function, and $\tilde{U}(s)$, which is the transform of the input $U(t)$, as in

$$Y(t) = L^{-1}\{G(s)\tilde{U}(s)\} \tag{F-22}$$

So, in the Laplace transform domain, the response is the *product* of the transfer function and the transform of the input.

F-14 Applying the Laplace Transform to the First-Order Model: The Impulse Response

As mentioned above, the behavior of the tank is described by the following differential equation

$$\tau \frac{dY}{dt} + Y = gU \tag{F-23}$$

After following the rule of replacing the derivative with the operator s, and assuming that initial values were all zero, Eq. (F-23) would become

$$s\tau\tilde{Y} + \tilde{Y} = g\tilde{U} \tag{F-24}$$

Specify that the time domain function $U(t)$ is a unit impulse or a Dirac delta function at time $t = 0$. Based on the above discussion of the Dirac delta function, having $U(t)$ be an impulse is a relatively difficult thing to conceive of physically, but bear with me. If, in fact, the input is an impulse then

$$\tilde{U}(s) = L\{U(t)\} = L\{\delta(t)\} = 1$$

and

$$s\tau\tilde{Y} + \tilde{Y} = g$$

$$\tilde{Y} = \frac{g}{\tau s + 1} \quad \text{(F-25)}$$

Note that we can also write

$$\tilde{Y} = \frac{g}{\tau s + 1} = G(s) \quad \text{(F-26)}$$

so, *the transfer function is also the transform of the impulse response*. Remember this because we will refer to it in App. I on the Z-transform.

Using Eq. (F-12), the inverse is obtained from Eq. (F-26) quickly as

$$Y(t) = \frac{g}{\tau} e^{-\frac{t}{\tau}} = I_m(t) \quad \text{(F-27)}$$

where the symbol $I_m(t)$ is used to indicate the impulse response (we will refer to it later on). Comparing Eqs. (F-26) and (F-27) we see again that the transfer function is the Laplace transform of the impulse response

$$L\{I_m(t)\} = G(s) \quad \text{(F-28)}$$

The response of the first-order process to a unit impulse at time $t = 0$ is an exponential decay from an initial value of g/τ. In other words, a spike in the input flow rate is manifested in an immediate height jump at $t = 0$ from zero to g/τ followed by a slow (or exponential) decay to zero.

Admittedly, it is difficult for the reader to conceptualize exactly how she would apply an impulse input to a process like the tank of liquid. However, don't let that bother you. Simply treat the impulse input as a handy mathematical idealization and stay tuned because it will raise its pretty head again.

F-15 Applying the Laplace Transform to the First-Order Model: The Step Response

In the previous sections the relationship between the input and output for the first-order model has repeatedly shown to be

$$\tilde{Y} = \frac{g}{\tau s+1}\tilde{U} \tag{F-29}$$

Assume that the time domain function $U(t)$ is a step function having a constant value of U_c for $t \geq 0$ and a value of zero for $t < 0$. The Laplace transform for U_c would be

$$\tilde{U} = \frac{U_c}{s}$$

So, Eq. (F-29) becomes

$$\tilde{Y} = \frac{g}{\tau s+1}\frac{U_c}{s} \tag{F-30}$$

To invert this transform to get $Y(t)$, Eq. (F-30) needs to be simplified to point where we can recognize a familiar Laplace transform and match it up with a time domain function. Partial fractions can be used to accomplish this.

Question F-2 What can the final value theorem and Eq. (F-30) tell you about whether $Y(t)$ settles down as $t \to \infty$?

Answer Applying the final value theorem to Eq. (F-30) gives

$$\lim_{s\to 0} s\frac{g}{\tau s+1}\frac{U_c}{s} = \lim_{s\to 0}\frac{gU_c}{\tau s+1} = gU_c$$

F-16 Partial Fraction Expansions Applied to Laplace Transforms: The First-Order Problem

In Sec. F-15 we arrived at the following Laplace transform while trying to solve the differential equation [Eq. (F-23)].

$$\tilde{Y} = \frac{g}{\tau s+1}\frac{U_c}{s} \tag{F-31}$$

where U_c is a constant. To get the time domain solution for $Y(t)$ we have to invert this transform. Finding a time domain function that has a transform like that in Eq. (F-31) would be nice. This transform looks somewhat familiar but it does not match up perfectly with the small group of transforms that we have already developed.

The tool of partial fractions, can be used to expand $\tilde{Y}$ into two terms

$$\tilde{Y} = \frac{\frac{g}{\tau}}{s+\frac{1}{\tau}} \frac{U_c}{s} = \frac{A}{s} + \frac{B}{s+\frac{1}{\tau}} \tag{F-32}$$

where the constants A and B are as yet undetermined. These two terms on the right-hand side of Eq. (F-32) should look familiar—they are the Laplace transforms of a constant (or step change at time zero) and an exponential. Before proceeding, the values of A and B need to be found. Upon cross-multiplying Eq. (F-32) by $s(1+1/\tau)$, the denominator of the transform in question, gives

$$\frac{gU_c}{\tau} = A\left(s+\frac{1}{\tau}\right) + Bs \tag{F-33}$$

Letting $s = 0$ in Eq. (F-33) yields a value for A

$$\frac{gU_c}{\tau} = \frac{A}{\tau} \qquad \text{or} \qquad A = gU_c$$

Then, letting $s = -1/\tau$ in Eq. (F-33) yields a value for B

$$\frac{gU_c}{\tau} = -\frac{B}{\tau} \qquad \text{or} \qquad B = -gU_c$$

So, the new expression for $\tilde{Y}$ is

$$\tilde{Y} = \frac{gU_c}{s} - \frac{gU_c}{s+\frac{1}{\tau}} = gU_c\left(\frac{1}{s} - \frac{1}{s+\frac{1}{\tau}}\right) \tag{F-34}$$

Thus, Eq. (F-31) has been expanded into two simpler terms, each of which is the Laplace transform of a known time domain function. From Eqs. (F-5) and (F-12) we can match the two transforms up with a constant and an exponential

$$L^{-1}\left\{gU_c\left(\frac{1}{s} - \frac{1}{s+\frac{1}{\tau}}\right)\right\} = gU_c\left(1-e^{-\frac{t}{\tau}}\right) \tag{F-35}$$

So, we have arrived at the solution to the first-order differential equation by a second method. Equation (F-35) is the classic *diminishing returns* curve. Comparing Eqs. (F-35) and (F-27), one sees that the response of our first-order process to a spike or impulse is quite different from that to a step (as you should suspect).

F-17 Partial Fraction Expansions Applied to Laplace Transforms: The Second-Order Problem

Later on in Chap. 3, where a second-order differential equation appears, the following transform is encountered in Eq. (3-39)

$$\tilde{Y} = \frac{gks + gI}{\tau s^2 + (1+gk)s + gI}\frac{S_c}{s} \tag{3-39}$$

Equation (3-39) has three poles, namely,

$$s_1, s_2 = -\frac{1+gk}{2\tau} \pm \frac{\sqrt{(1+gk)^2 - 4gI}}{2\tau} \qquad s_3 = 0$$

and one zero

$$s = -\frac{I}{k}$$

To expand Eq. (3-39) into partial fractions I resorted to the Internet because my algebraic bookkeeping skills are abysmal. Using Google, I searched "partial fractions" and ended up at a site called "QuickMath Automatic Math Solutions" (your search may turn up something different but keep trying unless you want to do the math by hand—ugh!). At this site I entered

$$(a + bx)/[x(x-c)(x-d)]$$

for

$$\frac{S_c}{\tau}(gI + gks)/[s(s - s_1)(s - s_2)] \tag{F-36}$$

in a little box on the site page. The site, using Mathematica, returned

$$\frac{a}{cdx} + \frac{a+bc}{c(c-d)(x-c)} + \frac{a+bd}{d(d-c)(x-d)} \tag{F-37}$$

So far, so good. The association between the variables, a, b, c, d, and x and the quantities in Eq. (3-39) should be apparent, especially the equivalence between x and s. With a little luck the reader should be able to deduce that the above three terms in the Laplace domain correspond to a constant and two exponentials in the time domain.

Still not trusting my algebraic abilities, I wrote a short Matlab script (using the symbolic toolbox) to make the substitutions and arrived at

$$Y(t) = \frac{gS_c}{\tau}\left(-\frac{k+Is_1}{s_1(s_2-s_1)}e^{s_1 t} + \frac{k+Is_2}{s_2(s_2-s_1)}e^{s_2 t} + \frac{I}{s_1 s_2}\right) \qquad \text{(F-38)}$$

As you might have perceived from this exercise, applying partial fractions is a busy algebraic exercise subject to many bookkeeping errors. For years I have pursued it with trepidation but now with the help of the Internet and Matlab I feel less dependent on my horrible bookkeeping skills. In any case, Eq. (F-38) shows the step response of a second-order system, namely a first-order process subject to proportional-integral feedback control. Note that there are two exponential terms and a constant. Also, note that neither s_1 or s_2 can be positive. Therefore, the exponential terms will die away and $Y(t)$ settle out at $IgS_c/(\tau s_1 s_2)$.

Question F-3 Can you simplify this a little further and obtain the final value?

Answer Using the expressions for s_1 and s_2, you should be able to show that $s_1 s_2 = gI/\tau$ so that the constant term becomes S_c. Since the other terms have exponentials that die away with time, the constant term shows that the process output reaches the set point.

F-18 A Precursor to the Convolution Theorem

Note that the common word "convoluted" means complicated, intricately involved, twisted, or coiled. That should be a clue as to the difficulty that we will encounter in this section.

We have been studying a first-order system based on a tank of liquid where the process variable Y is the tank height and the process input U is the input flow rate. At any time, $Y(t)$ is the result of past values of Y and past values of U. We can see this if we manipulate Eq. (F-23)

$$\tau\frac{dY}{dt} + Y = gU$$

$$\frac{dY}{dt} = \frac{-Y+gU}{\tau} \qquad \text{(F-39)}$$

$$dY = \frac{-Ydt + gUdt}{\tau}$$

Integrate both sides of Eq. (F-39) with respect to time and get

$$\int_0^t du \dot{Y}(u) = Y(t) - Y(0) = \frac{\int_0^t du(-Y(u) + gU(u))}{\tau}$$

so,

$$Y(t) = Y(0) + \frac{\int_0^t du[-Y(u) + gU(u)]}{\tau} \tag{F-40}$$

The integral is the sum of the area under the integrand. Thus, Eq. (F-40) shows that $Y(t)$ is a weighted sum of all the past values of $Y(t)$ and $U(t)$. This is not a particularly useful equation because it contains the response variable $Y(t)$ on both sides of the equation.

Question F-4 Can you use the Laplace transform to solve Eq. (F-40) for Y?

Answer Taking the Laplace transform of Eq. (F-40) yields

$$\tilde{Y} = \frac{Y(0)}{s} + \frac{g\tilde{U} - \tilde{Y}}{\tau s}$$

Here, Eqs. (F-3) and (F-19) were used. Solve for $\tilde{Y}$ and get

$$\tilde{Y}\left(1 + \frac{1}{\tau s}\right) = \frac{Y(0)}{\tau s} + \frac{g\tilde{U}}{\tau s}$$

$$\tilde{Y}(\tau s + 1) = Y(0) + g\tilde{U}$$

$$\tilde{Y} = \frac{Y(0) + g\tilde{U}}{\tau s + 1}$$

which, if $Y(0) = 0$, is the same as obtained in Eq. (F-29).

F-19 Using the Integrating Factor to Obtain the Convolution Integral

Consider yet another way to solve the first-order differential equation that describes the tank of liquid

$$\tau \frac{dY}{dt} + Y = gU(t)$$

or (F-41)

$$\frac{dY}{dt} + \frac{1}{\tau} Y = \frac{g}{\tau} U(t)$$

Note that the argument of U is added to emphasize that U may be a function of time rather than just a constant. Instead of using the Laplace transform or instead of trying an exponential time domain function as a solution, apply an integrating factor of $e^{t/\tau}$ to both sides of Eq. (F-41)

$$e^{\frac{t}{\tau}}\frac{dY}{dt}+e^{\frac{t}{\tau}}\frac{1}{\tau}Y=e^{\frac{t}{\tau}}\frac{g}{\tau}U \tag{F-42}$$

The left-hand side of Eq. (F-42) can be written as a derivative of the quantity $e^{t/\tau}Y$ because

$$\frac{d}{dt}\left(e^{\frac{t}{\tau}}Y\right)=e^{\frac{t}{\tau}}\frac{dY}{dt}+e^{\frac{t}{\tau}}\frac{1}{\tau}Y$$

so, Eq. (F-42) can be written as

$$\frac{d}{dt}\left(e^{\frac{t}{\tau}}Y\right)=e^{\frac{t}{\tau}}\frac{g}{\tau}U \tag{F-43}$$

Now, integrating across Eq. (F-43) gives

$$d\left(e^{\frac{t}{\tau}}Y\right)=e^{\frac{t}{\tau}}\frac{g}{\tau}Udt$$

$$\int_0^t d\left(e^{\frac{u}{\tau}}Y\right)=\int_0^t e^{\frac{u}{\tau}}\frac{g}{\tau}U(u)du$$

$$e^{\frac{u}{\tau}}Y\,\Big|_0^t=e^{\frac{t}{\tau}}Y(t)-Y(0)=\int_0^t e^{\frac{u}{\tau}}\frac{g}{\tau}U(u)du$$

After multiplying both sides by $e^{-(t/\tau)}$ and doing a little rearranging we get

$$Y(t)=Y(0)e^{-\frac{t}{\tau}}+e^{-\frac{t}{\tau}}\int_0^t e^{\frac{u}{\tau}}\frac{g}{\tau}U(u)du$$

If we place the factor $e^{-(t/\tau)}$ inside the integral (which we may do since it is a constant with respect to the dummy integration variable u) we get

$$\begin{aligned}Y(t)&=Y(0)e^{-\frac{t}{\tau}}+\int_0^t e^{-\frac{t}{\tau}}e^{\frac{u}{\tau}}\frac{g}{\tau}U(u)du\\&=Y(0)e^{-\frac{t}{\tau}}+\int_0^t e^{-\left(\frac{t-u}{\tau}\right)}\frac{g}{\tau}U(u)du\end{aligned} \tag{F-44}$$

Now comes the clever step! Remember the impulse response for this differential equation derived above and represented in Eq. (F-26)? Do you see anything in the integrand of Eq. (F-44) that resembles it? In fact, the impulse response is lurking there and we can rewrite Eq. (F-44) as

$$Y(t) = Y(0)e^{-\frac{t}{\tau}} + \frac{g}{\tau}\int_0^t I_m(t-u)U(u)du$$

$$I_m(t-u) = e^{-\left(\frac{t-u}{\tau}\right)} \tag{F-45}$$

This is the infamous convolution integral where the process input $U(t)$ is "convolved" with the impulse response $I_m(t)$.

Return to Eq. (F-44) and spend a few moments trying to figure out how you might evaluate the integral numerically.

$$Y(t) = Y(0)e^{-\frac{t}{\tau}} + \frac{g}{\tau}\int_0^t e^{-\left(\frac{t-u}{\tau}\right)}U(u)du$$

Suppose $g = 1$ and $\tau = 10$. The integrand is $e^{-[(t-u)/\tau]}\ U(u)$. Suppose that input U is a downward ramp described by $U(t) = 1 - 0.1t$. Thus, the input is initially 1.0 and then drops to zero at time $t = 10$. The varying factor in the impulse response is $e^{-(u/\tau)}$ and this initially equals 1.0 also and drops off slowly to a value of $e^{-1} = 0.368$ by time $t = 10$. However, the integrand contains the factor $e^{-[(t-u)/\tau]}$ which has a mirror image shape to that of $e^{-(u/\tau)}$. These different components are shown in Fig. F-1.

The response Y is seen to be a sum of the input U weighted by the impulse response that is "folded back" from the point of interest, namely,

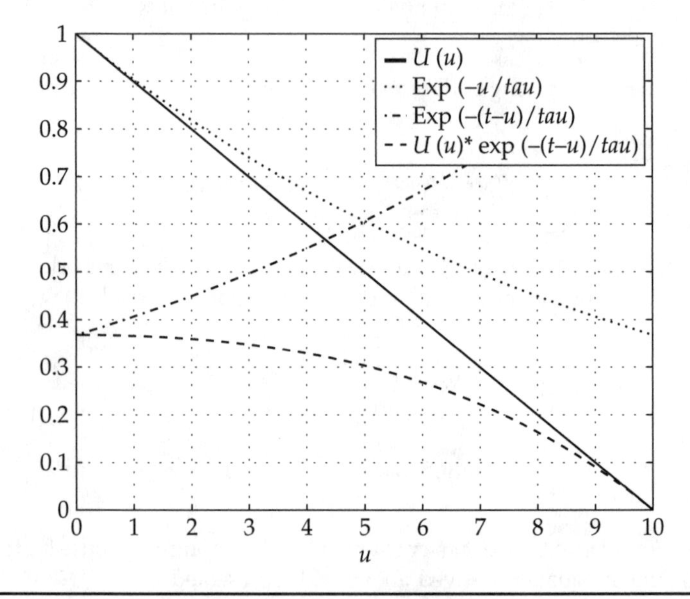

FIGURE F-1 Components of the convolution integrand.

time t. The Germans actually avoid the use of the word "convolution" and instead refer to the integral as the "folding" or "faltung" integral. The numerical evaluation of this integral is indeed convoluted.

Now, let's compare convolution in the time domain with another operation in the Laplace transform domain. We remember from Eq. (F-22) that the response $Y(t)$ is also the inverse of the product of the transfer function and the Laplace transform of the input, as in

$$Y(t) = L^{-1}\{G(s)U(s)\} = Y(0)e^{-\frac{t}{\tau}} + \frac{g}{\tau}\int_0^t e^{-\left(\frac{t-u}{\tau}\right)}U(u)du$$

So, we conclude that in the time domain we *convolve* and in the Laplace domain we *multiply*—which would you rather do?

F-20 Application of the Laplace Transform to a First-Order Partial Differential Equation

In Chap. 7 the following partial differential equation was derived to describe the behavior of a tubular heat exchanger.

$$\frac{\partial T}{\partial t} + v\frac{\partial T}{\partial z} = \frac{1}{\tau_T}[T_s - T(z,t)] \qquad \text{(F-46)}$$

The Laplace transform can be applied to remove the time dependence of Eq. (F-46). It could also be applied to remove the spatial dependence. Since we ultimately want to move from the time domain to the frequency domain, the Laplace transform will be applied to the time dependence first. The result, an ordinary differential equation, will then be solved by conventional means although the Laplace transform could be used again as an alternative.

We will apply the Laplace operator

$$\int_0^\infty dte^{-st}\ldots$$

to every term in Eq. (F-47) whether they are terms like $T_s - T$ or $\partial T / \partial t$ or $\partial T / \partial z$. The result is

$$\int_0^\infty dte^{-st}\frac{\partial T}{\partial t} + \int_0^\infty dte^{-st}v\frac{\partial T}{\partial z} = \int_0^\infty dte^{-st}\frac{T_s}{\tau_T} - \int_0^\infty dte^{-st}\frac{T(z,t)}{\tau_T} \qquad \text{(F-47)}$$

The first term on the left-hand side (LHS) is the Laplace transform of a derivative and yields

$$s\tilde{T}(z,s)$$

where the initial value $T(z,0)$, for all z, is assumed zero.

The next term on the LHS contains the partial derivative with respect to z so we exchange the order of the operators (we can do this since both operators are linear) and get

$$v\int_0^\infty dte^{-st}\frac{\partial T}{\partial z} = v\frac{\partial}{\partial z}\int_0^\infty dte^{-st}T = v\frac{\partial \tilde{T}}{\partial z}$$

The two terms on the right-hand side (RHS) of Eq. (F-47) are Laplace transforms of T and T_s, both divided by τ_T. Therefore, the result of applying the Laplace transform to the partial differential equation is

$$s\tilde{T} + v\frac{d\tilde{T}}{dz} = \frac{1}{\tau_T}(\tilde{T}_s - \tilde{T}) \qquad \text{(F-48)}$$

In review, $\partial T / \partial t$ was replaced by $s\tilde{T}$ (assuming for the time being that the initial value of T is zero). Next, $\partial T/\partial z$ was replaced by $d\tilde{T}/dz$ because we can exchange the order of the Laplace operator and the partial differentiation operator. Finally, $T_s - T$ became $\tilde{T}_s - \tilde{T}$. Note that $\tilde{T}$ is a function of s and z, so we could emphasize this by writing $\tilde{T}(z,s)$, but most of subsequent development will use $\tilde{T}$. The units of $\tilde{T}$ are °C sec.

Hopefully, you will admit that removing the time dependence from a partial differential equation is not that big of a deal. Thus, after the dust has settled, Eq. (F-48) is a first-order ordinary differential equation of the form

$$v\frac{d\tilde{T}}{dz} + \left(s + \frac{1}{\tau_T}\right)\tilde{T} = \frac{\tilde{T}_s}{\tau_T} \qquad \text{(F-49)}$$

where the Laplace variable s is just a parameter and where the independent variable is the position z. Remember that $\tilde{T}_s$ is the Laplace transform of the jacket temperature which we specified could be a function of time but not of axial position, that is, T_s or $\tilde{T}_s$ is not a function of z.

F-21 Solving the Transformed Partial Differential Equation

Now, how do we solve Eq. (F-49)? We could apply the Laplace transform again to remove the dependence on z with a different variable, say p, instead of s. Alternatively, we could look at the homogeneous part of Eq. (F-49)

$$v\frac{d\tilde{T}_h}{dz} + \left(s + \frac{1}{\tau_T}\right)\tilde{T}_h = 0$$

and try a solution of the form

$$\tilde{T}_h = Ce^{az}$$

This would yield

$$a = -\frac{1}{v}\left(s + \frac{1}{\tau_T}\right)$$

Note that the variable a has units of cm^{-1}.

The nonhomogeneous part of the equation (with the derivative equal to zero) would give us

$$\left(s + \frac{1}{\tau_T}\right)\tilde{T}_{nh} = \frac{\tilde{T}_s}{\tau_T}$$

$$\tilde{T}_{nh} = \frac{\tilde{T}_s}{\tau_T\left(s + \frac{1}{\tau_T}\right)} = \frac{\tilde{T}_s}{\tau_T s + 1}$$

Therefore, the total solution would be

$$\tilde{T}(z,s) = Ce^{az} + \frac{\tilde{T}_s(s)}{\tau_T s + 1} \tag{F-50}$$

To find the value of C, the condition at $z = 0$, the inlet condition of $T(0,t) = T_0(t)$, $t \geq 0$, is applied to Eq. (F-50). Since we are dealing with Laplace transforms, we write $\tilde{T}(0,s) = \tilde{T}_0(s)$ where $\tilde{T}_0(s)$ is the transform of the inlet temperature. Setting $z = 0$ in Eq. (F-50) gives

$$\tilde{T}(0,s) = \tilde{T}_0(s) = C + \frac{\tilde{T}_s(s)}{\tau_T s + 1}$$

$$C = \frac{\tilde{T}_0(s)(\tau_T s + 1) - \tilde{T}_s(s)}{\tau_T s + 1}$$

Inserting the value of C into Eq. (F-50) gives

$$\tilde{T}(z,s) = \tilde{T}_0(s)e^{az} + \tilde{T}_s(s)\frac{1 - e^{az}}{\tau_T s + 1}$$

$$a = -\frac{1}{v}\left(s + \frac{1}{\tau_T}\right) \tag{F-51}$$

Eq. (F-51) contains two transfer functions of interest. First, the transfer function showing how the inlet temperature affects the outlet temperature in the tube (at $z = L$), namely,

$$\frac{\tilde{T}(L,s)}{\tilde{T}_0(s)} = e^{aL} = e^{-\frac{L}{v}\left(s+\frac{1}{\tau_T}\right)} = e^{-s\frac{L}{v}} e^{-\frac{L}{v\tau_T}}$$
$$= e^{-st_D} e^{-\frac{t_D}{\tau_T}} \tag{F-52}$$

where $t_D = L/v$ is the average residence time or delay time for the tube.

Second, Eq. (F-51) yields the transfer function relating the steam jacket temperature to the outlet temperature.

$$\frac{\tilde{T}(L,s)}{\tilde{T}_s(s)} = \frac{1-e^{az}}{\tau_T s+1}$$
$$= \frac{1-e^{-st_D} e^{-\frac{t_D}{\tau_T}}}{\tau_T s+1} \tag{F-53}$$

To invert Eq. (F-53) for the case where T_s is a step change of size U_c, as in

$$\tilde{T}_s(s) = \frac{U_c}{s}$$

one would separate the terms as follows:

$$\tilde{T}(L,s) = \frac{1-e^{-st_D} e^{-\frac{t_D}{\tau_T}}}{\tau_T s+1} \frac{U_c}{s}$$
$$= \frac{1}{\tau_T s+1} \frac{U_c}{s} - \frac{U_c e^{-\frac{t_D}{\tau_T}}}{\tau_T s+1} \frac{e^{-st_D}}{s} \tag{F-54}$$

From Eqs. (F-31) and (F-34) we know that the first term on the RHS

$$\frac{1}{\tau_T s+1} \frac{U_c}{s}$$

has an inverse of

$$U_c\left(1-e^{-\frac{t}{\tau_T}}\right)$$

The second term on the RHS of Eq. (F-54) is the same as the first except for the constant coefficient of $e^{-(t_D/\tau_T)}$ and the factor e^{-st_D} which Eq. (F-6) shows is an indicator of a delayed unit step function $\hat{U}$. Therefore, the inverse of the second term on the RHS of Eq. (F-54) is

$$L^{-1}\left\{U_c e^{-\frac{t_D}{\tau_T}} \frac{1}{\tau_T s+1}\frac{e^{-st_D}}{s}\right\} = U_c e^{-\frac{t_D}{\tau_T}}\left(1-e^{-\frac{t-t_D}{\tau_T}}\right)\hat{U}(t-t_D) \qquad \text{(F-55)}$$

and the total solution is

$$T(L,t) = U_c\left(1-e^{-\frac{t}{\tau_T}}\right) - U_c e^{-\frac{t_D}{\tau_T}}\left(1-e^{-\frac{t-t_D}{\tau_D}}\right)\hat{U}(t-t_D) \qquad \text{(F-56)}$$

F-22 The Magnitude and Phase of the Transformed Partial Differential Equation

This section supplements Sec. 7-5 and assumes that the reader has read Chap. 4 which deals with moving from the Laplace s domain to the frequency domain. Start with the following transfer function which gives the response of the outlet tube temperature to the steam jacket temperature.

$$\frac{\tilde{T}(L,s)}{\tilde{T}_s(s)} = \frac{1-e^{-st_D}e^{-\frac{t_D}{\tau_T}}}{\tau_T s+1}$$

Make the usual substitution of $s \to j\omega$ getting

$$\frac{1-e^{-\frac{t_D}{\tau_T}}e^{-j\omega t_D}}{\tau_T j\omega+1} \qquad \text{(F-57)}$$

The denominator is an old friend and can be turned into magnitude and phase by inspection, as in

$$\tau_T j\omega + 1 = \sqrt{(\tau\omega)^2+1}\,e^{j\theta_d} \qquad \theta_d = \tan^{-1}\left(\frac{\tau_T j\omega}{1}\right)$$

If there is some question about this the reader can consult App. B.

The numerator is a little more involved. The complex exponential is replaced using Euler's formula.

$$\begin{aligned} 1-e^{-\frac{t_D}{\tau_T}}e^{-j\omega t_D} &= 1-e^{-\frac{t_D}{t_T}}(\cos(\omega t_D) - j\sin(\omega t_D)) \\ &= 1-e^{-\frac{t_D}{\tau_T}}\cos(\omega t_D) + je^{-\frac{t_D}{\tau_T}}\sin(\omega t_D) \end{aligned}$$

The magnitude is the square root of the sum of the squares of the real and imaginary parts, so the final expression for the numerator is

$$1-e^{-\frac{t_D}{\tau_T}}e^{-j\omega t_D} = \sqrt{\left(1-e^{-\frac{t_D}{t_T}}\cos(\omega t_D)\right)^2 + \left(e^{-\frac{t_D}{\tau_T}}\sin(\omega t_D)\right)^2}\, e^{j\theta_n}$$

$$= \sqrt{1-2e^{-\frac{t_D}{\tau_T}}\cos(\omega t_D)+e^{-2\frac{t_D}{\tau_T}}}\, e^{j\theta_n}$$

The phase of the numerator is

$$\theta_n = \tan^{-1}\left(\frac{e^{-\frac{t_D}{\tau_T}}\sin(\omega t_D)}{1-e^{-\frac{t_D}{\tau_T}}\cos(\omega t_D)}\right)$$

Therefore, the overall expression is

$$\frac{\tilde{T}(L,j\omega)}{\tilde{T}_s(j\omega)} = \frac{\sqrt{1-2e^{-\frac{t_D}{\tau_T}}\cos(\omega t_D)+e^{-2\frac{t_D}{\tau_T}}}\, e^{j\theta_n}}{\sqrt{(\tau\omega)^2+1}\, e^{j\theta_d}}$$

$$= \sqrt{\frac{1-2e^{-\frac{t_D}{\tau_T}}\cos(\omega t_D)+e^{-2\frac{t_D}{\tau_T}}}{(\tau\omega)^2+1}}\, e^{j\theta} \qquad \text{(F-58)}$$

$$\theta = \tan^{-1}\left(\frac{e^{-\frac{t_D}{\tau_T}}\sin(\omega t_D)}{1-e^{-\frac{t_D}{\tau_T}}\cos(\omega t_D)}\right) - \tan^{-1}(\tau_T j\omega)$$

F-23 A Brief History of the Laplace Transform

The transform is named after the Frenchman Pierre Simon Laplace but he did not actually come up with Eq. (F-1). There is some evidence that the Swiss mathematician Leonhard Euler may have used formulas of type in Eq. (F-1) in the 1700s. Probably unaware of Euler, the English engineer Oliver Heaviside, in the late 1800s, developed the recipe of replacing derivatives by the *s* operator (he used the symbol *p*). He then inverted by using partial fractions and tables of precomputed Laplace transforms. His recipe was largely empirical but quite useful. Mathematicians consistently criticized Heaviside for his lack of mathematical rigor. He is said to have replied, "I don't understand the digestive process but I still eat."

In the 1920s the German mathematician John I'Anson Bromwich developed a rigorous basis for Heaviside's recipe using equations similar to Eq. (F-1) but little was made of the connection in engineering. In 1937, Bromwich published a book *The Theory and Application of the Laplace Transform* in German. Also, in 1937 L. A- Pipes published an article in English showing how to use the Laplace transform for circuit analysis. In 1942, Gardner and Barnes published a classic text *Transients in Linear Systems: Studied by the Laplace Tranform* and the use of the transform in engineering quickly became wide spread. So, although Simon Laplace died in 1827, the transform named after him did not become an engineering tool until 110 years later.

F-24 Summary

Wow! This has been a long, painfully detailed appendix. Unfortunately, it is a subject that really has to be delved into. Just be happy it was not placed in the main body of the book.

The Laplace transform was introduced as a simple integral operation designed to move things to a new domain where there were not as many nasty differential equations. Transforms of several common and useful time domain functions were developed and catalogued. The Laplace transform was applied to the first-order differential equation to develop (1) the transfer function concept, (2) the impulse response, and (3) the step-change response. Partial fractions were shown to be a practical tool for breaking up a complicated transform into simple ones that could be associated with simple time domain functions. This tool was applied to solving first- and second-order differential equations.

The convolution integral was introduced to show that multiplication in the Laplace domain is equivalent to convolution in the time domain (and vice versa). Finally, the Laplace transform was applied to a simple partial differential equation which also described the behavior of the tubular energy exchanger discussed in Chap. 7.

The Laplace transform is the basic tool of control engineering so it would behoove the reader to be reasonably familiar with the material in this appendix—you may have to reread parts of it many times while wading through the rest of the book.

APPENDIX G

Vectors and Matrices

Matrices have a long history starting as early as 300 B.C. We will use them as a convenient and compact way of writing several equations as one. For example, in Chap. 4, equations describing the dynamics of the three-tank process were presented as

$$\begin{aligned}
\frac{dX_1}{dt} &= -\frac{X_1}{\tau_1} + \frac{R_1}{\tau_1}F_0 \\
\frac{dX_2}{dt} &= \frac{R_2}{R_1}\frac{X_1}{\tau_2} - \frac{X_2}{\tau_2} \\
\frac{dX_3}{dt} &= \frac{R_3}{R_2}\frac{X_2}{\tau_3} - \frac{X_3}{\tau_3} \\
Y &= X_3
\end{aligned} \tag{G-1}$$

These equations could also be written as

$$\begin{aligned}
\frac{d}{dt}\begin{pmatrix} X_1 \\ X_2 \\ X_3 \end{pmatrix} &= \begin{pmatrix} -\frac{1}{\tau_1} & 0 & 0 \\ \frac{R_2}{R_1}\frac{1}{\tau_2} & -\frac{1}{\tau_2} & 0 \\ 0 & \frac{R_3}{R_2}\frac{1}{\tau_3} & -\frac{1}{\tau_3} \end{pmatrix}\begin{pmatrix} X_1 \\ X_2 \\ X_3 \end{pmatrix} + \begin{pmatrix} \frac{R_1}{\tau_1} \\ 0 \\ 0 \end{pmatrix} F_0 \\
Y &= \begin{pmatrix} 0 & 0 & 1 \end{pmatrix}\begin{pmatrix} X_1 \\ X_2 \\ X_3 \end{pmatrix}
\end{aligned} \tag{G-2}$$

where

$$\begin{pmatrix} X_1 \\ X_2 \\ X_3 \end{pmatrix}$$

is a matrix with three rows and one column, called a size (3, 1) *column vector*, and

$$\begin{pmatrix} -\frac{1}{\tau_1} & 0 & 0 \\ \frac{R_2}{R_1}\frac{1}{\tau_2} & -\frac{1}{\tau_2} & 0 \\ 0 & \frac{R_3}{R_2}\frac{1}{\tau_3} & -\frac{1}{\tau_3} \end{pmatrix}$$

is a square matrix with three rows and three columns, as in a size (3, 3) matrix, sometimes called the coefficient matrix, and

$$(0 \quad 0 \quad 1)$$

is a matrix with one row and three columns, called a size (1, 3) *row vector*.

The matrices with either one row or one column are called *vectors* because they *could* represent the components of a vector in an appropriate dimensional space. For example, the column vector w

$$w = \begin{pmatrix} a \\ b \end{pmatrix}$$

could represent a vector having a magnitude of $\sqrt{a^2+b^2}$, an x-axis component of a and a y-axis component of b. Likewise, the vector u

$$u = \lambda \begin{pmatrix} a \\ b \end{pmatrix}$$

could represent a vector having the same direction as w but with a different magnitude, $\lambda\sqrt{a^2+b^2}$, because of the *scalar* multiplier λ. This expression used scalar multiplication which will be discussed in Sec. G-1. The elements of the example matrices have been one-dimensional quantities called scalars. They can be real, imaginary, or complex. In addition, the elements of matrices can be matrices themselves as we shall see.

G-1 Addition and Multiplication of Matrices

To make matrices useful there must be rules for addition, subtraction, and multiplication. Regarding addition, one adds matrices element by element:

$$\begin{pmatrix} a & b \\ c & d \end{pmatrix} + \begin{pmatrix} e & f \\ g & h \end{pmatrix} = \begin{pmatrix} a+e & b+f \\ c+g & d+h \end{pmatrix}$$

Multiplication is a little more subtle. When a matrix is multiplied by a scalar then all elements in the matrix are multiplied by the scalar, as in

$$a\begin{pmatrix} b & c \\ d & e \end{pmatrix} = \begin{pmatrix} ab & ac \\ ad & ae \end{pmatrix} = \begin{pmatrix} b & c \\ d & e \end{pmatrix} a$$

Note that this multiplication is commutative.

When matrices are multiplied together, one goes *across the rows* of the left matrix and *down the columns* of the right matrix and adds up the products of the appropriate elements as in

$$\begin{pmatrix} a & b \\ c & d \end{pmatrix}\begin{pmatrix} e & f \\ g & h \end{pmatrix} = \begin{pmatrix} ae+bg & af+bh \\ ce+dg & cf+dh \end{pmatrix}$$

$$\begin{pmatrix} e & f \\ g & h \end{pmatrix}\begin{pmatrix} a & b \\ c & d \end{pmatrix} = \begin{pmatrix} ea+fc & eb+fd \\ ga+hc & gb+hd \end{pmatrix}$$

For matrix multiplication to work, the number of columns of the left-hand matrix must equal the number of rows of the right-hand matrix. Unlike scalar multiplication, matrix multiplication is not commutative.

As an aid to following complicated equations containing many multiplications one often writes the number of rows and columns as in

$$(3,2)\ (2,3) = (3,3)$$

$$\begin{pmatrix} a & b \\ c & d \\ e & f \end{pmatrix}\begin{pmatrix} x & y & z \\ u & v & t \end{pmatrix} = \begin{pmatrix} ax+bu & ay+bv & az+bt \\ cx+du & cy+dv & cz+dt \\ ex+fu & ey+fv & ez+ft \end{pmatrix}$$

That is, the inside dimensions must be the same and can be crossed out leaving the outer two numbers as the size of the product matrix as in

$$(3,2)(2,3) = (3,3)$$

Sometimes it is useful to use subscripts and generalize a bit, as in

$$(n,m)(m,1)=(n,1)$$

$$\begin{pmatrix} a_{11} & a_{12} & \cdots & a_{1m} \\ a_{21} & a_{22} & m\ldots & a_{2m} \\ \cdots & \cdots & \cdots & \cdots \\ a_{n1} & a_{n2} & \cdots & a_{nm} \end{pmatrix} \begin{pmatrix} b_1 \\ b_2 \\ \cdots \\ b_m \end{pmatrix} = \begin{pmatrix} \sum_{k=1}^{m} a_{1k} b_k \\ \sum_{k=1}^{m} a_{2k} b_k \\ \cdots \\ \sum_{k=1}^{m} a_{nk} b_k \end{pmatrix}$$

Note that the number of columns m of the left-hand matrix must equal the number of rows of the right-hand matrix. The result of multiplying a size (n, m) matrix into a size (m, p) matrix is a size (n, p) matrix. Also note that for two-dimensional matrix elements the first subscript refers to the row and the second to the column where the element resides. Therefore, a_{ij} is the element lying in the ith row and the jth column.

A special case of multiplication is the "dot product" where a size $(1,n)$ row matrix (or vector) is multiplied into a size $(n,1)$ column matrix (or vector) producing a scalar, as in

$$(1,n)(n,1)=(1,1)$$

$$[a_1 \quad a_2 \quad \cdots \quad a_n] \begin{pmatrix} b_1 \\ b_2 \\ \cdots \\ b_n \end{pmatrix} = \sum_{k=1}^{n} a_k b_k$$

With these rules, hopefully the reader should be able to understand Eq. (G-2).

G-2 Partitioning

A matrix can be partitioned into submatrices as follows.

$$\begin{pmatrix} a & \begin{pmatrix} c & d \end{pmatrix} \\ \begin{pmatrix} e \\ f \end{pmatrix} & \begin{pmatrix} g & h \\ m & n \end{pmatrix} \end{pmatrix} = \begin{pmatrix} a & c & d \\ e & g & h \\ f & m & n \end{pmatrix}$$

The rules of multiplication still apply to the elements of the partititioned matrix, as in

$$\begin{pmatrix} a & (c \quad d) \\ \begin{pmatrix} e \\ f \end{pmatrix} & \begin{pmatrix} g & h \\ m & n \end{pmatrix} \end{pmatrix} \begin{pmatrix} x \\ \begin{pmatrix} y \\ z \end{pmatrix} \end{pmatrix} = \begin{pmatrix} ax + (c \quad d)\begin{pmatrix} y \\ z \end{pmatrix} \\ \begin{pmatrix} e \\ f \end{pmatrix} x + \begin{pmatrix} g & h \\ m & n \end{pmatrix}\begin{pmatrix} y \\ z \end{pmatrix} \end{pmatrix}$$

$$= \begin{pmatrix} ax + cy + zd \\ \begin{pmatrix} ex \\ fx \end{pmatrix} + \begin{pmatrix} gy + hz \\ my + nz \end{pmatrix} \end{pmatrix} = \begin{pmatrix} ax + cy + zd \\ \begin{pmatrix} ex + gy + hz \\ fx + my + nz \end{pmatrix} \end{pmatrix}$$

$$= \begin{pmatrix} ax + cy + zd \\ ex + gy + hz \\ fx + my + nz \end{pmatrix}$$

G-3 State-Space Equations and Laplace Transforms

Equation (G-2) can be further compacted as follows:

$$\frac{d}{dt}X = AX + BU$$

$$X = \begin{pmatrix} X_1 \\ X_2 \\ X_3 \end{pmatrix} \qquad A = \begin{pmatrix} -\frac{1}{\tau_1} & 0 & 0 \\ \frac{R_2}{R_1}\frac{1}{\tau_2} & -\frac{1}{\tau_2} & 0 \\ 0 & \frac{R_3}{R_2}\frac{1}{\tau_3} & -\frac{1}{\tau_3} \end{pmatrix} \qquad B = \begin{pmatrix} \frac{R_1}{\tau_1} \\ 0 \\ 0 \end{pmatrix} \qquad U = F_0 \tag{G-3}$$

$$Y = CX$$

$$C = (0 \quad 0 \quad 1)$$

Equation (G-3) shows the state-space way of writing the describing equations for processes. Note that the appearance of nonzero elements in the second row and first column (the nondiagonal positions) indicates a coupling between the first and second tank. This is discussed at length in Chap. 4.

The elements of a matrix can also be Laplace transforms. For example, Eq. (5-3) from Chap. 5 could also be written as

$$\begin{pmatrix} \rho A_1 s + \frac{1}{R_1} & 0 & 0 \\ -\frac{1}{R_1} & \rho A_2 s + \frac{1}{R_2} & 0 \\ 0 & -\frac{1}{R_2} & \rho A_3 s + \frac{1}{R_3} \end{pmatrix} \begin{pmatrix} \tilde{X}_1 \\ \tilde{X}_2 \\ \tilde{X}_3 \end{pmatrix} = \begin{pmatrix} 1 \\ 0 \\ 0 \end{pmatrix} \tilde{U}(s) \tag{G-4}$$

$$\tilde{Y}(s) = (0 \quad 0 \quad 1) \begin{pmatrix} \tilde{X}_1 \\ \tilde{X}_2 \\ \tilde{X}_3 \end{pmatrix}$$

or, after separating the Laplace s operator,

$$\left(\begin{pmatrix} \frac{1}{R_1} & 0 & 0 \\ -\frac{1}{R_1} & \frac{1}{R_2} & 0 \\ 0 & -\frac{1}{R_2} & \frac{1}{R_3} \end{pmatrix} + s \begin{pmatrix} \rho A_1 & 0 & 0 \\ 0 & \rho A_2 & 0 \\ 0 & 0 & \rho A_3 \end{pmatrix} \right) \begin{pmatrix} \tilde{X}_1 \\ \tilde{X}_2 \\ \tilde{X}_3 \end{pmatrix} = \begin{pmatrix} 1 \\ 0 \\ 0 \end{pmatrix} \tilde{U}(s) \tag{G-5}$$

$$\tilde{Y}(s) = (0 \quad 0 \quad 1) \begin{pmatrix} \tilde{X}_1 \\ \tilde{X}_2 \\ \tilde{X}_3 \end{pmatrix}$$

or

$$G(s)\tilde{X}(s) = B\tilde{U}$$

$$G(s) = \left(\begin{pmatrix} \frac{1}{R_1} & 0 & 0 \\ -\frac{1}{R_1} & \frac{1}{R_2} & 0 \\ 0 & -\frac{1}{R_2} & \frac{1}{R_3} \end{pmatrix} + s \begin{pmatrix} \rho A_1 & 0 & 0 \\ 0 & \rho A_2 & 0 \\ 0 & 0 & \rho A_3 \end{pmatrix} \right) \tag{G-6}$$

G-4 Transposes and Diagonal Matrices

The transpose of a matrix

$$A = \begin{pmatrix} a & b & c \\ d & e & f \\ g & h & i \end{pmatrix}$$

is

$$A^T = \begin{pmatrix} a & d & g \\ b & e & h \\ c & f & i \end{pmatrix}$$

That is, the off diagonal elements are exchanged but the diagonal elements are left unchanged. The transpose is denoted by a superscript "T" as in

$$A^T = \begin{pmatrix} a & b & c \\ d & e & f \\ g & h & i \end{pmatrix}^T = \begin{pmatrix} a & d & g \\ b & e & h \\ c & f & i \end{pmatrix}$$

A diagonal matrix has nonzero elements only on the diagonal, as in

$$\begin{pmatrix} a & 0 & 0 \\ 0 & b & 0 \\ 0 & 0 & c \end{pmatrix}$$

In the previous examples, the elements in the off-diagonal positions suggest coupling between the scalar equations.

In general, one could write the transpose in terms of the elements

$$(a_{ik})^T = (a_{ki})$$

Sometimes one needs to find the transpose of a matrix product as in

$$(AB)^T = B^T A^T$$

Therefore, taking the transpose of a product requires exchanging the order.

G-5 Determinants, Cofactors, and Adjoints of a Matrix

There are a variety of ways that the *determinant* of a matrix is introduced. In the simplest, usually given in high school, one goes diagonally down the matrix, computes the products of the elements and adds them up with plus signs for the forward diagonals and negative signs for the backward diagonals as in

$$\left|\begin{pmatrix} a & b & c \\ d & e & f \\ g & h & i \end{pmatrix}\right| = aei + bfg + chd - (ceg + bdi + ahf) \qquad \text{(G-7)}$$

An alternative more elegant definition uses *cofactors* which are the determinants constructed from the original matrix by deleting the *ij*th element and computing the determinant of the remainder with a plus or minus sign depending on whether the element indices add up to an even or odd number, as in

$$Cof(A)_{11} = \begin{vmatrix} e & f \\ h & i \end{vmatrix} \qquad Cof(A)_{12} = -\begin{vmatrix} d & f \\ g & i \end{vmatrix} \qquad Cof(A)_{23} = -\begin{vmatrix} a & b \\ g & h \end{vmatrix} \qquad \text{etc.}$$

Note that the cofactors are scalars.

The determinant of a matrix can then be defined as the sum of the cofactors times the deleted element from an arbitrarily chosen row or column, as in

$$|A| = \sum_{k=1}^{n} a_{ik} Cof(A)_{ik} \qquad i = 1,\ldots,n$$

where the elements of the matrix A are denoted as a_{ij} and where the above equation could also be written as

$$|A| = \sum_{i=1}^{n} a_{ik} Cof(A)_{ik} \qquad k = 1,\ldots,n$$

Finally, the *adjoint* of a matrix is constructed by placing the cofactors of the matrix in the place of the various elements of the matrix and transposing it, as in

$$A = (a_{ik})$$
$$Adj(A) = (Cof(A)_{ik})^T$$

Question G-1 Can you find the adjoint of the matrix in Eq. (G-7).

Answer

$$Adj(A) = \begin{pmatrix} ie - fh & -ib + ch & bf - ce \\ -id + fg & ia - cg & -af + cd \\ dh - eg & -ah + bg & ae - bd \end{pmatrix}$$

$$\frac{Adj(A)}{|A|} = \frac{1}{iae - afh - idb + dch + gbf - gce} \begin{pmatrix} ie - fh & -ib + ch & bf - ce \\ -id + fg & ia - cg & -af + cd \\ dh - eg & -ah + bg & ae - bd \end{pmatrix}$$

I used the symbolic toolbox in Matlab to verify that

$$A \frac{Adj(A)}{|A|} = I$$

where I represents the identity matrix; More on this in Sec. G-6.

G-6 The Inverse Matrix

Equation (G-6) is an algebraic equation in $\tilde{X}(s)$ so it would be nice if we could divide both sides of the equation by $G(s)$ and solve for $\tilde{X}(s)$. There is no rule for the division of one matrix by another but we can use the concept of the *inverse matrix* which in this case would be $[G(s)]^{-1}$ and which satisfies

$$G(s)G(s)^{-1} = G(s)^{-1}G(s) = I$$

$$I = \begin{pmatrix} 1 & 0 & 0 \\ 0 & 1 & 0 \\ 0 & 0 & 1 \end{pmatrix} \tag{G-8}$$

where I is the identity matrix, in this case for a (3, 3) matrix. Multiplying by the inverse is analogous to dividing by a factor in scalar algebra:

$$ax = 5$$
$$a^{-1}ax = a^{-1}5$$
$$x = a^{-1}5 = \frac{5}{a}$$

A formal definition of the inverse uses the adjoint matrix and the determinant, as in

$$A^{-1} = \frac{Adj(A)}{|A|} \tag{G-9}$$

Assuming for the time being that there is a way of obtaining the elements of $[G(s)]^{-1}$, the solution for $\tilde{X}(s)$ in Eq. (G-6) would be obtained by multiplying both sides of the equation *on the left* by $[G(s)]^{-1}$ as in.

$$\tilde{X}(s)=\left(\begin{pmatrix} \frac{1}{R_1} & 0 & 0 \\ -\frac{1}{R_1} & \frac{1}{R_2} & 0 \\ 0 & -\frac{1}{R_2} & \frac{1}{R_3} \end{pmatrix}+s\begin{pmatrix} \rho A_1 & 0 & 0 \\ 0 & \rho A_2 & 0 \\ 0 & 0 & \rho A_3 \end{pmatrix}\right)^{-1} B\tilde{U} \qquad \text{(G-10)}$$

The form of Eq. (G-10) is somewhat reminiscent of the first-order process:

$$\tilde{X}(s)=\frac{g}{\tau s+1}\tilde{U}=(\tau s+1)^{-1}g\tilde{U}$$

Finding the determinants and inverses of matrices is not trivial but almost every engineer has a package at her disposal that will compute them *numerically* (even Excel has an algorithm available). I used Matlab's Symbolic Toolbox to obtain the *symbolic*, rather than numeric, inverse of $G(s)$ in Eq. (G-10) which is

$$G(s)^{-1}=\begin{pmatrix} \frac{R_1}{\rho A_1 R_1 s+1} & 0 & 0 \\ \frac{R_2}{(\rho A_2 R_2 s+1)(\rho A_1 R_1 s+1)} & \frac{R_2}{\rho A_2 R_2 s+1} & 0 \\ \frac{R_3}{(\rho A_3 R_3 s+1)(\rho A_2 R_2 s+1)(\rho A_1 R_1 s+1)} & \frac{R_3}{(\rho A_3 R_3 s+1)(\rho A_2 R_2 s+1)} & \frac{R_3}{(\rho A_3 R_3 s+1)} \end{pmatrix}$$

The Matlab script to accomplish this is

```
syms R1 R2 R2 R3 s A1 A2 A3 rho A Ainv yt % declare the
variables as symbolic
A=[rho*A1*s+1/R1    0    0
   -1/R1       rho*A2*s+1/R2   0
   0    -1/R2    rho*A3*s+1/R3 ]; % define the matrix A
Ainv=inv(A)    % calculate and display the symbolic inverse
pretty(Ainv)   % display the results "prettier" form
```

The solution to Eq. (G-10) (still in the Laplace domain) for $\tilde{Y}(s)$ is obtained by multiplying the matrices:

$$\tilde{X}(s) = G(s)^{-1} B\tilde{U}$$

$$\tilde{X}(s) = \begin{pmatrix} \dfrac{R_1}{\rho A_1 R_1 s + 1} & 0 & 0 \\ \dfrac{R_2}{(\rho A_2 R_2 s + 1)(\rho A_1 R_1 s + 1)} & \dfrac{R_2}{\rho A_2 R_2 s + 1} & 0 \\ \dfrac{R_3}{(\rho A_3 R_3 s + 1)(\rho A_2 R_2 s + 1)(\rho A_1 R_1 s + 1)} & \dfrac{R_3}{(\rho A_3 R_3 s + 1)(\rho A_2 R_2 s + 1)} & \dfrac{R_3}{(\rho A_3 R_3 s + 1)} \end{pmatrix} \begin{pmatrix} 1 \\ 0 \\ 0 \end{pmatrix} \tilde{U}$$

$$= \begin{pmatrix} \dfrac{R_1 \tilde{U}}{\rho A_1 R_1 s + 1} \\ \dfrac{R_2 \tilde{U}}{(\rho A_2 R_2 s + 1)(\rho A_1 R_1 s + 1)} \\ \dfrac{R_3 \tilde{U}}{(\rho A_3 R_3 s + 1)(\rho A_2 R_2 s + 1)(\rho A_1 R_1 s + 1)} \end{pmatrix}$$

$$\tilde{Y}(s) = (0 \quad 0 \quad 1)\tilde{X}(s) = \frac{R_3 \tilde{U}(s)}{(\rho A_3 R_3 s + 1)(\rho A_2 R_2 s + 1)(\rho A_1 R_1 s + 1)}$$

which is seen to be identical to the result obtained in Chap. 5, Eq. (5-4).

As with transposes, the inverse of a matrix product is the product of the inverses in reverse order, as in

$$(AB)^{-1} = B^{-1} A^{-1}$$

Question G-2 What is the inverse of a diagonal matrix?

Answer Simply invert the nonzero diagonal elements, as in

$$\begin{pmatrix} a & 0 & 0 \\ 0 & b & 0 \\ 0 & 0 & c \end{pmatrix}^{-1} = \begin{pmatrix} \dfrac{1}{a} & 0 & 0 \\ 0 & \dfrac{1}{b} & 0 \\ 0 & 0 & \dfrac{1}{c} \end{pmatrix}$$

Verify this by multiplying the matrix by its inverse and see if you get the unity matrix.

G-7 Some Matrix Calculus

The derivative of a matrix is a matrix containing the derivatives of its elements.

$$\frac{d}{dt}\begin{pmatrix} a & b \\ c & g \end{pmatrix} = \begin{pmatrix} \frac{da}{dt} & \frac{db}{dt} \\ \frac{dc}{dt} & \frac{dg}{dt} \end{pmatrix}$$

Likewise, the integral of a matrix is a matrix containing the integrals of its elements. This means that the Laplace transform of a matrix can be the matrix of the transformed elements.

G-8 The Matrix Exponential Function and Infinite Series

Matrices can also occur as arguments of functions. For our book, the most important is the exponential.

$$A = \begin{pmatrix} a & b \\ c & g \end{pmatrix}$$

$$e^A = e^{\begin{pmatrix} a & b \\ c & g \end{pmatrix}}$$

Note that

$$e^{\begin{pmatrix} a & b \\ c & g \end{pmatrix}}$$

is a matrix of size 2 × 2. There are a variety of ways to evaluate the exponential of a matrix but we used the extension of the infinite series given in Eq. (D-4) in App. D.

$$e^A = I + A + \frac{1}{2!}AA + \frac{1}{3!}AAA + \text{h.o.t.}$$

For the size (2, 2) example, the first three terms of the infinite series would be

$$A = \begin{pmatrix} a & b \\ c & g \end{pmatrix}$$

$$e^A = e^{\begin{pmatrix} a & b \\ c & g \end{pmatrix}} = \begin{pmatrix} 1 & 0 \\ 0 & 1 \end{pmatrix} + \begin{pmatrix} a & b \\ c & g \end{pmatrix} + \frac{1}{2}\begin{pmatrix} a & b \\ c & g \end{pmatrix}\begin{pmatrix} a & b \\ c & g \end{pmatrix} + \text{h.o.t.}$$

$$= \begin{pmatrix} 1 & 0 \\ 0 & 1 \end{pmatrix} + \begin{pmatrix} a & b \\ c & g \end{pmatrix} + \frac{1}{2}\begin{pmatrix} a^2 + bc & ab + bg \\ ca + gc & cb + g^2 \end{pmatrix} + \text{h.o.t.}$$

$$= \begin{pmatrix} 1 + a + \frac{1}{2}(a^2 + bc) & b + \frac{1}{2}(ab + bg) \\ c + \frac{1}{2}(ca + gc) & 1 + g + \frac{1}{2}(cb + g^2) \end{pmatrix} + \text{h.o.t.}$$

Later on in this appendix, the Cayley-Hamilton theorem will provide a better way to evaluate the exponential of a matrix.

The derivative of and exponential with a matrix argument is similar to that in the scalar case, as in

$$\frac{d}{dt}e^{At} = Ae^{At}$$

G-9 Eigenvalues of Matrices

Although we will not use the concept in this book, a square matrix (one that has an equal number of rows and columns) is sometimes considered as an *operator* that can *rotate* a vector and produce a new vector with a different orientation. For example, in two dimensions, the following equation represents the rotation of a vector having a component only along the x-axis.

$$\begin{pmatrix} \cos\theta & -\sin\theta \\ \sin\theta & \cos\theta \end{pmatrix}\begin{pmatrix} 1 \\ 0 \end{pmatrix}$$

If the angle of rotation is 30° or $\pi/6$ radians then the rotation produces a new vector, as in

$$\begin{pmatrix} 0.8660 & -0.500 \\ 0.500 & 0.8660 \end{pmatrix}\begin{pmatrix} 1 \\ 0 \end{pmatrix} = \begin{pmatrix} 0.8660 \\ 0.5000 \end{pmatrix}$$

The "input" vector has a component along the x-axis and no component along the y-axis. The new "rotated" (or output) vector has an x-axis component of 0.8660 and a y-axis component of 0.500 and represents a vector that has a 30° angle with the x-axis along which the starting vector

$$\begin{pmatrix} 1 \\ 0 \end{pmatrix}$$

pointed. This example is a special case but, in general, the equation

$$Ax = y$$

can be considered as an example of a matrix A acting on a vector x and, as a result, rotating it into the vector y. After the rotation, the length (magnitude) and angle may have changed.

There is a special situation when the matrix A rotates an arbitrary vector x into a new vector that has the same direction, as in

$$Ax = \lambda x \tag{G-11}$$

where λ is a scalar multiplier or scaling factor indicating that the vector λx has the same direction (or the same ratio of components) as the x vector but can have a different magnitude. Hopefully, the reader can accept that the two-dimensional vector

$$\begin{pmatrix} a \\ b \end{pmatrix}$$

has the same direction as the vector

$$\lambda \begin{pmatrix} a \\ b \end{pmatrix}$$

The length of the former vector is $\sqrt{a^2+b^2}$ while the length of the latter vector is $\lambda\sqrt{a^2+b^2}$. At first blush, this idea of an operator generating something that is proportional to what it is operating on may seem nonsensical but it occurs many times in applied mathematics.

For example, if one considers d/dt as an operator, then when it operates on $e^{\lambda t}$ it generates something proportional to that which it operated on, as in

$$\frac{d}{dt}e^{\lambda t} = \lambda e^{\lambda t}$$

$$\left(\frac{d}{dt} - \lambda\right)e^{\lambda t} = 0$$

Note that $e^{j\lambda t}$ can also be an eigenvector of the d/dt operator.

Question G-3 Can you verify that $e^{j\lambda t}$ is also an eigenvector of the d/dt operator?

Answer The following should convince the reader:

$$\frac{d}{dt}e^{j\lambda t} = j\lambda e^{j\lambda t}$$

Using the identity matrix, Eq. (G-11) can be rewritten as

$$(A - \lambda I)x = 0$$

In two dimensions this equation would look like

$$\left(\begin{pmatrix} a & b \\ c & g \end{pmatrix} - \lambda\begin{pmatrix} 1 & 0 \\ 0 & 1 \end{pmatrix}\right)\begin{pmatrix} x_1 \\ x_2 \end{pmatrix} = \begin{pmatrix} 0 \\ 0 \end{pmatrix} \tag{G-12}$$

Without dredging up too much college algebra, Eq. (G-12) represents an example of a *homogeneous* system of equations. A *nonhomogeneous* version of Eq. (G-12) would be

$$\left(\begin{pmatrix} a & b \\ c & g \end{pmatrix} - \lambda \begin{pmatrix} 1 & 0 \\ 0 & 1 \end{pmatrix}\right)\begin{pmatrix} x_1 \\ x_2 \end{pmatrix} = \begin{pmatrix} h \\ r \end{pmatrix}$$

The obvious but trivial solution to Eq. (G-12) is

$$\begin{pmatrix} x_1 \\ x_2 \end{pmatrix} = \begin{pmatrix} 0 \\ 0 \end{pmatrix}$$

To have a nonzero solution to Eq. (G-12), the determinant of the coefficient matrix must equal zero (hopefully you remember this from high school or college algebra). That is, we need to find values of λ such that the following equation is satisfied

$$|A - \lambda I| = \left|\begin{pmatrix} a & b \\ c & g \end{pmatrix} - \lambda \begin{pmatrix} 1 & 0 \\ 0 & 1 \end{pmatrix}\right| = \left|\begin{pmatrix} a-\lambda & b \\ c & g-\lambda \end{pmatrix}\right| = 0 \qquad \text{(G-13)}$$

When the determinant in Eq. (G-13) is expanded, the following quadratic equation in λ is generated

$$\lambda^2 - (a+g)\lambda + ag - cb = 0$$

The two roots, λ_1 and λ_2, of this quadratic equation are called the *eigenvalues* of the matrix A. For each of these eigenvalues there will be a size (2, 1) vector satisfying Eq. (G-12). These two vectors, V_1 and V_2, are called the *eigenvectors* of the matrix A and they satisfy the following equations.

$$(A - \lambda_1 I)V_1 = 0$$

$$(A - \lambda_2 I)V_2 = 0$$

However, there is a problem in that the two equations generated by plugging the eigenvalue λ_1 into Eq. (G-12) are not independent. Therefore, the eigenvector V_1 associated with λ_1 will have a definite direction but an arbitrary magnitude. Usually, one makes the magnitude unity.

Consider the following numerical example. Find the eigenvalues and eigenvectors of the matrix

$$\begin{pmatrix} 1 & 2 \\ 3 & 5 \end{pmatrix}$$

The defining equation and the associated algebra for determining the eigenvalues and eigenvectors are given in the following

$$\left(\begin{pmatrix}1 & 2\\3 & 5\end{pmatrix} - \lambda\begin{pmatrix}1 & 0\\0 & 1\end{pmatrix}\right)\begin{pmatrix}v_1\\v_2\end{pmatrix} = \begin{pmatrix}0\\0\end{pmatrix}$$

$$\left|\begin{pmatrix}1 & 2\\3 & 5\end{pmatrix} - \lambda\begin{pmatrix}1 & 0\\0 & 1\end{pmatrix}\right| = 0$$

$$\left|\begin{pmatrix}1-\lambda & 2\\3 & 5-\lambda\end{pmatrix}\right| = 0$$

$$\lambda^2 - 6\lambda - 1 = 0$$

$$\lambda_1 = -0.1623 \qquad V_1 = \begin{pmatrix}-0.8646\\0.5025\end{pmatrix}$$

$$\lambda_2 = 6.1623 \qquad V_2 = \begin{pmatrix}-0.3613\\-0.9325\end{pmatrix} \tag{G-14}$$

where V_1 and V_2 are the two eigenvectors, the numerical origin of which will be explained shortly. These eigenvectors have been normalized to have unit magnitude.

Now, to the origin of these eigenvectors. When λ_1 is plugged into the first line of Eq. (G-14) one gets

$$(A - \lambda_1 I)V_1$$

or

$$\left(\begin{pmatrix}1 & 2\\3 & 5\end{pmatrix} + 0.1623\begin{pmatrix}1 & 0\\0 & 1\end{pmatrix}\right)\begin{pmatrix}v_{11}\\v_{21}\end{pmatrix} = \begin{pmatrix}0\\0\end{pmatrix}$$

or

$$\begin{pmatrix}1.1623 & 2.0\\3 & 5.1623\end{pmatrix}\begin{pmatrix}v_{11}\\v_{21}\end{pmatrix} = \begin{pmatrix}0\\0\end{pmatrix}$$

or

$$1.1623v_{11} + 2.0v_{21} = 0$$
$$3v_{11} + 5.1623v_{21} = 0$$

This represents two equations in two unknowns. If one were to multiply the first of the two equations by 2.4270 one would get the second equation. This is supported by the fact that the determinant of the coefficient matrix is zero

$$\left|\begin{pmatrix}1.1623 & 2.0\\3 & 5.1623\end{pmatrix}\right| = 0$$

This last equation basically says that the preceding two equations in v_{11} and v_{21} are not independent so we can only solve for v_{11} in terms of v_{21}. This is why we can find the direction of the eigenvectors but the magnitude is arbitrary—in fact, in this case, we arbitrarily choose the magnitude to be 1.0. For the sake of completeness, take one of the preceding two equations and solve for one of the components in terms of the other:

$$v_{11} = -\frac{2.0}{1.1623}v_{21} = -1.7207v_{21}$$

$$\begin{pmatrix} v_{11} \\ v_{21} \end{pmatrix} = \begin{pmatrix} -1.7207 \\ 1 \end{pmatrix} v_{21}$$

This is as far as we can go in getting numbers. To make this vector have unit magnitude we choose v_{21} to be the reciprocal of the magnitude of the vector, as follows.

$$v_{21} = \frac{1}{\sqrt{1.7207^2 + 1^2}} = \frac{1}{1.9902}$$

Now, the eigenvector has unit magnitude, as in

$$\begin{pmatrix} v_{11} \\ v_{21} \end{pmatrix} = \begin{pmatrix} -1.7207 \\ 1 \end{pmatrix} \frac{1}{1.9902} = \begin{pmatrix} -0.8646 \\ 0.5025 \end{pmatrix}$$

Note that the eigenvectors have been scaled to have unit magnitude, such that

$$V_1 V_1^T = I$$

You might want to multiply this out to confirm my contention.

G-10 Eigenvalues of Transposes

If the readers have access to Matlab, they should verify that the eigenvalues of a matrix A are the same as those of the transpose A^T.

G-11 More on Operators

At the risk of beating this eigenvalue thing to death, consider the operator

$$\frac{d^2}{dt^2}$$

The eigenvalue and eigenvector of this operator could be obtained from

$$\left(\frac{d^2}{dt^2}\right)y = \lambda\, y$$

as with many differential equations, let's try a solution of $y = Ce^{at}$ and see what happens.

$$\left(\frac{d^2}{dt^2}\right)Ce^{at} = Ca^2e^{\lambda t} = \lambda Ce^{\lambda t}$$

$$\lambda = a^2$$

Question G-4 What is the eigenvector result if a trial solution of e^{jat} is used for the d^2 / dt^2 operator?

Answer

$$\frac{d^2}{dt^2}e^{jat} = (ja)^2 e^{jat} = -a^2 e^{jat} = \lambda e^{jat}$$

So, e^{jat} is an eigenvector.

These last few paragraphs have probably been relatively painful but there is a reason for my madness and it has to do with solving the state-space equation for our process which is dealt with in App. H.

G-12 The Cayley-Hamilton Theorem

In general, the following equation that is used to solve for the eigenvalues, namely

$$|A - \lambda I| = 0 \tag{G-15}$$

is called the *characteristic equation* for the size (n, n) matrix A. In the case for the two-dimensional system, a polynomial of order two resulted. In general, a polynomial of order n in λ is generated, as in

$$\left|\begin{pmatrix} a_{11} & a_{12} & \cdots & a_{1n} \\ a_{21} & a_{22} & \cdots & a_{2n} \\ \cdots & \cdots & \cdots & \cdots \\ a_{n1} & a_{n2} & \cdots & a_{nn} \end{pmatrix} - \lambda \begin{pmatrix} 1 & 0 & \cdots & 0 \\ 0 & 1 & \cdots & 0 \\ \cdots & \cdots & \cdots & \cdots \\ 0 & 0 & \cdots & 1 \end{pmatrix}\right| = 0 \tag{G-16}$$

It is left to the reader to multiply Eq. (G-16) out and convince himself that this determinant generates an nth-order polynomial in λ that has n roots which are the eigenvalues of the matrix A.

The Cayley-Hamilton theorem says that each square matrix satisfies its own characteristic equation. Our two-dimensional example, introduced in Sec. G-9, gives

$$A = \begin{pmatrix} 1 & 2 \\ 3 & 5 \end{pmatrix}$$

$$(A - \lambda I) = 0$$

$$\left(\begin{pmatrix} 1 & 2 \\ 3 & 5 \end{pmatrix} - \lambda \begin{pmatrix} 1 & 0 \\ 0 & 1 \end{pmatrix} \right) \begin{pmatrix} v_1 \\ v_2 \end{pmatrix} = \begin{pmatrix} 0 \\ 0 \end{pmatrix}$$

$$\left| \begin{pmatrix} 1 & 2 \\ 3 & 5 \end{pmatrix} - \lambda \begin{pmatrix} 1 & 0 \\ 0 & 1 \end{pmatrix} \right| = 0$$

$$\left| \begin{pmatrix} 1-\lambda & 2 \\ 3 & 5-\lambda \end{pmatrix} \right| = 0$$

$$\lambda^2 - 6\lambda - 1 = 0$$

If A satisfies its characteristic equation, then

$$A^2 - 6A - I = 0? \tag{G-17}$$

To test this contention, plug the matrix A into the characteristic equation and get

$$\begin{pmatrix} 1 & 2 \\ 3 & 5 \end{pmatrix}\begin{pmatrix} 1 & 2 \\ 3 & 5 \end{pmatrix} - 6\begin{pmatrix} 1 & 2 \\ 3 & 5 \end{pmatrix} - 1\begin{pmatrix} 1 & 0 \\ 0 & 1 \end{pmatrix}$$

$$= \begin{pmatrix} 7 & 12 \\ 18 & 31 \end{pmatrix} - \begin{pmatrix} 6 & 12 \\ 18 & 30 \end{pmatrix} - \begin{pmatrix} 1 & 0 \\ 0 & 1 \end{pmatrix}$$

$$= \begin{pmatrix} 0 & 0 \\ 0 & 0 \end{pmatrix}$$

Another way of writing Eq. (G-17) is

$$A^2 - 6A - I = 0$$

or

$$A^2 = 6A + I \tag{G-18}$$

which suggests that higher powers of this (2, 2) matrix can be written in terms of A and the identity matrix I. For example,

$$\begin{aligned} A^3 &= AA^2 = A(6A+I) = 6A^2 + A \\ &= 6(6A+I) + A \\ &= 37A + 6I \end{aligned}$$

and

$$\begin{aligned} A^4 &= AA^3 = A(37A+6I) = 37A^2 + 6A \\ &= 37(6A+I) + 6A \\ &= 222A + 37I \end{aligned}$$

This is quite important because in control engineering there is a frequent need to deal with functions of a matrix like e^A. For this case, we know that

$$e^A = I + A + \frac{1}{2!}AA + \frac{1}{3!}AAA + \frac{1}{4!}A^4 + \cdots$$

This looks a bit awesome at first but from the Cayley-Hamilton theorem and Eq. (G-18) we know that all the powers of the (2, 2) matrix A can be condensed into a sum of two terms. It follows that e^A can be written as

$$e^A = h_0 I + h_1 A \tag{G-19}$$

Although I will not prove it here, a consequence of the Cayley Hamilton theorem is that Eq. (G-19) is also satisfied by the eigenvalues of A, as in

$$\begin{aligned} e^{\lambda_1} &= h_0 I + h_1 \lambda_1 \\ e^{\lambda_2} &= h_0 I + h_1 \lambda_2 \end{aligned} \tag{G-20}$$

These two equations can be solved for h_0 and h_1.

In general, functions of the (n, n) matrix A can be condensed into a sum of n terms containing up to the $n-1$th power of the matrix A. Hopefully, you will agree that the Cayley-Hamilton theorem is quite handy. In App. H, the Cayley-Hamilton theorem is applied to the development of the discrete time state-space equations.

G-13 Summary

Like the Laplace transform, matrices frequently occur in control engineering and are quite useful for condensing the mathematics, especially in the state-space approach. This appendix has covered only a few features of matrices such as the transpose, the inverse, the Cayley-Hamilton theory, and eigenvalues, which are used elsewhere in the book.

APPENDIX H

Solving the State-Space Equation

This appendix will address the solution of the state-space equations in two ways. The first, for the special case of a constant input, attempts to show the parallels between the matrix approach and the scalar approach in App. E and Chap. 3. Although it may be considered an interesting example of applying the matrix tricks presented in App. G, it can get quite involved so you may want to scan it first before diving in. The second approach uses the integrating factor, is more general, and may be a little easier to follow.

H-1 Solving the State-Space Equation in the Time Domain for a Constant Input

In Chap. 5 we introduced the state-space formulation, namely,

$$\frac{d}{dt}X = AX + BU$$

$$Y = CX$$

As with our scalar first-order differential equation, we break the state-space solution up into two parts: the homogeneous part and the particular part:

$$X = X_h + X_p$$

To make life simple, assume that the input is a constant vector U_c of size $(p, 1)$. This means that the vector B must have size (n, p). Start with the particular part of the state-space equation because in this case it is easier. Try a solution X_p, a size $(n, 1)$ vector which is constant since the input is a constant, as in

$$\frac{d}{dt}X_p = 0 = AX_p + BU_c$$

$$(n,\not{n})(\not{n},1)(n,\not{p})(\not{p},1)$$

$$0 = AX_p + BU_c$$

$$AX_p = -BU_c$$

$$(n,1)(n,\not{n})(\not{n},\not{p})(\not{p},1)$$

$$X_p = -A^{-1}BU_c$$

where we have used the inverse of the square, size (n, n) A matrix which we assume exists. This requirement will be satisfied if the determinant of A is nonzero.

The homogeneous part of the solution X_h comes next and it satisfies

$$\frac{d}{dt}X_h = AX_h$$

where the size of X_h is $(n, 1)$ and that of A is (n, n). Try a solution with an exponential form

$$X_h = Ce^{\lambda t}$$

where C is an unknown column vector of size $(n, 1)$ and λ is a scalar. The result of trying this solution is

$$\begin{aligned} C\lambda e^{\lambda t} &= ACe^{\lambda t} \\ (A - \lambda I)C &= 0 \end{aligned} \tag{H-1}$$

Note that both sides of Eq. (H-1) have been divided by $e^{\lambda t}$ because it is a scalar. However, we can not "divide" both sides of Eq. (H-1) by C because C is a column vector and does not have an inverse.

The solution of Eq. (H-1) will yield the eigenvalues of A, namely $\lambda_1, \lambda_2, \lambda_3, \ldots, \lambda_n$ and the eigenvectors of A, namely $C_1, C_2, C_3, \ldots, C_n$. As pointed out in App. G, the directions of these eigenvectors can be found, but the magnitudes are not determinable. The solution to the homogeneous differential equation therefore is a sum of the n exponential terms weighted by the eigenvectors, as in

$$
\begin{aligned}
X_h &= C_1 e^{\lambda_1 t} + C_2 e^{\lambda_2 t} + \cdots + C_n e^{\lambda_n t} \\
&= \sum_{k=1}^{n} C_k e^{\lambda_k t} \\
&= [C_1 \quad C_2 \quad \ldots \quad C_n] \begin{pmatrix} e^{\lambda_1 t} \\ e^{\lambda_2 t} \\ \ldots \\ e^{\lambda_N t} \end{pmatrix}
\end{aligned} \tag{H-2}
$$

Hopefully the reader can see parallels with the approach to solving first- and second-order differential equations presented earlier in App. E. You may want to think twice about continuing this section.

In Eq. (H-2) the size $(n, 1)$ X_h vector is the dot product between a row matrix containing the size $(n, 1)$ column vectors C_i, $i = 1, n$ and a size $(n, 1)$ column vector containing the scalar exponentials. This is a valid matrix multiplication because the number of columns in the left-hand row matrix (whose elements are column vectors) is n and the number of columns in the right-hand column vector, whose elements are scalars, is also n. Equation (H-2) is a n-dimensional extension of the kind of equations we used when solving the second-order differential equation.

Perhaps the reader remembers that only the directions of the eigenvectors are known so that we could write them as $C_i = b_i B_i$ where B_i is a known size $(n, 1)$ vector with a magnitude of unity that contains the direction of C_i and b_i is the unknown scalar magnitude of the ith eigenvector.

This means that Eq. (H-2) could be rewritten as

$$
\begin{aligned}
X_h &= b_1 B_1 e^{\lambda_1 t} + b_2 B_2 e^{\lambda_2 t} + \cdots + b_n B_n e^{\lambda_n t} \\
&= \sum_{k=1}^{n} b_k B_k e^{\lambda_k t} \\
&= [B_1 \quad B_1 \quad \ldots \quad B_n] \begin{pmatrix} b_1 e^{\lambda_1 t} \\ b_2 e^{\lambda_2 t} \\ \ldots \\ b_n e^{\lambda_N t} \end{pmatrix}
\end{aligned} \tag{H-3}
$$

These last equations could be a bit off-putting. The reader is encouraged to wade through them. Remember that the leftmost row vector on the right-hand side of Eq. (H-3), namely,

$$[B_1 \quad B_1 \quad \ldots \quad B_n]$$

actually contains n size $(n, 1)$ vectors and therefore can be considered of size (n, n) or if we look just at the elements of the matrix then it is a row matrix of size $(1, n)$. That is, it can be considered a square matrix which we will denote as $\hat{B}$ so as to keep it from being confused with the input matrix B.

$$\hat{B} = [B_1 \quad B_1 \quad \dots \quad B_n]$$

As with the first-order differential equation that we solved in Chap. 3, we need to combine the particular solution with the homogeneous solution before we can apply the initial conditions and find the values of the b_i's. The sum of the two parts of the solution is

$$\begin{aligned} X &= X_h + X_p \\ &= \hat{B}\begin{pmatrix} b_1 e^{\lambda_1 t} \\ b_2 e^{\lambda_2 t} \\ \dots \\ b_n e^{\lambda_N t} \end{pmatrix} - A^{-1}BU_c \end{aligned} \tag{H-4}$$

Assume that the initial value of the X vector is known as X_0 which is also a size $(n, 1)$ vector. Then, applying this condition to Eq. (H-4), remembering that the elements in the vector of exponentials are all unity when $t = 0$, gives

$$X_0 = \hat{B}\begin{pmatrix} b_1 \\ b_2 \\ \dots \\ b_n \end{pmatrix} - A^{-1}BU_c \tag{H-5}$$

The only unknown in Eq. (H-5) is the column vector containing the b_i's. To solve for that column vector we multiply both sides of Eq. (H-5) by the inverse of $\hat{B}$.

$$\hat{B}^{-1}X_0 = \hat{B}^{-1}\hat{B}\begin{pmatrix} b_1 \\ b_2 \\ \dots \\ b_n \end{pmatrix} - \hat{B}^{-1}A^{-1}BU_c \tag{H-6}$$

Since $\hat{B}^{-1}\hat{B} = I$, Eq. (H-6) can be solved for the column vector containing the b_i's.

$$\begin{pmatrix} b_1 \\ b_2 \\ \dots \\ b_n \end{pmatrix} = \hat{B}^{-1}X_0 + \hat{B}^{-1}A^{-1}BU_c$$

Therefore, the solution to the state-space equation for the special case where the input is a constant (as in a step change at time zero) is

$$x = -A^{-1}BU_c + \hat{B}\begin{pmatrix} b_1 e^{\lambda_1 t} \\ b_2 e^{\lambda_2 t} \\ \cdots \\ b_n e^{\lambda_N t} \end{pmatrix} \tag{H-7}$$

For the sake of comparison to what we did with the first-order process model in the book, we could try to expand Eq. (H-7). However, it would get us unnecessarily deeper into matrix manipulations so we will pass on to the next section which is a little cleaner and more general. The reader might want to redo Eqs. (H-1) through (H-7) with $n=1$ and $p=1$ and see how those equations collapse into the equations derived early for one dimension.

After all the considerable matrix dust has settled, the reader should see that (1) the solution of the state-space equation depends on the eigenvalues and eigenvectors of the A matrix and that (2) the eigenvalues, which end up as the arguments of exponentials, should all have negative real parts if *stability* is to be obtained. The reader had to navigate through some serious matrix manipulations that should stand him in good stead for the other parts of this book...or not.

Question H-1 Can you check that Eq. (H-7) is a multidimensional version of the scalar equation that we derived earlier, namely, Eq. (E-6)?

Answer Make the following pairings

$$\frac{d}{dt}y = -\frac{1}{\tau}y + \frac{g}{\tau}u \qquad \frac{d}{dt}x = Ax + Bu$$

$$y_h = Ce^{at} \Leftrightarrow x_h = Ce^{\lambda t}$$

$$(\tau a + 1)Ce^{at} = 0 \Leftrightarrow (A - \lambda I)Ce^{\lambda t} = 0$$

$$a = -\frac{1}{\tau} \Leftrightarrow |A - \lambda I| = 0$$

$$1 \Leftrightarrow n \qquad 1 \Leftrightarrow p$$

$$y = Ce^{-\frac{t}{\tau}} \Leftrightarrow x = \sum_{k=1}^{n} b_k B_k e^{\lambda_k t}$$

$$C \Leftrightarrow (b_1 B_1 \quad b_1 B_1 \quad \ldots \quad b_n B_n)$$

$$C = Y_0 - gU_c \Leftrightarrow \begin{pmatrix} b_1 \\ b_2 \\ \cdots \\ b_n \end{pmatrix} = \hat{B}^{-1}X_0 + \hat{B}^{-1}A^{-1}BU_c$$

$$Y = Y_0 e^{-\frac{t}{\tau}} + gU_c(1 - e^{-\frac{t}{\tau}}) \Leftrightarrow x = -A^{-1}BU_c + \hat{B}\begin{pmatrix} b_1 e^{\lambda_1 t} \\ b_2 e^{\lambda_2 t} \\ \cdots \\ b_n e^{\lambda_N t} \end{pmatrix}$$

So, the manipulations for the simple first-order differential equation are in many ways parallel to those for the multidimensional state space. It is worthwhile to wade through the n, p dimensional approach and collapse it to one dimension.

Question H-2 What is a difference between the unity matrix

$$I = \begin{pmatrix} 1 & 0 & \ldots & 0 \\ 0 & 1 & \ldots & 0 \\ \ldots & \ldots & \ldots & \ldots \\ 0 & 0 & \ldots & 1 \end{pmatrix}$$

and a column vector of ones

$$\begin{pmatrix} 1 \\ 1 \\ \ldots \\ 1 \end{pmatrix}?$$

Answer The unity matrix multiplies any matrix or vector of correct dimension and produces that same matrix, as in

$$IA = A = AI \qquad \text{or} \qquad Ix = x = xI$$

However, a column of ones generates a new vector as in

$$\begin{pmatrix} a & b \\ c & d \end{pmatrix}\begin{pmatrix} 1 \\ 1 \end{pmatrix} = \begin{pmatrix} a+b \\ c+d \end{pmatrix}$$

or

$$\begin{pmatrix} 1 \\ 1 \end{pmatrix}\lfloor a \quad b \rfloor = \begin{pmatrix} a & b \\ a & b \end{pmatrix}$$

This subsection has been tough sledding and relatively unimportant. The next subsection on the integrating factor is more important and easier to follow.

H-2 Solution of the State-Space Equation Using the Integrating Factor

The matrix approach in the time domain has some interesting and perhaps elegant features but the reader can see that if the process input were nonconstant then the approach given in Sec. H-1 would be insufficient. However, it is possible to extend the integrating factor that was used for the first-order differential equation (in App. F) into n dimensions.

The state-space differential equation can be written as

$$\frac{d}{dt}x - Ax = BU \tag{H-8}$$

Upon applying an integrating factor of e^{-At}, which is a matrix of size (n, n), Eq. (H-8) becomes

$$e^{-At}\frac{d}{dt}x - Axe^{-At} = e^{-At}Bu$$

The left-hand side is an exact differential, as in

$$e^{-At}\frac{d}{dt}x - Axe^{-At} = \frac{d}{dt}(e^{-At}x) = e^{-At}BU$$

Therefore, the following manipulations should be relatively straightforward.

First, change the differential equation slightly to enable integration

$$d(e^{-At}x) = e^{-At}BUdt$$

$$\int_0^t d(e^{-At}x) = \int_0^t e^{-Av}BU(v)dv$$

Carry out the integration on the left-hand side by observation

$$e^{-At}x - x_0 = \int_0^t e^{-Av}BU(v)dv$$

Multiply both sides by e^{At} which is a square matrix with the same dimensions as the matrix A.

$$\begin{aligned} x &= e^{At}x_0 + \int_0^t e^{At}e^{-Av}BU(v)dv \\ x &= e^{At}x_0 + \int_0^t e^{A(t-v)}BU(v)dv \end{aligned} \tag{H-9}$$

Note that the inverse of the integrating factor is straightforward.

$$(e^{-At})^{-1} = e^{At}$$
$$e^{-At}e^{At} = I$$

Everything here is in parallel with the development of Eq. (F-43) in App. F which was for the first-order scalar process model. There are, however, a couple of interesting differences. First, the integrating factor is a matrix e^{-At} of size (n, n). Second, we will give the inverse of the integrating factor a name, namely, the *fundamental* matrix (also sometimes called the *transition* matrix), $\Phi(t) = e^{At}$, so Eq. (H-9) becomes

$$x = \Phi(t)x_0 + \int_0^t \Phi(t-v)BU(v)dv \tag{H-10}$$

The fundamental matrix is analogous to the impulse response in the convolution integral of Eq. (F-44). Now, if the process input U is variable, in principle, the process response can be found using Eq. (H-10).

H-3 Solving the State-Space Equation in the Laplace Transform Domain

You must have been wondering how long it would take me to conjure up the Laplace transform, once we got cracking with the state-space equations. The procedure is relatively straightforward as shown in the following.

$$\frac{d}{dt}x = Ax + BU$$
$$s\tilde{x}(s) - x(0) = A\tilde{x}(s) + B\tilde{U}(s)$$
$$(sI - A)\tilde{x}(s) = x(0) + B\tilde{U}(s)$$
$$\tilde{x}(s) = (sI - A)^{-1}x(0) + (sI - A)^{-1}B\tilde{U}(s)$$

The inverse Laplace transform gives $x(t)$

$$x(t) = L^{-1}\{(sI - A)^{-1}x(0)\} + L^{-1}\{(sI - A)^{-1}B\tilde{U}(s)\} \tag{H-11}$$

From Eq. (H-9) we had

$$x = e^{At}x_0 + \int_0^t e^{A(t-v)}BU(v)dv$$

so, comparing the two equations suggests the following

$$L^{-1}\{(sI - A)^{-1}x(0)\} = e^{At}x_0$$

and

$$L^{-1}\{(sI-A)^{-1}B\tilde{U}(s)\} = \int_0^t e^{A(t-v)}BU(v)dv$$

This further suggests that

$$L^{-1}\{(sI-A)^{-1}\} = e^{At} \tag{H-12}$$

From App. G, the inverse of a matrix is given by

$$A^{-1} = \frac{Adj(A)}{|A|}$$

Therefore, Eq. (H-12) can be written

$$L^{-1}\{(sI-A)^{-1}\} = e^{At} = \frac{Adj(sI-A)}{|sI-A|} \tag{H-13}$$

The inverse is the matrix $Adj(sI-A)$ divided by the scalar determinant $|sI-A|$.

This is analogous to inverting the scalar Laplace transforms in App. F where the roots of the denominator of the Laplace transform in question (the poles of the Laplace transform in question) were directly related to exponential terms. Also, the real part of those roots had to be in the left-hand side of the s-plane for there to be stability.

The determinant $|sI-A|$ generates a polynomial in s and the roots of that polynomial are the eigenvalues of the matrix A. These roots, s_1, $s_2,\ldots,s_n$, correspond to e^{s_it} terms in the time domain solution for x.

H-4 The Discrete Time State-Space Equation

In Chap. 9, a transition from the continuous time domain to the discrete time domain was made via step-change inputs to the scalar first-order model.

$$\tau\frac{dy}{dt} + y = gU(t)$$

$$y(h) = y_0 e^{-\frac{h}{\tau}} + gU_h\left(1-e^{-\frac{h}{\tau}}\right)$$

$$y_i = y_{i-1}e^{-\frac{h}{\tau}} + gU_{i-1}\left(1-e^{-\frac{h}{\tau}}\right) \qquad i = 0,1,2,\ldots$$

Earlier in this Appendix, the continuous time state-space equation was solved as follows:

$$\frac{d}{dt}X = AX + BU$$
$$Y = CX \qquad \text{(H-14)}$$
$$x(t) = e^{At}x_0 + \int_0^t e^{A(t-v)}BU(v)dv$$

When the process input is constant at the value U_i over a time interval of size h between t_i and t_{i+1} Eq. (H-14) can be written as

$$x_{i+1} = e^{Ah}x_i + \int_{t_i}^{t_{i+1}} e^{A(t-v)}BU_i\,dv$$
$$= e^{Ah}x_i - A^{-1}(I - e^{Ah})BU_i \qquad \text{(H-15)}$$

where the order of the multiplication of the matrices is important. For the sake of brevity and ease of bookkeeping, Eq. (H-15) will be written as

$$x_{i+1} = e^{Ah}x_i + A^{-1}(e^{Ah} - I)BU_i$$
$$x_{i+1} = \Phi x_i + \Gamma U_i \qquad \text{(H-16)}$$
$$\Phi = e^{Ah} \qquad \Gamma = A^{-1}(e^{Ah} - I)B$$

Equation (H-16) shows that in the multidimensional case, the state and process input satisfy a simple discrete time indexed matrix equation.

For the case of the underdamped process covered in Chap. 6, we have in the continuous time domain

$$\frac{d}{dt}\begin{pmatrix} x_1 \\ x_2 \end{pmatrix} = \begin{pmatrix} 0 & 1 \\ -\omega_n^2 & -2\zeta\omega_n \end{pmatrix}\begin{pmatrix} x_1 \\ x_2 \end{pmatrix} + \begin{pmatrix} 0 \\ g\omega_n^2 \end{pmatrix}U$$
$$A = \begin{pmatrix} 0 & 1 \\ -\omega_n^2 & -2\zeta\omega_n \end{pmatrix} \qquad \text{(H-17)}$$
$$B = \begin{pmatrix} 0 \\ g\omega_n^2 \end{pmatrix}$$
$$y = (1 \quad 0)\begin{pmatrix} x_1 \\ x_2 \end{pmatrix} \qquad H = (1 \quad 0)$$

By combining knowledge of the A and B matrices from Eq. (H-17) with knowledge of the Cayley-Hamilton theorem in App. G, one could compute numerical values for the size (2, 2) Φ and Γ matrices relatively straightforwardly. For example, with $\zeta = 0.1$, $h = 0.5$, and the other parameters scaled to unity, one can calculate the following

$$\Phi = e^{Ah} = \begin{pmatrix} 0.88154 & 0.45624 \\ -0.45624 & 0.79029 \end{pmatrix}$$

$$\Gamma = A^{-1}(I - e^{Ah})B = \begin{pmatrix} 0.11845 \\ 0.45624 \end{pmatrix}$$

At this point, one is ready to use the discrete time form of the state-space equation for the underdamped process.

If you or your control engineer associate is familiar with the software tool, Matlab, then the above can be done quite quickly with the following Matlab script:

```
zeta=.7;
omega=1;
h=1.;
g=1;
A=[0 1;-omega^2 -2*zeta*omega];
B=[0;g*omega^2];
GsysD = c2d(ss(A,B,[1 0],0),h,'zoh');
[AD,BD,CD,DD] = ssdata(GsysD);
AD
BD
```

The built-in functions `c2d` and `ssdata` do the calculations seamlessly. The last two lines simply ask for the numerical values of the matrices `AD` and `BD` to be displayed. As with many mature software packages, there are several ways to carry out the same calculation other than that shown in the above script.

H-5 Summary

The continuous time state-space equation was solved in two ways. The first approach, for the case of a constant input, showed parallels with the approach for the scalar first-order model. The second approach used the integrating factor and was both more general and easier to develop.

The discrete time state-space equation was developed from the continuous time domain equations in a manner similar to that done in Chap. 9. A simple Matlab script showed how one might generate numerical values for the discrete time version if one has information about the continuous time version.

APPENDIX I

The Z-Transform

In Chap. 9 we uncovered the Z-transform by starting with the time domain solution of the first-order model in response to a step change in the process input. The process input was divided into a series of steps of different value separated by a constant time interval h. The time domain solution for the process output was appropriately modified. The back shift operator z^{-1} was introduced and the time domain solution was converted to a Z-transform. This "backdoor" approach is similar to that in Chap. 3 (and App. F) where we used the Heaviside operator p (or s) to replace derivatives and generate algebraic equations; a technique that led to the Laplace transform.

The act of breaking the process input into a series of contiguous steps of constant value separated in time by a constant interval is, in effect, the same as repeatedly sampling the process input every h seconds and holding it at that sampled value over the subsequent interval lasting h seconds. To develop a rationale for the Z-transform, the sampling and the holding process have to be quantified. In this appendix the Laplace transform will provide a starting place.

I-1 The Sampling Process and the Laplace Transform of a Sampler

Sampling a continuous signal at a constant time interval can be accomplished by modulating the signal with an infinite sequence of pulses. If the continuous signal to be sampled is $y(t)$, then the sampling is equivalent to multiplying the signal by an infinite string of Dirac deltas (see App. F), as in

$$\begin{aligned} y^*(t) &= \sum_{k=0}^{\infty} y(t)\delta(t-kh) \\ &= y(t)\delta(t) + y(t)\delta(t-h) + y(t)\delta(t-2h) + \cdots \end{aligned} \tag{I-1}$$

where $y^*(t)$ denotes the sampled signal. As mentioned in App. F, the Dirac delta $\delta(t)$ is defined as a pulse

$$\delta(t) = 0 \qquad \text{for } t \neq 0$$

undefined for $t = 0$

but
$$\int_0^\infty dt\,\delta(t) = 1$$

The last expression with the integral specifies that the Dirac delta has unit area. The definition says nothing about the height or width of the pulse. Also, as part of its definition, the Dirac delta can "pluck" the integrand from an integral, as in

$$\int_0^\infty dt\, f(t)\delta(t-\tau) = f(\tau)$$

This definition suggests that Eq. (I-1) might yield the following values:

$$\begin{aligned} y^*(h/2) &= \sum_{k=0}^\infty y(h/2)\delta(h/2 - kh) = 0 \\ y^*(2h) &= \sum_{k=0}^\infty y(2h)\delta(2h - kh) = y(2h)\delta(0) =^? y(2h) \end{aligned} \tag{I-2}$$

Equation (I-2) for $y^*(2h)$ is a little bit shaky (hence the ? mark) because of the rather nebulous definition of the Dirac delta function. Nowhere in the definition of the Dirac delta do we specify that $\delta(0) = 1$ which Eq. (I-2) implies. We will not pursue this here; consider it a mathematical slight of hand that many respected authors tend to gloss over and let it go. Instead, we will move immediately to the Laplace s-domain where the discomfort in Eq. (I-2) will perhaps be ameliorated.

The Laplace transform of the sampled signal can be written as

$$\begin{aligned} L\{y^*(t)\} = \tilde{y}^*(s) &= \int_0^\infty dt e^{-st} y^*(t) \\ &= \int_0^\infty dt e^{-st} \sum_{k=0}^\infty y(t)\delta(t-kh) \\ &= \sum_{k=0}^\infty \int_0^\infty dt e^{-st} y(t)\delta(t-kh) \\ &= \sum_{k=0}^\infty y(kh) e^{-skh} \end{aligned} \tag{I-3}$$

In the third line of Eq. (I-3) the order of the summation and the integration is exchanged. In going from line three to line four, the "plucking" feature of the Dirac delta function has been applied (does this truly make Eq. (I-2) more bearable?). Frankly, you might have to look at Eq. (I-2) as an artificial starting point, chosen because it leads to a useful result.

The simple change of variable $z = e^{sh}$ converts Eq. (I-3) into

$$\sum_{k=0}^{\infty} y(kh)z^{-k}$$

which is the Z-transform of $y(t)$ or $\hat{y}(z)$, as in

$$Z\{y(t)\} = \hat{y}(z) = \sum_{k=0}^{\infty} y(kh)z^{-k} \tag{I-4}$$

Therefore, the Z-transform is a somewhat cunning result of applying the Laplace transform to a sampled signal (a signal modulated by an infinite train of impulses). By the way, remember that change of variable, $z = e^{sh}$; we will refer to it later in this appendix.

I-2 The Zero-Order Hold

Chapter 9 introduced the zero-order hold by modifying the time domain solution of the first-order process model when the process input is a contiguous series of steps. The backshift operator was inserted into the modified equation and the result was called a Z-transform. We need to quantify this operation by finding the Laplace transform transfer function of the zero-order hold. The term "transfer function" is used because the zero-order hold operates on an input and generates an output. In App. F the step response of a process described by the transfer function $G(s)$ was shown to be

$$L^{-1}\left\{\frac{G(s)}{s}\right\}$$

where $1/s$ represents the Laplace transform of a unit step at time zero. Likewise, the impulse response of a process described by $G(s)$ was shown to be

$$L^{-1}\{G(s)\,1\} = L^{-1}\{G(s)\}$$

where 1 represents the Laplace transform of a unit impulse at time zero.

Therefore, the transfer function of the zero-order hold will be developed by starting with its impulse response in the time domain and working backward. If the input is a unit pulse at time zero, then the zero hold should generate an output consisting of a step that lasts for h seconds during the interval $0 \le t < h$, as in

$$\Pi_h(t) = \hat{U}(t) - \hat{U}(t-h) \tag{I-5}$$

where $\hat{U}(t)$ is the unit step function at time zero and $\Pi_h(t)$ is the symbol denoting the zero-order hold having an interval of h. The Laplace transform of $\Pi_h(t)$ can be obtained from Eq. (I-5) by taking the Laplace transform of $\hat{U}(t)$ and subtracting the transform of $\hat{U}(t-h)$. Referring to App. F, if necessary, one finds that the Laplace transform of the time domain function in Eq. (I-5) is

$$\begin{aligned} L\{\Pi_h(t)\} = \tilde{\Pi}_h(s) &= \frac{1}{s} - \frac{e^{-sh}}{s} \\ &= \frac{1-e^{-sh}}{s} \end{aligned} \tag{I-6}$$

Before applying Eq. (I-6), a small repertoire of Z-transforms will be developed.

I-3 Z-Transform of the Constant (Step Change)

Consider the transform of a constant C.

$$\begin{aligned} Z\{C\} &= \sum_{k=0}^{\infty} Cz^{-k} = C\sum_{k=0}^{\infty} z^{-k} = C(1+z^{-1}+z^{-2}+z^{-3}+\cdots) \\ &= C\frac{1}{1-z^{-1}} \end{aligned} \tag{I-7}$$

where the reader can verify the second line of Eq. (I-6) by long division. For the long division to be valid, the infinite series must converge which is ensured if $|z^{-1}| < 1$ or $|z| > 1$. Since transformed quantities are assumed to be zero for $t < 0$, Eq. (I-7) is also the Z-transform of a step change at $t = 0$. More formally, we can write

$$Z\{\hat{U}(t)\} = \frac{1}{1-z^{-1}} = \frac{z}{z-1} \tag{I-8}$$

By the way, do not confuse the rounded hat of the Z-transforms, as in $\hat{Y}(z)$, with the sharp hat of the unit step change, as in $\hat{U}(t)$.

Question I-1 What is the Z-transform of a step change starting at $t = nh$?

Answer

$$Z\{\hat{U}(t-nh)\} = \sum_{k=0}^{\infty} \hat{U}(t_k - nh) z^{-k} = (0 + \cdots + z^{-n} + z^{-n-1} + z^{-n-3} + \cdots)$$

$$= z^{-n}(1 + z^{-1} + z^{-2} + \cdots)$$

$$= \frac{z^{-n}}{1 - z^{-1}}$$

I-4 Z-Transform of the Exponential Function

The exponential function e^{-at} in the continuous time domain becomes $e^{-aih}, i = 0, 1, 2, \ldots$ in the discrete time domain. The Z-transform is

$$Z\{e^{-aih}\} = \sum_{k=0}^{\infty} e^{-akh} z^{-k} = \sum_{k=0}^{\infty} (e^{-ah} z^{-1})^k$$

$$= \frac{1}{1 - e^{-ah} z^{-1}} \tag{I-9}$$

$$= \frac{z}{z - e^{-ah}}$$

As with Eq. (I-7), long division has been invoked and convergence of Eq. (I-9) requires that $\left|e^{-ah} z^{-1}\right| < 1$ or $|z| > e^{-ah}$.

I-5 The Kronecker Delta and Its Z-Transform

In the discrete time domain, the unit pulse or Kronecker delta $\hat{\delta}(k-n)$ is simply an isolated spike of unit magnitude at time $t_n = nh$ (not a Dirac delta function $\delta(t)$ which in this book has no sharp hat) as in

$$\hat{\delta}(k-n) = 0 \qquad k \neq n$$

$$= 1 \qquad k = n$$

Analogously to the Dirac delta function, the Kronecker delta can also "pluck" a value, not from an integral but from a sum, as in

$$\sum_{k=0}^{\infty} y(k) \hat{\delta}(k-n) = y(n)$$

The Z-transform of the Kronecker delta is

$$Z\{\hat{\delta}(k-n)\} = \sum_{k=0}^{\infty} \hat{\delta}(k-n)z^{-k} = \hat{\delta}(0-n)1 + \hat{\delta}(1-n)z^{-1}$$

$$+ \hat{\delta}(2-n)z^{-1} + \cdots + \hat{\delta}(n-n)z^{-n} + \cdots$$

$$= 0 + 0 + 0 + \cdots + z^{-n} + \cdots$$

$$= z^{-n}$$

For the special case of a Kronecker delta at time zero, the Z-transform is simply

$$Z\{\delta(k)\} = \sum_{k=0}^{\infty} \delta(k)z^{-k} = \delta(0)1 + \delta(1)z^{-1} + \delta(2)z^{-1} + \cdots$$

$$= 1$$

I-6 Some Complex Algebra and the Unit Circle in the z-Plane

The Laplace transform variable s was shown to be complex in App. F. Its domain was the complex plane and we found that poles of a transfer function had to occur in the left-hand side of the complex plane for Laplace transforms to represent stable functions. The Z-transform variable z is also complex and Z-transforms also have poles. The Z-transform of the exponential function in Eq. (I-9) has a pole at a value of z that causes the denominator of the Z-transform to vanish, that is, that satisfies

$$1 - e^{-ah}z^{-1} = 0 \qquad \text{or} \qquad z - e^{-ah}$$

or

$$e^{-ah}z^{-1} = 1$$

or

$$z = e^{-ah}$$

Since both a and h are real and positive, the pole is real, and lies on the positive real axis. Furthermore, it lies inside the unit circle in the complex z-plane, defined by $|z| = 1$ because $|e^{-ah}| < 1$.

The idea that $|z| = 1$ defines a unit circle can be understood as follows. Since z is a complex number it can be written as a phasor or vector $z = |z|e^{j\theta}$ where $|z|$ is the magnitude of the vector and θ is the angle of the vector with the x-axis (see App. B). If the magnitude is constant at unity and the angle is allowed to vary from 0 to 2π, a circle with unit radius is described in the complex z-plane. The pole of the Z-transform for the exponential function $e^{-aih}, i = 0, 1, 2, \ldots$, lies inside the unit circle at e^{-ah} and, as long as $a > 0$, the function is bounded.

Had we been working with e^{at} or $e^{aih}, i = 0, 1, 2, \ldots$, where $a > 0$, we could formally show that the Z-transform would look like

$$Z\{e^{aih}\} = \sum_{k=0}^{\infty} e^{akh} z^{-k} = \sum_{k=0}^{\infty} (e^{ah} z^{-1})^k$$
$$= \frac{1}{1 - e^{ah} z^{-1}} \quad ? \tag{I-10}$$

But this is, in fact, a formality because we can conclude simply by observation that this infinite series will not converge. We also see that the pole of Eq. (I-10) lies at $z = e^{ah}$ which is outside the unit circle on the positive real axis in the z-plane because $|e^{ah}| > 1$. This suggests that the unit circle in the z-plane plays an analogous role to the imaginary axis in the s-plane. More about this later in this appendix.

I-7 A Partial Summary

So far we have developed three Z-transforms and we know, from App. F and Chap. 3, the associated Laplace transforms. The following table summarizes this.

The Z-transform for the zero-order hold will be developed in the following section.

Function	Laplace Transform	Z-Transform
Dirac delta or Kronecker delta	$L\{\delta(t-a)\} = e^{-sa}$	$Z\{\hat{\delta}(k-n)\} = z^{-n}$
Step Change at $t = L = Nh$	$\frac{e^{-sL}}{s}$	$\frac{z^{-N}}{1 - z^{-1}}$
Exponential Function $e^{-at} = e^{-iah}$	$\frac{1}{s-a}$	$\frac{z}{z - e^{-ah}}$

Table I-1 Laplace and Z-transforms for Three Functions

I-8 Developing Z-Transform Transfer Functions from Laplace Tranforms with Holds

If the process model is described by $G(s)$ and if there is a sampler/zero-order hold applied to the process input, what is the Z-transform transfer function that can be used to find the process output? In Chap. 9 we arrived at an answer by developing the time domain solution for a piecewise stepped process input and then applied the backshift operator. Here, the following must be evaluated:

$$Z\{\tilde{\Pi}_h(s)G(s)\}$$

Start with the first-order process where

$$G(s) = \frac{g}{\tau s + 1}$$

Remember that the zero-order hold $\Pi_h(t)$ has the Laplace transform of

$$\frac{1 - e^{-sh}}{s}$$

So, now we must evaluate the following:

$$\frac{\hat{y}(z)}{\hat{U}(z)} = Z\left\{\frac{1 - e^{-sh}}{s}\frac{g}{\tau s + 1}\right\}$$

It is simplest to manipulate the expression a little and use partial fractions.

$$\begin{aligned} Z\left\{\frac{1 - e^{-sh}}{s}\frac{g}{\tau s + 1}\right\} &= Z\left\{g(1 - e^{-sh})\frac{1}{s(\tau s + 1)}\right\} \\ &= Z\left\{g(1 - e^{-sh})\left(\frac{1}{s} - \frac{1}{s + \frac{1}{\tau}}\right)\right\} \end{aligned} \tag{I-11}$$

where partial fractions were used to expand $1/[s(\tau s + 1)]$ (see App. F for the algebraic manipulations). Using the table given in Sec. I-7, we can write the Z-transforms for $1/s$ and $1/(s + 1/\tau)$ immediately.

Furthermore, we know that z^{-1} corresponds to e^{-sh}. Therefore, Eq. (I-11), by inspection, becomes

$$\frac{\hat{y}(z)}{\hat{U}(z)} = Z\left\{ g(1-e^{-sh})\left(\frac{1}{s} - \frac{1}{s+\frac{1}{\tau}}\right)\right\} = g(1-z^{-1})\left(\frac{1}{1-z^{-1}} - \frac{1}{1-e^{-\frac{h}{\tau}}z^{-1}}\right) \quad \text{(I-12)}$$

To make sense out of Eq. (I-12) one simply collects coefficients of the backshift operator z^{-1} and after a little algebra, one obtains

$$\hat{y}(z)\left(1-e^{-\frac{h}{\tau}}z^{-1}\right) = g\left(1-e^{-\frac{h}{\tau}}\right)z^{-1}\hat{U}(z)$$

or

$$\hat{y}(z) = e^{-\frac{h}{\tau}}z^{-1}\hat{y}(z) + g\left(1-e^{-\frac{h}{\tau}}\right)z^{-1}\hat{U}(z) \quad \text{(I-13)}$$

which is the same as Eq. (9-6) which is

$$y_i = y_{i-1}e^{-\frac{h}{\tau}} + g\left(1-e^{-\frac{h}{\tau}}\right)U_{i-1} \qquad i = 0, 1, 2, \ldots \quad \text{(I-14)}$$

Therefore, we have shown the effect of the zero-order hold in both the time and the Z-domains.

I-9 Poles and Associated Time Domain Terms

In Sec. I-7 we hinted at a general feature of the Z-transform where poles in the Z-domain correspond to terms in the time domain containing Cr^k where C is a coefficient, k is the sample index, and r is a pole of the Z-transform. To illustrate this concept, consider Eq. (I-13) which can also be written as

$$\hat{y}(z)\left(1-e^{-\frac{h}{\tau}}z^{-1}\right) = g\left(1-e^{-\frac{h}{\tau}}\right)z^{-1}\hat{U}(z)$$

$$\frac{\hat{y}(z)}{\hat{U}(z)} = \frac{g\left(1-e^{-\frac{h}{\tau}}\right)z^{-1}}{1-e^{-\frac{h}{\tau}}z^{-1}}$$

$$= \frac{g\left(1-e^{-\frac{h}{\tau}}\right)}{z-e^{-\frac{h}{\tau}}}$$

If the process input is a step or a constant with a value of U_c, then

$$\hat{U}(z) = \frac{U_c z}{z-1}$$

so, the process output can be written as

$$\hat{y}(z) = \frac{g\left(1-e^{-\frac{h}{\tau}}\right)}{z-e^{-\frac{h}{\tau}}} \frac{U_c z}{z-1} \tag{I-15}$$

The same approach used for inverting Laplace transforms will be used here. First, partial fractions can be used to expand Eq. (I-15), as in

$$\hat{y}(z) = gU_c\left(\frac{z}{z-1} - \frac{z}{z-e^{-\frac{h}{\tau}}}\right) \tag{I-16}$$

Second, from the above developments, we can pick off the two time domain functions associated with the two terms in Eq. (I-16), as in

$$y(k) = gU_c\left(1-(e^{-\frac{h}{\tau}})^k\right) \tag{I-17}$$

There should be nothing startling about Eq. (I-17) but I show it because it points to the fact that the two poles in Eq. (I-15), at $e^{-h/\tau}$ and 1.0, lead to two terms in the time domain of the form

$$\begin{aligned} y(t) &= C_1 r_1^k + C_2 r_2^k \\ r_1 &= 1 \qquad r_2 = e^{-\frac{h}{\tau}} \\ C_1 &= gU_c \qquad C_2 = -gU_c \end{aligned} \tag{I-18}$$

Therefore, one might induce a general rule that the poles in the z-plane must lie inside or on the unit circle for there to be stability. Furthermore, a pole at $z = r$ corresponds to a time domain term of r^k and a pole at $z = 1$ leads to a constant. Finally, as the position of the pole moves toward the origin of the z-plane, which is also the center of the unit circle, the transient will have shorter duration. For example, in Eqs. (I-17) and (I-18) one can see that, as the time constant τ decreases, the pole location r_2 moves toward the origin and the transient becomes shorter.

Before leaving this section we need to point out the correspondence between the z-plane poles of a Z-transform and the s-plane poles of a Laplace transform. Remember that

$$g(1-e^{-sh})\left(\frac{1}{s}-\frac{1}{s+\frac{1}{\tau}}\right) \tag{I-19}$$

corresponds to

$$g(1-z^{-1})\left(\frac{1}{1-z^{-1}}-\frac{1}{1-e^{-\frac{h}{\tau}}z^{-1}}\right) \tag{I-20}$$

By inspection, Eq. (I-19) has poles at $s=0$ and $s=-1/\tau$. Those two poles in the s-plane correspond to the two poles in Eq. (I-20) located at $z=1$ and $z=\exp(-h/\tau)$, respectively. These two equivalences are special cases of the general relationship between the poles of a Laplace transform and a Z-transform given in

$$\boxed{z=e^{sh}} \tag{I-21}$$

which was just the variable substitution made in the development of Eq. (I-4).

I-10 Final Value Theorem

Fittingly, we conclude the appendix with a handy trick called the *final value theorem* which we will present without derivation. The final value of $y(t)$, given the Z-transform $\widehat{Y}(z)$, can be obtained the following operation:

$$\boxed{\begin{aligned}\lim_{t\to\infty} y(t) &= \lim_{z\to 1}(1-z^{-1})\widehat{Y}(z)\\ &= \lim_{z\to 1}\frac{z-1}{z}\widehat{Y}(z)\end{aligned}} \tag{I-22}$$

Applied to Eq. (I-15), the final value theorem gives

$$\lim_{t\to\infty} y(t) = \lim_{z\to 1}(1-z^{-1})\widehat{Y}(z) = \lim_{z\to 1}\frac{z-1}{z}\frac{g\left(1-e^{-\frac{h}{\tau}}\right)}{z-e^{-\frac{h}{\tau}}}\frac{U_c z}{z-1}$$

$$= \lim_{z\to 1}\frac{g\left(1-e^{-\frac{h}{\tau}}\right)}{z-e^{-\frac{h}{\tau}}}U_c = gU_c$$

which makes sense because in response to a unit step change the process output of a first-order model should settle out to the gain multiplied by the value of the input step.

By the way, remember the final value theorem for the Laplace transform?

$$\lim_{s\to 0} s\tilde{Y}(s) = \lim_{t\to\infty} y(t)$$

What is the connection? In the Laplace domain, s is an operator that causes differentiation. In the Z-transform domain, $1-z^{-1}$ is an operator that causes differencing.

I-11 Summary

In Chap. 9, we used the backshift operator as a means of familiarizing ourselves with the Z-transform. In this appendix we took a more rigorous approach using the Laplace transform as a starting point.

With this alternative approach in hand we developed the Z-transform of the zero-order hold and a couple of common time domain functions. The Kronecker delta was introduced and shown to be analogous to the Dirac delta.

The poles of a Z-transform were discussed in a manner similar to that used with the Laplace transform. An important equivalence between the poles of a Laplace transform and Z-transform was discussed. Finally, the final value theorem was presented.

APPENDIX J

A Brief Exposure to Matlab

Back in the early 1980s I got my first copy of Matlab. The slim manual began with, "If you feel you can't bother with this manual, start here." In effect, Matlab was presented using "backward chaining. " The manual quickly showed you what it could do and motivated you to dig into the details to figure out how you could avail yourself of such awesome computing power. Encountering Matlab was a mind-blowing experience (remember, this was circa 1984 and most of us were using BASIC, Quick BASIC, and Fortran).

Matlab mfiles consist of "scripts" which you write in a BASIC-like language and which allow you to call many built-in incredibly powerful routines or functions to carry out calculations. For example, the function `eig` calculates eigenvalues of a matrix. You can use the Matlab editor to look at the `eig` function code and find out that "it computes the generalized eigenvalues of A and B using the Cholesky factorization of B. " Fortunately, you do not have to understand the Cholesky factorization to use the function—it is completely transparent.

This appendix will show an example script with some comments (anything after a % is a comment) and let you take it from there.

```
% Matlab Example script
close all % close all existing graphs from previous
% sessions
clear   % clear all variables
% make up a (3,3) numerical matrix and display it
Am=[1 2 4;-4 2 1;0 9 2];
disp('starting matrix ')
Am
Aminv=inv(Am); %numerically calculate the matrix inverse
and display
disp(['inverse ' ])
Aminv
disp(['determinant = ' num2str(det(Am))]) % determinant
disp('eigenvalues ')  % calculate the eigenvalues
eig(Am)
```

```
% do some symbolic math
syms R1 R2 R2 R3 s A1 A2 A3 rho A Ainv yt % declare the
variables as symbolic
% make up a matrix symbolically
A=[rho*A1*s+1/R1   0   0
  -1/R1     rho*A2*s+1/R2  0
  0   -1/R2   rho*A3*s+1/R3 ];
disp('starting matrix')
pretty(A)
disp('inverse')
Ainv=inv(A);  % invert the matrix symbolically
pretty(Ainv)
% invert a laplace transform
disp('Laplace transform')
pretty(1/(rho*A1*s+1/R1)) % a simple first order transform
yt=ilaplace(1/(rho*A1*s+1/R1)) % invert the transform
disp('inverse Laplace transform')
pretty(yt)
% generate a test sinusoid and plot it
N=1000;
t=0:N-1;
y=sin(2*pi*t/50);
figure(1)
plot(t,y),grid  % plot the sinusoid on a grid
title('A Test Sinusoid')
xlabel('time')
ylabel('y')
% put the signal through a filter
tau=20; % filter time constant
a=exp(-1/tau); % filter parameters
b=1-a;
n=[0 b]; % filter numerator
d=[1 -a]; % filter denominator
yf=filter(n,d,y); % apply the filter to y
figure(gcf+1)  % set up the next graph
plot(t,y,t,yf),grid % plot the sine and the filtered
signal
xlabel('time')
legend('y','{\ity}_{\itf}') % put subscripts in the legend
ylabel('filter input & output')
% develop the FFT of the filtered signal (1000 pts)
YF=fft(yf); % (YF is a complex number)
h=1; % assume sampling at 1 sec intervals
fs=1/h; % sampling frequency
f1=1/N; % fundamental frequency
fNY=fs/2; % Nyquist frequency
f=0:f1:fNY; % generate the frequencies in the Nyquist
interval
Nf=length(f); % number of frequencies
mag=abs(YF); % calculate the power
magplot=mag(1:Nf); % pick only the magnitudes in the
Nyquist interval
figure(gcf+1)
set(gcf,'DefaultLineLineWidth',1.5) % choose a thicker
```

```
% line
plot(f,magplot),grid
xlabel('frequency')
ylabel('Magnitude')
title('Power Spectrum of Filter Output')
```

Use the editor to save this script and give it a name, as in `test.m`. Then run in by entering the name at the Matlab prompt, as in

```
>>test
```

Use the help function to get information on the built-in functions, as in

```
>> help filter
```

FILTER One-dimensional digital filter.

Y = FILTER(B,A,X) filters the data in vector X with the filter described by vectors A and B to create the filtered data Y. The filter is a "Direct Form II Transposed" implementation of the standard difference equation:

```
a(1)*y(n) = b(1)*x(n) + b(2)*x(n-1) + ··· + b(nb+1)*x(n-nb)
                      - a(2)*y(n-1) - ··· - a(na+1)*y(n-na)
```

If a(1) is not equal to 1, FILTER normalizes the filter coefficients by a(1).

FILTER always operates along the first non-singleton dimension, namely dimension 1 for column vectors and nontrivial matrices, and dimension 2 for row vectors.

[Y,Zf] = FILTER(B,A,X,Zi) gives access to initial and final conditions, Zi and Zf, of the delays. Zi is a vector of length MAX (LENGTH(A),LENGTH(B))-1 or an array of such vectors, one for each column of X.

FILTER(B,A,X,[],DIM) or FILTER(B,A,X,Zi,DIM) operates along the dimension DIM.

See also FILTER2 and, in the Signal Processing Toolbox, FILTFILT.

Overloaded methods

```
    help par/filter.m
    help dfilt/filter.m
    help cas/filter.m
>>
```

I also use Matlab's Simulink extensively but I will leave it to the reader to figure it out, other than to say that it is equally friendly and powerful. Aside from the basic Matab and Simulink packages, I use the following toolboxes in the book: control systems, signal processing, symbolic and system identification.

Index

E

F

G

H

I

K

L

M

N

O

P

Q

R

S

T

U

V

W

Z